# Heat Transfer Enhancement Techniques

**Scrivener Publishing**
100 Cummings Center, Suite 541J
Beverly, MA 01915-6106

*Publishers at Scrivener*
Martin Scrivener (martin@scrivenerpublishing.com)
Phillip Carmical (pcarmical@scrivenerpublishing.com)

# Heat Transfer Enhancement Techniques

## Thermal Performance, Optimization and Applications

Edited by
**Ashwani Kumar**
**Nitesh Dutt**
and
**Mukesh Kumar Awasthi**

Scrivener Publishing

WILEY

This edition first published 2025 by John Wiley & Sons, Inc., 111 River Street, Hoboken, NJ 07030, USA and Scrivener Publishing LLC, 100 Cummings Center, Suite 541J, Beverly, MA 01915, USA

For more information about Scrivener publications please visit www.scrivenerpublishing.com.



**Wiley Global Headquarters**
111 River Street, Hoboken, NJ 07030, USA

For details of our global editorial offices, customer services, and more information about Wiley products visit us at www.wiley.com.



***Library of Congress Cataloging-in-Publication Data***

ISBN 978-1-394-27096-5

Front cover images supplied by Pixabay.com
Cover design by Russell Richardson

Set in size of 11pt and Minion Pro by Manila Typesetting Company, Makati, Philippines

Printed in the USA

10 9 8 7 6 5 4 3 2 1

# Contents

**Aim and Scope** xvii

**Preface** xix

**Acknowledgement** xxiii

**1 Recent Innovation in Heat Transfer Enhancement Techniques** 1
*Ashwani Kumar, Mukesh Kumar Awasthi, Nitesh Dutt and Varun Pratap Singh*
1.1 Introduction 2
1.1.1 Industrial Application of Heat Transfer Enhancement Techniques 3
1.1.2 Standards and Regulations 5
1.2 Important Heat Transfer Enhancement Techniques and Their Effect 7
1.2.1 Effect of Fins and Extended Surfaces 7
1.2.2 Effect of Phase Change Materials (PCMs) 11
1.2.3 Effect of Heat Exchangers 13
1.2.4 Effect of Microchannels 13
1.2.5 Effect of Nanofluids 16
1.2.5.1 Analytical Approaches to Understanding Physical Characteristics of Nanofluids 17
1.2.6 Effect of Porous Media 20
1.2.6.1 Effect of Porosity 21
1.2.7 Effect of Jet Impingement 22
1.2.8 Effect of Heat Pipes 22
1.2.9 Effect of Vortex Generators 23
1.2.10 Effect of Ribbed Surfaces 25
1.2.11 Effect of Artificial Roughness-Based Turbulence 25
1.3 Numerical Analysis of Heat Transfer Problem 27
1.4 Conclusion 30
References 31

**2 Renewable Thermal Energy Systems: Sustainable, Modern and Reliable Energy** **39**
*Bipasa B. Patra and Pratik Sharad Chirmade*
2.1 Introduction 39
2.2 Sustainable Development Goals (SDG) 42
2.3 Discussion 54
References 56

**3 HVAC System Efficiency Improvement Through Heat Transfer Enhancement Techniques** **63**
*Md Naim Hossain and Arijit Kundu*
3.1 Introduction 64
3.2 Passive Heat Transfer Enhancement Techniques 66
3.2.1 Surfactants 68
3.2.2 Straight Microfins and Helical Microfins Tubes 69
3.2.3 Herringbone Tube 72
3.2.4 Twisted Tapes Insert Tube 74
3.2.5 Wired Coils 76
3.2.6 Dimpled Tubes 77
3.3 Electro-Passive Heat Transfer Enhancement Techniques 78
3.4 Conclusion 79
References 80

**4 Indoor Thermal Performance Enhancement of Sustainable Buildings** **87**
*D.B. Jani*
List of Nomenclature 87
4.1 Introduction 88
4.2 Background of the Present Study 90
4.3 System Operation 91
4.4 Comparison of Desiccant Cooling with Traditional VCR Cooling 96
4.5 Conclusions 99
References 100

**5 Eco-Friendly Paint for Sustainable Building Applications to Enhance Thermal Life Comfort** **105**
*Vikas Chaubey, Atul Kumar, Aakash Singh and Shekhar Yadav*
5.1 Introduction 106
5.2 Advantages of Vedic Plaster Over Conventional Plaster 108
5.3 Need for Vedic Paints 110
5.4 Types of Vedic Paints 112

5.5 Chemical Properties of Vedic Paints 113
5.6 Factors Increasing Comfort 114
5.7 Conclusion 116
5.8 Future Outlook 116
References 117

**6 Augmentation of Solar, Geothermal, and Earth-Air Heat Exchanger in Sustainable Buildings 119**
***Varun Pratap Singh, Ashwani Kumar and Mukesh Kumar Awasthi***
6.1 Introduction 120
6.2 Current State of Renewable Energy Technologies 121
6.3 Solar Augmentation Strategies 122
6.3.1 Advanced Solar Technologies 124
6.3.2 Integration Into Building Design 125
6.3.3 Energy Efficiency and Environmental Impact 127
6.3.3.1 Energy Efficiency 127
6.3.3.2 Environmental Impact 128
6.4 Geothermal Energy in Building Systems 128
6.4.1 Harnessing Subsurface Heat 130
6.4.2 Applications in Space Heating, Cooling, and Power Generation 131
6.4.3 Innovative Geothermal Solutions 133
6.5 Earth-Air Heat Exchangers: Passive and Active Cooling 135
6.5.1 Principles and Functionality of Earth-Air Heat Exchangers (EAHE) 136
6.5.2 Benefits of Earth-Air Heat Exchanger (EAHE) Systems 139
6.5.3 Practical Implementation Techniques for Earth-Air Heat Exchangers (EAHE) 140
6.6 Combined Augmentation Strategies for Sustainable Buildings 141
6.6.1 Synergies Among Solar, Geothermal, and Earth-Air Systems 142
6.6.2 Challenges and Considerations in the Integration of Solar, Geothermal, and Earth-Air Systems for Sustainable Buildings 143
6.6.3 Future Trends and Prospects in the Integration of Solar, Geothermal, and Earth-Air Systems for Sustainable Buildings 146
6.7 Conclusion 148
6.7.1 Implications for Sustainable Building Practices 150

6.7.2 Recommendations for Future Research 151
References 152

**7 CFD Numerical Investigation of Thermal Performance of Diamond Shape Micro Rectangular Heat Exchanger 159**
***Jaideep, Pritosh Tomar and Ashwani Kumar***
7.1 Introduction 160
7.1.1 Heat Transfer Enhancement Method 163
7.1.2 Microchannel Heat Sink Method 164
7.1.3 Geometry of the Advanced Heat Sink Channel 164
7.2 Objective and Methodology 165
7.2.1 Methodology of the Microchannel Heat Exchanger Rectangular Channel 165
7.2.2 Methodology of Microchannel Diamond Fin Heat Sink 165
7.3 Parameters of Microchannel Fin Heat Sink 166
7.4 Governing Equation Used in Microchannel 168
7.4.1 K-e Model Used in Microchannel Heat Sink 169
7.4.2 Continuity Equation Applied 169
7.4.3 Momentum Equation 169
7.4.4 Energy Equation 170
7.4.5 Assumption for the Microchannel Heat Sink 170
7.5 Material Properties and Boundary Conditions 171
7.5.1 Thermophysical Properties of Copper-Inserted Diamond Fin Heat Sink 171
7.5.2 Boundary Conditions Applied in Diamond Microchannel 172
7.5.3 Mesh Generation of the Microchannel Heat Sink with Diamond Fin Shape 172
7.5.4 Validation for the Microplate Fin 174
7.6 Result and Discussion 174
7.6.1 The Smooth Microchannel and Dittus Boelter Equation 175
7.6.2 Smooth Microchannel and Blasius Friction Equation 176
7.6.3 Results and Effects of Microchannel with Diamond Fin Heat Sink 177
7.6.4 Results and Effects of the Velocity Contour 177
7.6.5 Results and Effects of the Temperature Contour 178
7.6.6 Effects of Inserted Diamond Fin with Reynolds Number 180
7.6.7 Friction Factor in the Roughened Microchannel Heat Sink 183
7.7 Thermal Hydraulic Efficiency of Diamond Shape Heat Exchanger Sink 186

7.8 Conclusion 187
References 187

**8 Particle Swarm Optimization Technique for Determining Optimal Process Parameters for Counter Flow Double Pipe Heat Exchanger 193**
***Sridharan M.***
Nomenclature 193
Abbreviations 194
8.1 Introduction 194
8.1.1 Need for Optimization in Heat Exchangers 195
8.2 Experimental Setup 201
8.2.1 Data Reduction 203
8.3 Mathematical Model 204
8.3.1 Formulation of Objective Function 204
8.3.2 Process Parameters 205
8.3.2.1 Hot Fluid Outlet Temperature 205
8.3.2.2 Mass Flow Rate of Hot Fluid 207
8.3.2.3 Cold Fluid Outlet Temperature 207
8.3.2.4 Mass Flow Rate of Cold Fluid 207
8.4 Implementation of Multi-Objective Type Optimization Technique [MOTOT] 207
8.4.1 Particle Swarm Optimization 208
8.4.1.1 Condition 1 208
8.4.1.2 Condition 2 208
8.4.2 Merits of PSO 209
8.4.3 Algorithm 209
8.4.4 Parameters of PSO 209
8.4.5 Numerical Illustration of PSO 210
8.4.5.1 Calculation of $m_h$ 210
8.4.5.2 Calculation of $m_c$ 210
8.4.5.3 Calculation of $T_2$ 210
8.4.5.4 Calculation of $t_2$ 210
8.4.5.5 Calculation of Objective Function 212
8.4.5.6 Calculation of Particle Best Value 212
8.4.5.7 Calculation of Global Best Value 212
8.4.6 Computational Result of PSO 212
8.5 Confirmation Experiments 212
8.6 Results and Discussion 212
8.6.1 Initial Experiments 214
8.6.2 The PSO Analysis 214

8.6.3 The Validation Experiments 216
8.7 Conclusions 217
References 218

**9 Application of Geothermal Energy-Based Earth-Air Heat Exchanger in Sustainable Buildings 221**
***Arijit Kundu***
9.1 Introduction to Sustainable Building 221
9.2 System Approach for Complex System Study 222
9.3 Earth-to-Air Heat Exchanger for Sustainable Buildings 223
9.4 EAHE Performance Evaluation: Numerical Method 227
9.5 Discussion 229
References 230

**10 Numerical Study of Solar Air Heater with Semi-Cylindrical Tube Roughness 233**
***Ankush Hedau and S. K. Singal***
Nomenclature 233
Abbreviations 234
10.1 Introduction 234
10.2 Numerical Simulation 236
10.2.1 Preprocessing 237
10.2.2 Processing 239
10.2.3 Postprocessing 240
10.2.4 Data Reduction 241
10.3 Validation 241
10.4 Results and Discussions 242
10.4.1 Effect of Roughness Height Ratio ($e_r$/H) 242
10.4.2 Effect of Tube Pitch Ratio (P/H) 244
10.4.3 Thermohydraulic Performance (THP) 246
10.5 Conclusions 247
Declaration of Competing Interest 248
Data Availability 248
Acknowledgement 248
References 248

**11 Design and Analysis of Solar Tracking System for PV Thermal Performance Enhancement 251**
***Bhupender Singh, Preet Kaur, Ashok Kumar Yadav, Mukesh Kumar Awasthi and Ashwani Kumar***
11.1 Introduction 252
11.2 Background and Motivation 256

11.2.1 Types of Solar Tracking System 256
11.3 Fundamentals of Arduino-Based Solar Tracking System 257
11.3.1 Components of a Sun Tracking Solar Panel System 257
11.3.2 Working Procedure of Solar Tracking System 258
11.3.3 Development of Model 258
11.4 Benefits and Challenges 260
11.5 Conclusions 260
References 261

**12 An Overview on Thermal Characterization of Lithium-Ion Batteries for Enhancing the Durability 269**
***Vikas Chaubey, Atul Kumar, Shailendra Sinha and Rakesh Verma***
12.1 Introduction 270
12.1.1 Li-Ion Batteries 270
12.2 Thermal Behavior of Li-Ion Battery 271
12.3 Heat Generation Mechanism and Thermal Modeling 272
12.3.1 Ohmic Heating 272
12.3.2 Faradaic Heating 272
12.3.3 Concentration Polarization 273
12.3.4 Side Reactions 273
12.3.5 1D, 2D, or 3D Thermal Models 273
12.3.5.1 Multiphysics Coupling 273
12.3.5.2 Thermal Runaway Prediction Models 273
12.4 The Effect of Temperature on Li-Ion Batteries 274
12.4.1 Optimal Operating Temperature Range 274
12.4.2 Capacity and Power Output 275
12.4.3 Degradation and Aging 275
12.4.4 Safety Concerns 275
12.4.5 Thermal Management 275
12.4.6 Charging and Discharging Behavior 275
12.5 Thermal Runway Modeling and Safety Tests 276
12.5.1 Thermal Runaway Modeling 276
12.5.1.1 Multiphysics Simulations 276
12.5.1.2 Reaction Kinetics 276
12.5.1.3 Catastrophic Failure Prediction 277
12.5.2 Safety Tests 277
12.5.2.1 Accelerated Rate Calorimetry (ARC) 277
12.5.2.2 Differential Scanning Calorimetry (DSC) 277
12.5.2.3 Thermal Abuse Testing 277

12.5.2.4 Thermal Imaging and *In-Situ* Measurements 277
12.6 Interior Electrode Modifications 278
12.6.1 Nanostructured Materials 278
12.6.2 Coating and Additives 278
12.6.3 Conductive Networks 278
12.6.4 Porosity and Structure Control 278
12.6.5 Thermal Conductive Materials 278
12.6.6 Advanced Composite Structures 279
12.7 Exterior Thermal Management System 279
12.7.1 Liquid Cooling Systems 279
12.7.2 Air Cooling Systems 279
12.7.3 Phase Change Materials (PCMs) 279
12.7.4 Thermal Insulation 280
12.7.5 Active Thermal Control Systems 280
12.7.6 Thermal Interface Materials 280
12.7.7 Design Optimization 280
12.8 Safety Management Strategy 280
12.8.1 Thermal Monitoring Systems 281
12.8.2 Temperature Thresholds and Alarms 281
12.8.3 Thermal Shutdown Mechanisms 281
12.8.4 Ventilation and Containment 282
12.8.5 Safety Protocols and Training 282
12.8.6 Battery Management Systems (BMS) 282
12.8.7 Rapid Response Plans 282
12.8.8 Regulatory Compliance 282
12.9 Thermal Analysis of Lithium-Ion Batteries 283
12.9.1 Calorimetry Techniques 283
12.9.2 Thermal Imaging 283
12.9.3 Thermal Modeling 283
12.9.4 Electrochemical Impedance Spectroscopy (EIS) 283
12.9.5 Thermal Conductivity Measurements 284
12.9.6 Cycling and Aging Studies 284
12.9.7 Thermal Runaway Assessment 284
12.10 Failures in Lithium-Ion Batteries Pack 284
12.10.1 Thermal Runaway 284
12.10.2 Capacity Fade 285
12.10.3 Internal Resistance Increase 285
12.10.4 Electrode Degradation 285
12.10.5 Electrolyte Breakdown 285
12.10.6 Thermal Stress 285

12.10.7 External Damage 285
12.11 Conclusion and Suggestions 286
12.11.1 Advanced Thermal Analysis 286
12.11.2 Materials Innovation 286
12.11.3 Optimized Thermal Management 286
12.11.4 Safety Protocols and Training 286
12.11.5 Integrated Design Approach 287
References 287

**13 An In-Depth Introduction to State of Health Estimation Methods of Li-Ion Batteries 291**
***Prateek Verma***
13.1 Introduction 292
13.2 State of Health 293
13.2.1 Direct Measurement 295
13.2.2 Indirect Measurement 295
13.2.2.1 Model-Based Estimation Method 295
13.2.2.2 Data-Driven Model 301
13.3 Conclusion 305
References 306

**14 Heat and Mass Transportation Enhancement of Casson Cu-AA7075-AA7072/Methanol Tri-Hybrid Nanofluid Flow Past A Porous Spinning Disk: A Computational Assessment 311**
***Bhagyashri Patgiri and Ashish Paul***
Nomenclature 312
14.1 Introduction 313
14.2 Problem Formulation 314
14.3 Numerical Method and Validation 319
14.4 Results and Discussion 319
14.5 Conclusions 327
References 327

**15 Thermal Performance of MXene ($Ti_3C_2$) Nanoparticles in Blood Flow Over a Curved Region: A Biomedical Application 331**
***Niraj Rathore and N. Sandeep***
15.1 Introduction 332
15.2 Characteristics of MXene Nanomaterials 334
15.2.1 Drug Delivery 335
15.2.2 Imaging 335
15.2.3 Photothermal Therapy 335
15.2.4 Biosensing 335

15.2.5 Antibacterial Applications 335
15.2.6 Tissue Engineering 336
15.2.7 Neural Interfaces 336
15.2.8 Anti-Inflammatory Applications 336
15.3 Problem Description 337
15.4 Flow Nature for Different Thermal Conductivity Models 341
15.5 Outcomes and Discourse of Results 344
15.5.1 Velocity and Temperature Variance for MXene ($Ti_3C_2$) Emerged Blood Flow 345
15.5.2 Variation in Heat Transfer Rate and Drag of the Flow 352
15.6 Conclusion 354
References 355

**16 Strong Magnetic Shock Wave Propagation in a Dusty Gas 359**
***Akmal Husain, S. A. Haider, M. K. Shukla, Mohd Miyan and A. Taqvi***
16.1 Introduction 360
16.2 Fundamental Set of Equations 362
16.3 Rankine-Hugoniot Jump Boundary Conditions for Strong Shocks 364
16.4 Closed Form Solution for Strong Shocks 366
16.5 Results and Conclusions 370
References 373

**17 The Effect of Casson Fluid Flow on a Stagnation Point Over a Porous Stretching Sheet with Thermal Radiation 375**
***Wajeeha K., Sushma M. N., U.S. Mahabaleshwar, Mahesh R. and Dhananjay Yadav***
17.1 Introduction 376
17.2 Mathematical Formulation 378
17.3 Result and Discussion 381
17.4 Conclusion 386
References 386

**18 Emerging Trends in Smart Green Building Technologies 391**
***Gongutri Borah***
18.1 Introduction 392
18.2 Environmental Challenges and the Need for Innovation 393
18.3 The Urgency for Adopting Smart Green Building Technologies 394

18.4 Innovative Architectural Designs That Prioritize Energy Efficiency 395
18.5 Passive Design Principles and Their Impact on Building Performance 397
18.6 Exploration of Eco-Friendly and Sustainable Construction Materials 398
18.7 Case Studies Showcasing the Use of Advanced Materials in Real-World Projects 399
18.7.1 Case Study: The Edge, Amsterdam—Innovative Use of Recycled Materials 399
18.7.2 Case Study: One Central Park, Sydney—Living Façade with Green Wall Technology 399
18.7.3 Case Study: Bosco Verticale, Milan—Vertical Forest Towers 400
18.7.4 Case Study: The Crystal, London—Sustainable Building with Photovoltaic Integration 400
18.7.5 Case Study: The Edge, New York—High-Performance Façade 400
18.7.6 Case Study: King Abdullah Petroleum Studies and Research Center (KAPSARC), Riyadh—Advanced Building Envelope 400
18.8 The Intersection of IoT and Smart Green Buildings 401
18.9 Artificial Intelligence in Smart Buildings 402
18.10 The Role of Solar and Wind Energy in Achieving Net-Zero Energy Buildings 403
18.11 Energy Storage Solutions in Buildings for Balancing Intermittent Renewable Sources 404
18.12 Technologies Enhancing Occupant Well-Being and Productivity 405
18.13 The Impact of a Human-Centric Approach on Building Design 406
18.14 Smart Green Building Policies and Certifications 407
18.15 The Influence of Regulations on Industry Adoption of Smart Green Technologies 408
18.16 Speculations on the Future Trajectory of Smart Green Building Technologies 409
18.16.1 Incorporation of Artificial Intelligence (AI) and Machine Learning (ML) 409
18.16.2 Advanced Building Materials and Energy Harvesting 409

18.16.3 IoT-Driven Smart Building Ecosystems 410
18.16.4 Circular Economy and Sustainable Construction Practices 410
18.16.5 Human-Centric Design and Well-Being Focus 410
18.17 Conclusion: Socio-Economic Impact and Community Resilience 410
Acknowledgement 411
References 412

**About the Editors** **417**

**Index** **419**

# Aim and Scope

The book *Heat Transfer Enhancement Techniques: Thermal Performance, Optimization and Applications* is designed to consider the cutting-edge heat transfer enhancement techniques that will obliterate the barriers of traditional thinking. It defies delving into the realms of the next generation, igniting a relentless pursuit of new devices and ground-breaking technologies. This book aims to bridge the gap between conventional methodologies and innovative strategies in the realm of heat transfer. This book will serve as a beacon for researchers, practitioners, and scholars seeking a deeper comprehension of cutting-edge techniques that go beyond conventional boundaries, aiming to revolutionize the field of heat transfer.

The practical guidelines provided by the authors can help engineers select the appropriate enhancement technique based on the specific requirements of the application, *i.e.*, sustainable buildings, electronics industry, and batteries.

Furthermore, the book's overarching objective is to integrate theoretical insights with practical implications, catering to a diverse audience invested in various industries. Its comprehensive approach not only focuses on elucidating complex heat transfer phenomena but also accentuates the practical implementation of these techniques across multiple sectors. By delving into interdisciplinary aspects, this book endeavors to offer a holistic understanding of heat transfer enhancement, thereby empowering professionals and researchers to leverage these advancements across diverse fields such as energy, electronics, construction, and automotive industries.

Top of form. The book has dedicated industrial application of heat transfer enhancement techniques, which makes it ideal for students and industry professionals since it helps them comprehend the background, current, and prospective future of the subject under discussion. The research content will work as a reference book for post-graduate students, Ph.D. researchers, heat transfer scientists, mechanical engineers, budding researchers, production engineers, design engineers, and energy study professionals. The applicability of this book covers a wide range of industries such as heat transfer, production sector, fluid dynamics, research and development engineer, automotive industry, electronics industry, civil engineers, and researchers in nuclear heat transfer sectors, and it will help the audience conducting research in these industries.

Editors

# Preface

The book *Heat Transfer Enhancement Techniques: Thermal Performance, Optimization and Applications* have eighteen chapters covering from introduction to advanced heat transfer enhancement techniques.

**Chapter 1** provides a concise overview of the recent advancements in heat transfer enhancement methods, encompassing both experimental and computational approaches. It also covers recent technological advancements which enable the development of intelligent thermal management systems capable of adapting to dynamic operating conditions, thereby enhancing overall system efficiency. In continuation, **Chapter 2** covers sustainable development goals and highlights the role of different governments, organizations, and communities worldwide for implementing renewable energy policies, investing in research and development, and promoting the adoption of clean energy technologies.

**Chapter 3** deals with the application portion of heat transfer enhancement techniques in heating, ventilation, and air conditioning (HVAC) system. The chapter describes a concise overview of the fundamental methods that are employed to optimize the HVAC system performance through heat transfer enhancement. The enhancement of HVAC system efficiency through heat transfer enhancement techniques may offer multiple feasible design solutions that can be used by suitable trade-offs to meet specific building requirements and local climate conditions. **Chapter 4** highlights the inventive strategy for sustainable cooling that can be achieved by the use of desiccant-assisted dehumidification and cooling systems in place of contemporary air conditioning systems to save energy and employ green cooling techniques for building comfort. To meet the energy demands in sustainable buildings, **Chapter 5** highlights the use of eco-friendly paints. The primary objective of this chapter is to highlight the potential of eco-friendly paint as a sustainable option for smart buildings. The future of the study is concluded for the construction sector to develop energy efficient paint materials with better sustainability by incorporating natural waste. **Chapter 6** promotes renewable energy application in green buildings to meet out the energy demands. The chapter presents an in-depth exploration of the integration and enhancement of renewable energy technologies in sustainable building practices. The chapter systematically dissects the current state of renewable energy technologies, focusing on solar energy, geothermal energy, and Earth-air heat exchangers (EAHE).

**Chapter 7** present numerical approaches to examine the thermo-hydraulic behavior of micropin fin heat sink with a different shape of pin fins (diamond), constant surface area, and air at 4% volumetric concentration with Reynolds number variations from 1500 to 4000 for further optimization of heat transfer enhancement in solar-based systems. In continuation, **Chapter 8** deals with the particle swarm optimization (PSO) to find the optimal operating parameters for maximizing the heat transfer rate in the double pipe heat exchanger (DPHE). In this study, the PSO-based process parameter optimization of the counter flow double pipe heat exchanger [CF-DPHE] is tested. Initially, real-time experiments are conducted with the CP-DPHE rig at the heat and mass transfer laboratory, Saranathan college of Engineering, A.P., India.

**Chapter 9** discusses the application of geothermal energy-based Earth-air heat exchanger (EAHE) systems within the context of sustainable building designs and operations. It is possible to draw the conclusion that effective usage of EAHE systems in conjunction with environmentally friendly energy sources and cutting-edge technology will be crucial in reducing energy consumption and protecting the environment not just in India but globally.

**Chapter 10** presents a numerical study of solar air heater for drying application. The chapter presents a numerical investigation at various design parameters of roughness elements such as tube pitch ratios (P/H) of 1 to 4, roughness height ratios ($e_r/H$) of 0.2 to 0.5, under various values of Re from 3000 to 19000. This research could prove useful for further studies and industries to develop efficient SAH for drying applications. **Chapter 11** deals with the development and assessment of a solar tracking system capable of optimizing the solar panel's power output is the primary goal of this chapter. The solar tracking system prototype was constructed and put to the test. The data results are collated, and a graphical method is used to compare the performance of a static solar panel with a solar tracking system. As a result, tracking solar panels are able to produce greater current, voltage, and power than stationary solar panels.

Moving towards the second industrial application of heat transfer enhancement techniques, **Chapter 12** provides a thorough overview of the thermal characterization of lithium-ion batteries (LIBs). The chapter serves as a valuable resource for researchers, engineers, and practitioners seeking a comprehensive understanding of thermal characterization techniques applied to lithium-ion batteries, particularly concerning the promotion of more durable, reliable, and safer lithium-ion batteries. The advanced techniques employed in battery state of health (SOH) prediction are discussed in **Chapter 13**. The SOH of a lithium-ion (Li-ion) battery is a measure of its overall health and performance relative to its original or designed specifications. **Chapter 14** focuses on the computational analysis of mass and heat relocation upon the flow of Casson Cu-AA7075-AA7072/Methanol tri-hybrid nanofluids across a spinning disk under thermal radiation, slip condition, and chemical reaction. Cu nanoparticles are utilized as coolant in disk drives, computer monitors, televisions, and in many other electronic equipment. **Chapter 15** investigates the thermal conduction and rate of heat transmission due to MXene (Ti3C2) nanoparticles flowing in blood arteries. For this purpose, mathematical modeling for blood flow on the curved region of the artery is developed using the Navier-Stokes equation.

**Chapter 16** aims to create a precise mathematical solution for the movement of a shockwave in a gas that is not perfect and contains tiny solid particles. Additionally, the study will take into account the impact of a magnetic field. **Chapter 17** investigates the influence of magnetite ($Fe_3O_4$) nanoparticles mixed with water ($H_2O$) on a Casson fluid with a stagnation point on a porous stretched sheet.

**Chapter 18** concludes the book with the idea of green building technologies. The chapter conducts a thorough examination of recent innovations, methodologies, and practices driving the evolution of smart green buildings. The socio-economic impact of smart green buildings is also investigated, highlighting their potential contribution to creating sustainable and resilient communities. The chapter concludes by envisioning the future trajectory of thermal renewable energy application in smart green buildings, foreseeing their continued integration into urban planning and the construction industry.

The above short summary of each chapter states the book's nature, which makes it ideal for usage in graduate students curricula and makes it useful for research and academic purposes. It is helpful for mechanical-electrical-electronics-automobile engineers, policy makers, energy scientists, and industry professionals to study the subject under discussion.

Editors
Dr. Ashwani Kumar
Dr. Nitesh Dutt
Dr. Mukesh Kumar Awasthi

# Acknowledgement

We express our gratitude to **Wiley Scrivener Publishing** and the editorial team for their suggestions and support during the completion of this book. We are grateful to all contributors and reviewers for their illuminating views on each book chapter presented in the book *Heat Transfer Enhancement Techniques: Thermal Performance, Optimization and Applications.*

"*This book is dedicated to all engineers, researchers, and academicians.*"

# 1

# Recent Innovation in Heat Transfer Enhancement Techniques

**Ashwani Kumar[1]*, Mukesh Kumar Awasthi[2], Nitesh Dutt[3] and Varun Pratap Singh[4,5]**

[1]*Department of Mechanical Engineering, Technical Education Department, Uttar Pradesh, Kanpur, India*
[2]*Department of Mathematics, Babasaheb Bhimarao Ambedkar University, Lucknow, India*
[3]*Department of Mechanical Engineering, COER University, Roorkee Uttarakhand, India*
[4]*Solar Thermal Energy Research Group (STERG), Department of Mechanical Engineering and Mechatronics Engineering, University of Stellenbosch, Stellenbosch, South Africa*
[5]*Department of Mechanical Engineering, School of Advanced Engineering, UPES, Bidholi, Dehradun, Uttarakhand, India*

## *Abstract*

Heat transfer plays a critical role in various industrial processes, electronic cooling systems, battery thermal management and renewable energy applications. In recent years, significant strides have been made in the field of heat transfer enhancement techniques to improve efficiency and optimize thermal management. This abstract provides a concise overview of the recent advancements in heat transfer enhancement methods, encompassing both experimental and computational approaches. Innovative materials, such as advanced nanomaterials and metamaterials, have emerged as key contributors to enhanced heat transfer. Nanofluids, which consist of nanoparticles dispersed in traditional heat transfer fluids, exhibit improved thermal conductivity, leading to more efficient heat transfer in various applications.

**Corresponding author*: drashwanikumardte@gmail.com
Ashwani Kumar: ORCID: 0000-0003-4099-935X
Mukesh Kumar Awasthi: ORCID: 0000-0002-6706-5226
Nitesh Dutt: ORCID: 0000-0002-0405-3850
Varun Pratap Singh: ORCID: 0000-0002-2148-5604

Ashwani Kumar, Nitesh Dutt and Mukesh Kumar Awasthi (eds.) Heat Transfer Enhancement Techniques: Thermal Performance, Optimization and Applications, (1–38) © 2025 Scrivener Publishing LLC

Additionally, the design and fabrication of structured surfaces with micro and nano-scale features, known as surface modifications, have proven effective in augmenting convective heat transfer. Additive manufacturing technologies have revolutionized the fabrication of intricate heat exchanger geometries, offering a high degree of customization and geometric complexity. This has led to the creation of compact and efficient heat exchangers with improved thermal performance. The integration of artificial intelligence and machine learning algorithms in heat transfer research has opened new avenues for optimization and control. These technologies enable the development of intelligent thermal management systems capable of adapting to dynamic operating conditions, thereby enhancing overall system efficiency. Integration of novel materials, innovative surface modifications, advanced computational simulations, additive manufacturing, and the incorporation of artificial intelligence plays a crucial role in heat transport enhancement.

***Keywords:*** Heat transfer, innovative techniques, artificial intelligence, advance materials, heat exchangers

## 1.1 Introduction

Heat transfer is a fundamental process that plays a crucial role in the exchange of thermal energy between systems and their surroundings. Three distinct mechanisms govern heat transfer: conduction, convection, and radiation. Each method operates in unique ways, influencing the transfer of heat through different mediums and under diverse conditions. Understanding these mechanisms is essential for comprehending the thermal behavior of materials and systems in various environments, paving the way for advancements in fields such as engineering, physics, and materials science. Figures 1.1 and 1.2 show brief descriptions of methods of heat

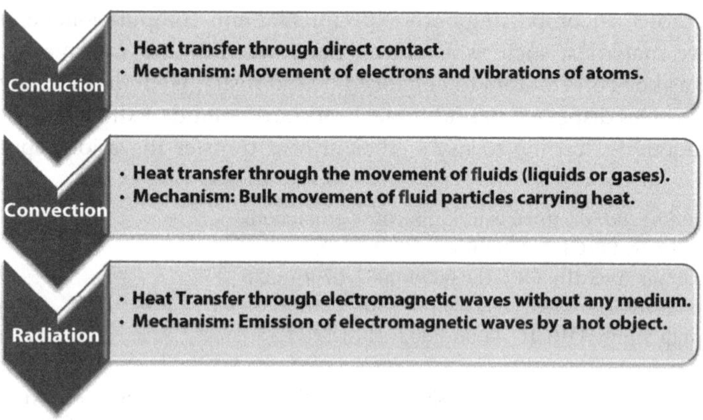


**Figure 1.1** Methods of heat transfer.

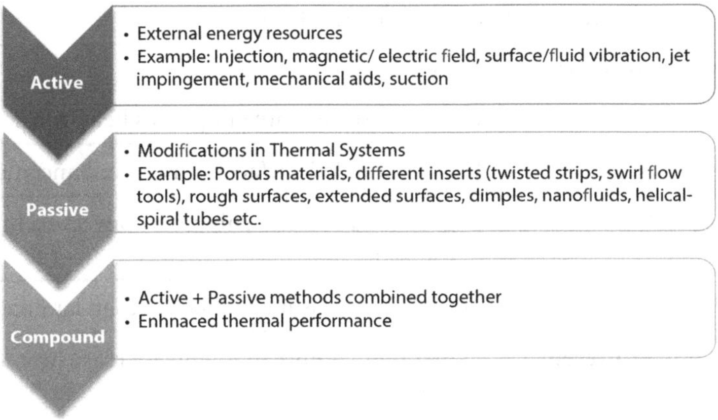


**Figure 1.2** Classifications of heat transfer methods [1–3].

transfer. Table 1.1 highlights the factors and parameters that collectively contribute to the effectiveness of heat transport enhancement in various applications such as cooling systems, heat exchangers, and thermal management devices.

Conduction, convection, and radiation are three fundamental methods of heat transfer. Conduction involves the direct transfer of heat through a material, typically a solid, by the collision of particles. In convection, heat is transferred through the movement of fluid (liquid or gas) particles, creating a cycle of warmer and cooler regions. This method is commonly observed in fluids where warmer portions rise and cooler portions sink, establishing a continuous circulation. Radiation, on the other hand, is the transfer of heat through electromagnetic waves that can propagate through a vacuum. Unlike conduction and convection, radiation does not require a medium and can occur in the absence of material substance. These three methods are crucial in various scientific and engineering applications, ranging from designing efficient heat exchangers to explaining the thermal dynamics of natural phenomena [1, 2].

### 1.1.1 Industrial Application of Heat Transfer Enhancement Techniques

Table 1.2 and Figure 1.3 highlight the key factors and different industrial applications of the heat transport enhancement techniques. Heat transfer enhancement techniques are often utilized in combination based on specific industrial requirements. Power generation, particularly in thermal power plants, the use of advanced heat exchanger designs and surfaces

**Table 1.1** Important factors and their effect [1–4].

| Important parameters | Effect |
|---|---|
| **Surface area** | Increased surface area improves heat transfer. |
| | Use of extended surfaces, fins, or enhancements. |
| **Fluid velocity** | Higher fluid velocity enhances convective heat transfer. |
| | Turbulent flow is more effective than laminar flow. |
| **Temperature difference** | Larger temperature gradients increase heat transfer. |
| | Maintaining a large ΔT across surfaces. |
| **Thermal conductivity** | Greater thermal conductivity materials transfer heat more effectively. |
| | Selection of materials in heat exchanger design. |
| **Fluid properties** | Heat transfer coefficient varies with fluid properties. |
| | Consideration of specific heat, viscosity, density, etc. |
| **Heat exchanger design** | Efficient design improves heat transfer performance. |
| | Optimization of geometry and arrangement of surfaces. |
| **Surface roughness** | Controlled roughness can enhance convective heat transfer. |
| | Impact on boundary layer and flow characteristics. |
| **Phase change** | Latent heat transfer during phase change (*e.g.*, boiling, condensation). |
| | Utilizing phase change for enhanced heat transfer. |
| **Flow regimes** | Transitioning from laminar to turbulent flow enhances heat transfer. |
| | Understanding and controlling different flow regimes. |

(*Continued*)

**Table 1.1** Important factors and their effect [1–4]. (*Continued*)

| Important parameters | Effect |
|---|---|
| **Agitation/mixing** | Stirring or mixing enhances heat transfer in fluids. |
| | Promotes better heat distribution within the fluid. |
| **Thermal resistance** | Reducing thermal resistance in the system improves heat transfer. |
| | Minimizing resistances in series and parallel paths. |
| **Heat transfer fluid** | Selection of appropriate fluid for the specific application. |
| | Consideration of thermal properties and stability. |

with enhanced thermal conductivity improves overall energy conversion efficiency. In the aerospace industry, where thermal management is critical, techniques such as finned surfaces and phase change materials aid in dissipating excess heat generated during flight. Additionally, in electronics and microelectronics manufacturing, heat transfer enhancement is vital to prevent overheating and ensure the reliability of electronic components. Furthermore, in the field of chemical processing, innovative heat exchanger configurations and materials facilitate more efficient heat transfer, reducing energy consumption and enhancing overall process productivity. Overall, the integration of heat transfer enhancement techniques is pivotal across industries, contributing to increased energy efficiency, improved system reliability, and enhanced performance in a wide range of applications.

### 1.1.2 Standards and Regulations

International standards and regulations pertaining to heat transfer are crucial for ensuring safety, efficiency, and environmental sustainability in various industries. Organizations such as the International Organization for Standardization (ISO) play a significant role in developing and maintaining these standards. ISO 22007 establishes guidelines for determining thermal conductivity, while ISO 6946 addresses heat transfer through building elements. Some general standards and organizations that are relevant to heat transfer and related fields are listed in Table 1.3. The applicability of these standards may vary depending on the industry and specific applications.

**Table 1.2** Techniques to enhance heat transfer and their industrial applications [5, 6].

| Heat transfer enhancement technique | Applications in industrial processes |
|---|---|
| **Fins and extended surfaces** | Cooling of electronic components |
| | Heat exchangers for HVAC systems |
| | Radiators in automotive engines |
| **Heat exchangers** | Chemical processing |
| | Oil and gas refining |
| | Power generation |
| **Phase change materials (PCMs)** | Thermal energy storage |
| | Solar water heaters |
| | Food processing |
| **Nanofluids** | Electronics cooling |
| | Solar collectors |
| | Biomedical applications |
| **Jet impingement** | Electronics cooling |
| | Aerospace applications |
| | Metal quenching |
| **Fluidized bed heat transfer** | Chemical and petrochemical processes |
| | Biomass and coal combustion |
| | Pharmaceutical manufacturing |
| **Thermosyphon** | Solar water heaters |
| | Refrigeration systems |
| | Heat recovery systems |
| **Vortex generators** | Heat exchangers |
| | Automotive radiators |
| | HVAC systems |

*(Continued)*

**Table 1.2** Techniques to enhance heat transfer and their industrial applications [5, 6]. (*Continued*)

| Heat transfer enhancement technique | Applications in industrial processes |
|---|---|
| **Porous media** | Electronics cooling |
| | Nuclear reactors |
| | Heat exchangers |
| **Ribbed surfaces** | Air-cooled heat exchangers |
| | Gas turbine blades |
| | Compact heat exchangers |

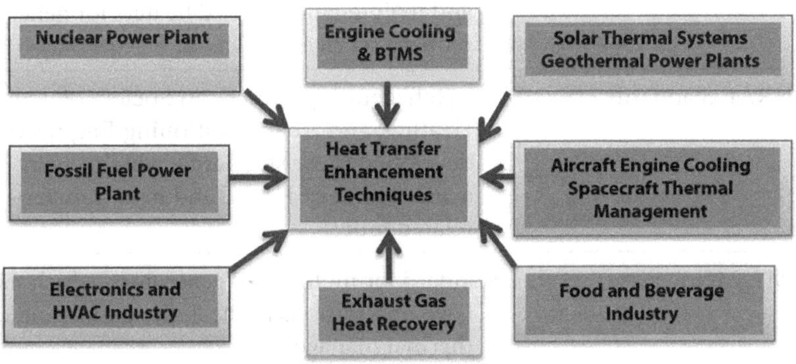


**Figure 1.3** Different industrial applications of heat transfer enhancement techniques.

## 1.2 Important Heat Transfer Enhancement Techniques and Their Effect

### 1.2.1 Effect of Fins and Extended Surfaces

The effect of fins and extended surfaces on heat transfer is a fundamental concept in thermal engineering and heat transfer. Fins are protrusions or extended surfaces attached to a heat-conducting base, such as a heat exchanger or a cooling system, designed to enhance the rate of heat transfer. The primary purpose of fins is to increase the surface area available for heat exchange, thereby improving the overall heat dissipation efficiency. This increased surface area promotes greater convection and radiation, facilitating the transfer of thermal energy from the solid base to the surrounding fluid

**Table 1.3** International standards and regulations related to heat transfer [1, 2].

| Standards and regulations | Details |
|---|---|
| ASME Boiler and Pressure Vessel Code | Contains standards for the design, fabrication, and inspection of boilers and pressure vessels, including heat exchangers. |
| ASTM International | Creates and disseminates voluntary consensus technical standards for a variety of goods, services, systems, and materials, including heat transfer-related ones. |
| ISO 22000:2018 | Lays out specifications for a food safety management system, which might take heat transfer in food processing into account. |
| API Standard 560 | Covers the design, fabrication, installation, and maintenance of fired heaters for general refinery service. |
| ASHRAE Standards | The Refrigerating, American Society of Heating, and Air-Conditioning Engineers developed these standards, which cover a wide range of HVAC and heat transfer-related topics. |
| NEMA MG 1 | Standard for motors and generators, which may have implications for heat dissipation and thermal management. |
| European Pressure Equipment Directive (PED) | Provides requirements for the design, manufacturing, and conformity assessment of pressure equipment, including heat exchangers. |
| ASCE 7 | Thermal loads are one of the minimum design loads and related requirements for buildings and other structures. |
| ANSI/IEEE C62.41 | Provides guidelines for the selection of surge arresters used in protecting electrical equipment from over-voltage transient events, which can affect heat transfer in electrical systems. |

or environment (Table 1.4). Fins are especially useful in applications where efficient heat dissipation is crucial, such as in electronic devices, radiators, and various industrial heat exchangers. The design and arrangement of fins play a significant role in optimizing heat transfer performance (Table 1.5), and engineers carefully consider factors such as fin geometry, material properties, and operating conditions to achieve the desired thermal performance.

**Table 1.4** Details of fin and extended surfaces [7, 8].

| Aspect | Fins | Extended surfaces |
|---|---|---|
| **Purpose** | Increase surface area. Length, thickness, and spacing are crucial parameters.<br>Pin fins, the diameter is an additional parameter. | Increase surface area |
| Heat transfer mechanism | Convection and radiation | Conduction, convection, and radiation |
| Primary application | Cooling electronic devices, heat exchangers | Heat exchangers, fins in engines, air-cooled systems |
| Heat transfer rate | Increases | Increases (dependent on material conductivity) |
| Surface area | Greater plane area for the dissipation of heat | Greater plane area for the dissipation of heat |
| Material selection | Good thermal conductivity (*e.g.*, aluminum) | Good thermal conductivity (*e.g.*, metals) |
| Geometry | Various shapes (Straight, pin, annular, louvered, serrated) fins<br>**Louvered fins** focus on improving airflow around the surface.<br>**Serrated fins** introduce turbulence to enhance heat transfer | Typically planar, but various shapes possible |
| Effectiveness | Highly effective in enhancing heat transfer | Effective, especially in applications with limited airflow |

**Table 1.5** Geometrical details of fins and their effects [9–12].

| Geometry | Shape | Effect |
|---|---|---|
| **Fins** | Rectangular | Higher surface area improves heat dissipation |
| | Triangular | Efficient for certain applications, enhances convective heat transfer |
| | Circular | Suitable for cylindrical structures, improves heat dissipation |
| | Annular | Effective for certain compact designs |
| **Pin fins** | Long, slender structures perpendicular to base | Enhances heat transfer in compact spaces, improves convective heat transfer |
| **Plate fins** | Flat surfaces attached to the base material | Effective for applications with limited space, enhances convective heat transfer |
| **Louvered fins** | Fins with spaced openings or slots | Improves airflow and convective heat transfer, suitable for forced convection |
| **Serrated fins** | Fins with serrations or notches | Enhances heat transfer by disrupting boundary layer, effective in forced convection |
| **Wavy fins** | Fins with a wavy or sinusoidal profile | Increases surface area and disrupts boundary layer, enhances convective heat transfer |
| **Extended tubes** | Tubes with external fins or ribbed surfaces | Increases heat transfer area, improves convective heat transfer in tubes |
| **Microfins** | Extremely small fins on a microscale | Enhances heat transfer in microchannels, suitable for miniaturized devices |
| **Spine fin** | Central spine with fins extending radially | Enhances heat transfer in both radial and axial directions |
| **Swirl fins** | Fins with a swirling or helical pattern | Enhances heat transfer through increased turbulence, suitable for certain applications |

### 1.2.2 Effect of Phase Change Materials (PCMs)

Phase change materials (PCMs) show unique thermal and mechanical properties that make them integral in various heat transfer applications (Table 1.6). Phase change materials' thermal properties involve their ability to absorb or release latent heat during phase transitions, providing a stable temperature within a defined range. The mechanical properties of PCMs, such as solid-liquid transition, enable them to undergo reversible phase changes without significant degradation. In heat transfer enhancement, PCMs play a crucial role by efficiently storing and releasing thermal energy, acting as thermal buffers to regulate temperature variations (Table 1.7). This is particularly beneficial in maintaining a constant temperature in applications like building insulation, electronics cooling, and renewable energy systems. The industrial applications of PCMs extend to diverse fields, including energy storage, heating, ventilation, and air

**Table 1.6** PCM thermal and mechanical properties.

| PCM | Thermal properties | Mechanical properties |
|---|---|---|
| Paraffin waxes | Melting range: 20°C to 70°C | Soft and pliable at room temperature |
| Bio-based PCMs | Variable depending on the specific material | Mechanical properties depend on the source and process |
| Salt hydrates | Melting range: 0°C to 100°C | Brittle at low temperatures, can be encapsulated |
| Metallic PCMs | High thermal conductivity, variable melting temperatures | Mechanical properties depend on the specific metal |
| Eutectic alloys | Sharp melting point, good thermal stability | Mechanical properties depend on the specific alloy |
| Polymer-based PCMs | Variable depending on the polymer used | Mechanical properties depend on the specific polymer |
| Inorganic PCMs | Variable depending on the specific material | Mechanical properties vary; some may be brittle |
| Carbon-based PCMs | High thermal conductivity, tunable properties | Mechanical properties depend on the carbon structure |

**Table 1.7** Role of PCM in heat transfer enhancement [12, 70–75].

| Role | Details |
|---|---|
| Thermal energy storage | PCM absorbs and releases heat energy through phase change (solidification and melting), providing a high latent heat capacity for effective energy storage. |
| Temperature regulation | PCM helps regulate temperatures by maintaining a nearly constant temperature during phase change, preventing temperature spikes. |
| Heat transfer enhancement | PCM improves heat transfer efficiency by storing and releasing energy at a constant temperature, leading to a more uniform and controlled thermal profile. |
| Thermal mass | PCM adds thermal mass to systems, which helps in stabilizing temperature variations and reducing temperature fluctuations. |
| Energy conservation | PCM aids in conserving energy by storing excess heat during peak periods and releasing it during low-demand or off-peak periods. |
| Thermal comfort improvement | In buildings and textiles, PCM can contribute to improved thermal comfort by absorbing and releasing heat to maintain a comfortable environment. |
| Increased efficiency in cooling | PCM can enhance the efficiency of cooling systems by absorbing heat during the day and releasing it during the night when ambient temperatures are lower. |
| Reduced energy consumption | The use of PCM can lead to reduced energy consumption in heating and cooling systems, as it allows for better utilization of energy and reduces peak energy demands. |
| Temperature smoothing | PCM helps in smoothing out temperature variations by absorbing and releasing heat, contributing to a more stable and controlled thermal environment. |
| Passive cooling applications | PCM finds applications in passive cooling systems, such as building materials and textiles, where it helps maintain comfortable temperatures without active energy input. |

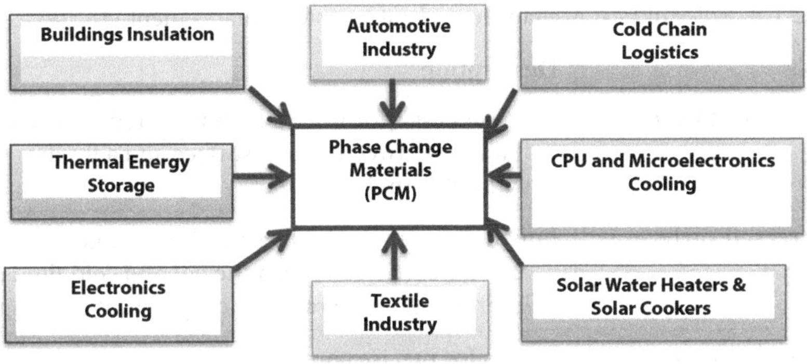


**Figure 1.4** Different industrial applications of phase change materials.

conditioning (HVAC) systems, and thermal management in electronics, showcasing their versatility and significance in addressing heat transfer challenges across various sectors (Figure 1.4).

### 1.2.3 Effect of Heat Exchangers

Heat exchangers are devices designed to efficiently transfer thermal energy between two or more fluids, and their effectiveness is crucial for optimizing processes that involve heat exchange. These devices play a vital role in improving the overall efficiency of heat exchange systems. Heat exchangers facilitate the transfer of heat from one fluid to another, either for the purpose of heating or cooling, and their design and functionality can be tailored to enhance heat transfer rates. This enhancement is achieved through features such as increased surface area, improved fluid flow patterns, and the incorporation of advanced materials (Table 1.8). By employing heat exchangers with enhanced heat transfer capabilities, industries can achieve better energy efficiency, reduced operating costs, and improved overall performance in various thermal processes.

### 1.2.4 Effect of Microchannels

For improving heat transport in a variety of engineering applications, microchannels are vital. Heat exchangers, electronic cooling systems, and other thermal management devices use these channels, which usually have dimensions in the micrometer range [14]. There are various reasons why microchannels have such a large impact on heat transfer enhancements:

1. Increased surface area: because of their high plane area to volume ratio, microchannels help the working fluid and the

**Table 1.8** Roles of heat exchangers in heat transfer enhancement [13, 79].

| Role | Description |
|---|---|
| **Transfer of heat** | Makes it easier for heat to be transferred between fluids or between a solid surface and a fluid without combining them. |
| **Increased surface area** | Enhances heat transfer by providing a larger surface area for contact between the hot and cold fluids, improving the overall efficiency of heat exchange. |
| **Temperature control** [71, 72] | Allows for precise control of temperatures in industrial processes, HVAC systems, and various applications by regulating the flow rates of the hot and cold fluids. |
| **Energy efficiency** | Improves the overall energy efficiency of systems by recovering and utilizing waste heat, reducing the need for additional energy input. |
| **Phase change applications** | Used in applications involving phase changes (*e.g.*, condensation or evaporation), contributing to efficient heat transfer in processes such as refrigeration and power generation. |
| **Heat recovery** | Captures and reuses heat that would otherwise be lost, minimizing energy waste and reducing operational costs. Common in industrial processes and power generation. |
| **Thermal integration** | Integrates with other systems for efficient heat management, allowing for the optimization of processes and reducing environmental impact. |
| **Preventing fouling and scaling** | Helps prevent the buildup of deposits, fouling, and scaling on heat exchange surfaces, ensuring continuous and efficient heat transfer over time. |
| **Enhanced heat transfer surfaces** | Incorporates advanced surface geometries, materials, or coatings to increase heat transfer rates, promoting efficient thermal performance. |
| **Compact design** | Enables the design of compact systems, reducing space requirements while maintaining effective heat transfer, crucial in applications with limited space availability. |

(*Continued*)

**Table 1.8** Roles of heat exchangers in heat transfer enhancement [13, 79]. (*Continued*)

| Role | Description |
|---|---|
| **Fluid mixing control** | Aids in controlling and minimizing undesirable mixing of fluids, maintaining the integrity of each fluid stream during heat exchange. |
| **Maintenance ease** | Designed to be easily maintainable, allowing for inspection, cleaning, and repairs to ensure long-term and reliable operation. |
| **Versatility** | Adaptable to a wide range of industries and applications, including chemical processing, power generation, HVAC, automotive, and more. |
| **Temperature homogenization** | Contributes to achieving uniform temperatures within a fluid or system, important in processes where temperature consistency is critical. |
| **Safety** | Plays a role in maintaining safe operating temperatures in various systems, preventing overheating or excessive cooling. |

channel walls make better contact. Higher heat transfer rates are encouraged by the more effective heat exchange made possible by the larger surface area.

2. Reduced thermal resistance: microchannels have less thermal resistance than larger channels or traditional heat exchangers because of their smaller dimensions. A lower thermal resistance results in faster heat dissipation and better heat transfer efficiency overall.
3. Enhanced convective heat transfer: forced convection is encouraged by microchannels, and fluid flow displays special properties at the microscale, such as higher heat transport coefficients. The fluid's high velocity within the microchannels facilitates convective heat transfer, increasing the process's efficiency.
4. Increased heat transfer coefficient: the heat transport coefficients of microchannels are generally higher than those of macroscale channels. Increased surface area, decreased thermal resistance, and improved convective heat transfer work together to produce the intensified heat transfer coefficients.

5. Phase change heat transfer: microchannels are also utilized in applications relating phase change heat transfer, such as condensation and boiling. The small dimensions of microchannels facilitate rapid phase change processes, leading to improved heat transfer rates.
6. Compact design and space efficiency: microchannels enable the design of compact heat exchangers, making them suitable for applications with limited space. The compact design also reduces the thermal inertia of the system, allowing for faster response to changes in heat load.

In summary, microchannels contribute to heat transfer enhancements through their ability to increase surface area, reduce thermal resistance, promote convective heat transfer, and facilitate efficient phase change processes. These advantages make microchannels valuable in a variety of applications where effective thermal management is critical [15–17].

### 1.2.5 Effect of Nanofluids

The amazing heat transfer capabilities of nanofluids—engineered colloidal suspensions of nanoparticles in a base fluid—have attracted a lot of attention. When compared to their base fluids, these nanofluids have higher convective heat transport coefficients and thermal conductivities. Improved heat transfer performance is the outcome of changing the thermal characteristics of the base fluid by adding nanoparticles, such as metal oxides, carbon-based compounds, or other nanoparticles [19, 20]. Because of their enhanced thermal conductivity, nanofluids are good candidates for a variety of heat transfer applications because of their ability to carry heat efficiently [21–23]. Colloidal combinations of metallic or ceramic particles of nano-metric sizes suspended in a base fluid, like oil, ethylene glycol, or water, are known as nanofluids. In recent years, a great deal of research has been done on the application of nanofluids in heat transfer improvement strategies [24–26].

Particle composition, size, shape, concentration, and type of base fluid are among the variables that affect how well heat transfer is enhanced by nanofluids [27]. The degree of the thermal conductivity increase is dependent on the size of the nanoparticles; the largest increase was noted in nanofluids containing high aspect ratio nanoparticles [25]. To further optimize the nanofluid, a mixture of particles can be used to increase thermal conductivity without producing excessive viscosity. According to the results that are currently available in the literature, raising the heat exchanger's flow rate could improve heat transfer more than adding nanoparticles could [24]. Furthermore, a standard for impartially assessing the impact of adding nanoparticles to working fluids is needed for the use of nanofluids in heat transport enhancement approaches.

One of the significant advantages of nanofluids in heat transport techniques is their ability to improve heat transfer rates. Due to their enhanced thermal conductivity, nanofluids can effectively dissipate heat, leading to better thermal performance in heat exchangers, refrigeration systems, and cooling applications. The nanoparticles in nanofluids can significantly modify the convective heat transport characteristics, promoting better heat transfer efficiency. Their use can contribute to more compact and efficient heat exchange systems, leading to improved energy efficiency and reduced operational costs in various industries.

However, despite their promising potential, challenges exist in the practical implementation of nanofluids. Issues related to nanoparticle stability, aggregation, and the cost of producing stable nanofluids at large scales are areas that require further research and development. Nonetheless, ongoing studies continue to explore and optimize nanofluid formulations and application techniques to harness their potential benefits for advancing heat transfer technologies.

It has been discovered that nanofluids have greater heat transfer coefficients than their base fluids, which makes them more efficient cooling agents in a range of circumstances. However, there are many drawbacks to using nanofluids in real-world heat exchangers, such as agglomeration, reduced heat transfer, and the expensive price of traditional nanoparticles. The type of base fluid, concentration, size, shape, and composition of the particles all affect how well heat transfer is enhanced by nanofluids. A higher order of magnitude improvement in heat transfer could be achieved by increasing the flow rate inside the heat exchanger than by adding nanoparticles.

### *1.2.5.1 Analytical Approaches to Understanding Physical Characteristics of Nanofluids*

#### 1.2.5.1.1 Density

The density $\rho_n$ of nanofluids depends upon the base fluid's density and nanoparticle's density of nanoparticles. The expression of density can be represented as (Awasthi *et al.* [28, 29], Agarwal *et al.* [30]):

$$\rho_n = \phi \rho_p + (1-\phi)\rho_f \tag{1.1}$$

In this context, $\rho_f$ denotes the base fluid's density, $\rho_p$ signifies the density of the metal particles, and $\phi$ is the volume fraction of nanoparticle.

To assess the accuracy of Equation 1.1, experimental investigations were conducted separately by Pak and Cho [31] as well as by Ho *et al.* [32] to determine the room temperature's density of $Al_2O_3$–water nanofluids. Their findings showcased remarkable conformity between the experimental data and the predictions made through Equation 1.1.

Engineering computations heavily rely on the physical properties observed at various temperatures. Ho *et al.* [32] conducted density measurements of $Al_2O_3$–water nanofluid across different temperature ranges and varying nanoparticle volume fractions. In this study, a correlation for the density of $Al_2O_3$–water nanofluid was established, utilizing the experimental data from Ho *et al.* [32], factoring in temperature and nanoparticle volume fraction. The newly formulated correlation can be denoted as

$$\rho_n = 1001.064 + 2738.6191\phi - 0.2095T;\ \ 0.04 \geq \phi \geq 0, \quad 40°C \geq T \geq 5°C \tag{1.2}$$

With a regression $R^2$ of 99.97% and a maximum relative error of 0.22%, this correlation exhibits high precision.

#### 1.2.5.1.2 Viscosity

The expression of viscosity of nanofluids can be written as

$$\mu_n = \mu_{ef}\left(1 - \frac{\phi_{ag}}{\phi_m}\right)^{-[\delta]\phi_m} \tag{1.3}$$

Here, $\mu_{ef}$ denotes the base fluid's density, $\phi_m$ the maximum volume fraction that can be attained for spheroidal nanoparticles, and $\delta$ the shape parameter of the nanoparticles. According to Agarwal *et al.* [30], the equivalent volume fractions of the nanoparticles, represented as $\phi_{ag}$, and the fractal index $\phi_{mod}$ can be used to express the aggregate volume fraction of the nanoparticles, denoted as $d$.

$$\phi_{ag} = \phi_{\text{mod}}\left(\frac{r_a}{r}\right)^{3-d} \tag{1.4}$$

The equivalent volume fractions of the nanoparticles, represented by the $\phi_{mod}$, can be expressed in this context in terms of the volume fraction ($\phi$), the thickness of the interfacial layer ($\gamma$), and the lengths of the semi-major and semi-minor axes ($b$ and $a$), respectively.

$$\phi_{\text{mod}} = \phi\left(1+\frac{\gamma}{a}\right)\left(1+\frac{\gamma}{b}\right)^2 \tag{1.5}$$

### 1.2.5.1.3 Heat Capacity Of Nanofluids

The determination of nanofluid's specific heat $c_n$ involves presuming a state of thermal balance between the nanoparticles and the base fluid phase.

$$c_n = \frac{\phi\rho_p c_p + (1-\phi)\rho_f c_f}{\rho_n} \tag{1.6}$$

While some of the researchers are using the following expression:

$$c_n = \phi c_p + (1-\phi)c_f \tag{1.7}$$

Here $c_f$ denotes the specific heat of the base fluid, $c_p$ signifies the specific of the metal particles.

Using both equations to compare the specific heat of the $Al_2O_3$–water nanofluid at room temperature to the experimental results of Zhou and Ni [33] at various volume fractions x, it is clear that the specific heat of the nanofluid falls with increasing volume fraction of nanoparticles, in accordance with the patterns shown in Equations 1.6 and 1.7.

### 1.2.5.1.4 Natural Heat Convection In Nanofluids

There are contradictory reports in the literature regarding the use of nanofluids to improve natural convection heat transfer. The results obtained from analytical and experimental research continue to differ. In contrast to experimental results, analytical studies indicate that heat transfer increases as the volume fraction of nanoparticles rises. The ratio between the Rayleigh numbers of the nanofluid and the base fluid is provided because of the significance of the Rayleigh number in natural convection, which represents the ratio of buoyancy to viscous forces.

$$\frac{Ra_n}{Ra_f} = \frac{\beta_n}{\beta_f}\frac{\nu_f}{\nu_n}\frac{\alpha_f}{\alpha_n} \tag{1.8}$$

### 1.2.6 Effect of Porous Media

Porous media refer to materials that possess a complex structure, consisting of a solid matrix with pores or voids. These pores can be natural or artificial and can vary in size, shape, and distribution within the medium. In many different applications, such as power plants, heat and mass transfer devices, and heat pumps, porous media—which are frequently found in nature, such as soil, rock, and biological tissues—are employed [34]. Here is a summary of how porous media affects heat transfer:

1. Thermal dispersion coefficient: the presence of pores and the porous complex structure lead to thermal dispersion, which is the spreading of heat due to the tortuous flow paths within the medium [35]. The thermal dispersion coefficient is a key parameter for understanding heat transfer in porous media.
2. Porosity: because it influences both the surface area available for heat exchange and the flow resistance, the porosity of the medium is an important factor in heat transfer. Lower flow resistance and greater heat transfer are typically the outcomes of higher porosity. Convective heat transfer coefficient: measuring the efficiency of convective heat transfer within the medium, the internal convective heat transfer coefficient is a crucial parameter in the study of heat transfer through porous media [37–39]. Cryogenic conductors are useful in enclosures and can enhance heat transfer.
3. Multiscale processes: heat and mass transfer in porous media involve multiscale processes that involve momentum and energy [35]. These processes can be complex and depend on various factors, such as the pore size, shape, and distribution, as well as the thermal conductivity of the medium.
4. Geometrical shapes: numerous geometrical forms, including cavities, vertical plates, and cylindrical shapes, can be used to study heat transport in porous media [36]. When examining heat transport in porous media, these shapes should be taken into account as they have the potential to affect the process.
5. Computational fluid dynamics: porous media conditions have been incorporated into computational fluid dynamics

simulations to predict heat transfer in porous media. These simulations take into account the porosity of the medium and the time-derivative term in the scalar transport equation to model heat transfer enhancement in porous media.

Because of their intricate structure, which affects convective heat transfer, thermal dispersion, porosity, and multiscale processes, porous media can have a substantial impact on heat transfer. For the purpose of maximizing the efficiency of a variety of applications, including power plants, heat and mass transfer devices, and heat pumps, it is imperative to comprehend these factors and how they affect heat transfer.

#### *1.2.6.1 Effect of Porosity*

The porosity of a medium significantly affects heat transfer due to its impact on various factors, such as thermal conductivity, flow resistance, and surface area available for heat exchange. Based on the search results, the following points highlight how porosity influences heat transfer:

1. Apparent thermal conductivity: a medium's apparent thermal conductivity rises with decreasing porosity, increasing the rate of heat transfer [40]. This is so that more heat can pass through the medium because of the decreased porosity, which causes less heat to be absorbed by the medium.
2. Fabric porosity: fabric porosity plays a major role in the heat transfer process in the context of porous materials, which in turn affects moisture transfer [35]. This implies that the efficiency of heat transfer is highly dependent on the structure and characteristics of the medium, including the pore size distribution and shape.
3. Heat exchanger performance: a heat exchanger's performance may vary if a porous medium is present because this could impact the heat transfer process [37]. By lengthening the heat exchanger's wavelength and decreasing its wave amplitude, a medium with a higher porosity can enhance heat transfer.

In summary, a medium's porosity affects thermal conductivity, flow resistance, and the amount of surface area that can be used for heat exchange, all of which are critical components of heat transfer. It is crucial to comprehend how porosity affects heat transfer in order to maximize the efficiency of a variety of applications, including heat pumps, power plants, and mass and heat transfer devices.

### 1.2.7 Effect of Jet Impingement

A high-velocity fluid jet is used in the heat transfer process known as "jet impingement" to transfer heat to a target surface [18]. Numerous engineering applications, including the cooling of tempering glass, hot steel plates, turbine blades, electronic components of turbines, and the drying of film and paper, make extensive use of this technique. In jet impingement, the heat transfer coefficient is commonly expressed as $h(x) = q/(T_w - T_j)$, where ($q$) represents the wall heat flux, ($T_w$) denotes the local wall temperature, and ($T_j$) represents the jet temperature [41–46].

By confined flow parallel to the cooled surface, the impinging jet generates heat transfer coefficients that are significantly higher than those obtained by traditional convection cooling. This is due to the fact that the impingement boundary layers are substantially thinner, and the higher heat transfer rates are caused by the spent flow that remains after the impingement and turbulences the surrounding fluid [45, 46].The heat transfer from a surface to jet flow is highly variable, and the properties of jet impingement heat transfer are extremely complex. The flow condition at the nozzle exit, flow Reynolds number, jet type (conventional/swirled), jet-to-plate spacing, jet inclination, and surface type are some of the factors that affect the heat transfer coefficient [46]. Because of the direct impact of free-stream turbulence on the boundary layer as well as the faster mixing of the fluid close to the surface, turbulent jets have higher heat transfer coefficients. Liquid jet impingement is appealing because of its high heat transfer coefficients in circumstances where a high heat load needs to be eliminated while preserving a minimum temperature or temperature differential within the system [47].

With its high local and average heat transfer coefficients, jet impingement is an efficient heat transfer method that can be applied to a variety of engineering applications. Many factors affect the heat transfer coefficient in jet impingement, and the method is especially appealing in circumstances where it is necessary to remove large heat loads while preserving low surface temperatures.

### 1.2.8 Effect of Heat Pipes

Heat pipes consist of a sealed, evacuated tube containing a working fluid, typically in a liquid state. When one end of the heat pipe is subjected to heat, the fluid evaporates, forming vapor that travels to the cooler end of the pipe. At the cooler end, the vapor condenses back into liquid, releasing latent heat. This phase change process facilitates rapid heat transfer, as it involves the absorption and release of energy during the phase transitions. Heat pipes effectively transport heat over long distances with minimal temperature gradients, making them highly efficient in dissipating heat from hot spots [50]. Their ability to operate passively, without the need for external power, makes them versatile and widely used in electronics, aerospace, and various industrial applications for optimizing thermal management and improving overall system performance (Table 1.9).

### 1.2.9 Effect of Vortex Generators

Vortex generators are devices designed to enhance heat transfer in various engineering applications, such as heat exchangers, cooling systems, and aerodynamics. They work by inducing swirling vortices in the fluid flow, which can lead to improved heat transfer rates [51]. The effects of vortex generators on heat transfer enhancements can be summarized as follows:

1. Turbulence enhancement: vortex generators generate controlled turbulence in the fluid flow, which disrupts the boundary layer. Increased turbulence enhances the mixing of fluid layers with different temperatures, promoting better heat transfer between the fluid and the solid surface.
2. Boundary layer disruption: vortex generators disrupt the boundary layer, which is the thin layer of fluid near the solid surface with a slower flow than the bulk fluid. Disrupting the boundary layer helps prevent the formation of a stagnant layer of fluid, promoting better convective heat transfer.
3. Heat transfer enhancement: the induced vortices can lead to increased convective heat transfer coefficients, resulting in more effective cooling or heating of the surface. The enhanced heat transfer can be particularly beneficial in applications where efficient thermal management is crucial.
4. Delay of flow separation: vortex generators can delay flow separation, which is the detachment of the boundary layer from the surface. Delaying separation helps maintain attached flow, improving the efficiency of heat transfer and reducing the likelihood of heat transfer deterioration.

**Table 1.9** Roles of heat pipes in heat transfer enhancement [48, 49].

| Parameter | Role of heat pipes |
|---|---|
| **Mechanism** | Heat pipes operate on the principle of phase change. They use the evaporation and condensation of a working fluid. |
| **Heat conduction** | Heat pipes provide highly efficient heat conduction, often surpassing traditional methods like solid conductors. |
| **Thermal conductivity** | Heat pipes exhibit high effective thermal conductivity, allowing for rapid heat transfer over long distances. |
| **Temperature control** | Heat pipes help in maintaining uniform temperatures by quickly distributing heat throughout the system. |
| **Cooling efficiency** | They enhance cooling efficiency by efficiently transporting heat away from the source, preventing overheating. |
| **Passive operation** | Heat pipes are passive devices, requiring no external power. They rely on natural convection and gravity for operation. |
| **Reduced weight** | Compared to traditional cooling methods, heat pipes are lightweight, making them advantageous in weight-sensitive applications. |
| **Versatility** | Heat pipes can be shaped and oriented as needed, offering flexibility in design for various applications. |
| **Thermal management** | They play a crucial role in thermal management, especially in electronic devices and aerospace applications. |
| **Reliability** | Heat pipes have no moving parts, contributing to their reliability and long operational life. |
| **Heat recovery** | Heat pipes can be used for heat recovery applications, transferring waste heat to areas where it can be utilized. |
| **Space applications** | Heat pipes are commonly used in space applications due to their ability to function in microgravity environments. |

5. Improved mixing: vortices generated by vortex generators facilitate the mixing of fluid at different temperatures, promoting a more uniform temperature distribution. Uniform temperature distribution is essential in heat exchangers and other systems to avoid hot spots and improve overall efficiency.
6. Reduced drag: in aerodynamic applications, vortex generators can be strategically placed to reduce drag and improve the overall performance of the system. The reduction in drag can contribute to more efficient heat transfer in aircraft, automobiles, and other vehicles.
7. Optimization challenges: designing an effective vortex generator configuration requires careful consideration of factors such as size, shape, and placement. Optimization is essential to achieve the desired balance between enhanced heat transfer and the associated pressure drop.

In summary, vortex generators can significantly enhance heat transfer by promoting turbulence, disrupting the boundary layer, delaying flow separation, and improving mixing. The application of vortex generators is diverse, ranging from heat exchangers and cooling systems to aerodynamic surfaces in vehicles and aircraft. However, achieving optimal performance requires careful design and consideration of the specific requirements of the system in question [51–53].

### 1.2.10 Effect of Ribbed Surfaces

The presence of ribs on a surface creates turbulence in the fluid flow, leading to increased heat transfer rates. This is attributed to the disruption of the boundary layer, which reduces thermal resistance. The ribs act as turbulators, intensifying the convective heat transfer process by promoting the formation of vortices and enhancing fluid mixing. Additionally, ribbed surfaces often result in increased surface area, facilitating greater contact between the fluid and the solid surface, further augmenting heat transfer (Table 1.10).

### 1.2.11 Effect of Artificial Roughness-Based Turbulence

The use of artificial roughness to enhance heat transfer is a well-studied aspect in fluid dynamics and thermal engineering. The effects of artificial

**Table 1.10** Roles of ribbed surfaces in heat transfer enhancement [51–57].

| Parameter | Details |
|---|---|
| Geometry | Ribbed surfaces introduce protrusions or depressions on the heat transfer surface, altering its topography. Common geometries include rectangular, triangular, or sinusoidal ribs. |
| Turbulence | Ribs disrupt the flow of the fluid, promoting turbulence. Increased turbulence enhances heat transfer by improving the mixing of fluid layers and reducing the thickness of the thermal boundary layer. |
| Heat transfer coefficient | Ribbed surfaces generally lead to an increase in the overall heat transfer coefficient. The enhanced turbulence and increased surface area contribute to improved convective heat transfer. |
| Pressure drop | The introduction of ribs may result in higher pressure drops due to increased fluid resistance. Balancing heat transfer improvement with pressure drop considerations is crucial in practical applications. |
| Reynolds number | Heat transfer enhancement is often more pronounced at higher Reynolds numbers. Below a certain Reynolds number, the benefits of ribbed surfaces may be limited. |
| Thermal performance factor | This factor compares the enhancement in heat transfer to the increase in pressure drop. A higher thermal performance factor indicates a more efficient heat transfer enhancement. |
| Applications | Ribbed surfaces are commonly employed in heat exchangers, cooling channels, and electronic devices to enhance heat dissipation in Industry 4.0. The choice of rib geometry depends on the specific application and flow conditions [76–80]. |
| Challenges | Challenges associated with ribbed surfaces include manufacturing complexities, increased pressure drop, and the potential for fouling in certain applications. |

(*Continued*)

**Table 1.10** Roles of ribbed surfaces in heat transfer enhancement [51–57]. (*Continued*)

| Parameter | Details |
|---|---|
| Optimization | Optimal rib design involves a trade-off between heat transfer enhancement and pressure drop. Computational fluid dynamics (CFD) simulations and experimental studies are often conducted to optimize rib parameters. |
| Materials | The material of the ribbed surface plays a role in heat transfer performance. Materials with high thermal conductivity facilitate efficient heat transfer [81–85]. |
| Research trends | Ongoing research explores advanced rib configurations, materials, and surface treatments to further enhance heat transfer while minimizing associated drawbacks. |

roughness on heat transfer enhancement depend on various parameters, including the type of roughness, Reynolds number, pitch, height, and other geometrical and flow-related parameters [90] (Table 1.11). Here is a tabulated form summarizing some of these effects:

## 1.3 Numerical Analysis of Heat Transfer Problem

The simulation of heat transfer problems through numerical methods often involves the use of various software tools. Here is a tabulated list of some commonly used software for solving heat transfer problems through numerical simulation (Table 1.12).

The choice of software depends on the specific requirements of the heat transfer problem, the user's familiarity with the software, and the computational resources available. Different software tools may have varying levels of complexity, simulation capabilities, and user interfaces.

**Table 1.11** Effect of artificial roughness-based turbulence on heat transfer [58–62].

| Parameters | Details |
|---|---|
| Type of roughness | Different types (*e.g.*, ribbed, dimpled, vortex generators) have varied effects on heat transfer. |
| Reynolds number | Higher Reynolds numbers generally lead to increased heat transfer enhancement due to improved mixing and turbulence. |
| Roughness height | Increased roughness height generally results in higher heat transfer enhancement, but there is an optimal range. |
| Pitch | Pitch affects the flow between roughness elements; optimal spacing is critical for maximizing heat transfer enhancement. |
| Orientation of roughness | The orientation of roughness elements influences the flow patterns and heat transfer; optimal orientations vary. |
| Flow inlet conditions | Inlet velocity and temperature conditions affect the onset and intensity of heat transfer enhancement. |
| Fluid properties | Viscosity and thermal conductivity of the working fluid influence the heat transfer performance. |
| Surface material | Manufacturing Industry, The thermal conductivity and surface characteristics of the material used for roughness elements impact heat transfer [86–89]. |
| Surface roughness distribution | Uniform or non-uniform distribution of roughness elements affects the overall heat transfer enhancement. |
| Surface curvature | Curved surfaces with artificial roughness may exhibit different heat transfer characteristics compared to flat surfaces |
| Aspect ratio roughness | The ratio of the height to the characteristic length influences the heat transfer enhancement; optimal ratios exist. |
| Turbulence intensity | Higher turbulence levels generally enhance heat transfer, and artificial roughness can contribute to increased turbulence. |

**Table 1.12** List of numerical approaches used for heat transfer analysis [62–70].

| Software package | Details | Features | Application |
|---|---|---|---|
| **COMSOL Multiphysics** | Multiphysics simulation software that includes modules for heat transfer, fluid flow, and structural mechanics. | Finite element method (FEM), heat transfer module, CFD module | General-purpose heat transfer simulations, including fluid flow and structural interactions. |
| **ANSYS Fluent** | CFD software for simulating fluid flow and heat transfer. | Finite volume method (FVM), conjugate heat transfer (CHT) | Applications in fluid dynamics and heat transfer in various engineering domains. |
| **OpenFOAM** | Open-source CFD software with a focus on flexibility and extensibility. | Finite volume method (FVM) | Widely used for research and industrial applications in fluid dynamics and heat transfer. |
| **Abaqus** | Finite element analysis (FEA) software for structural and thermal simulations. | Finite element method (FEM), heat transfer module | Suitable for coupled thermal-structural analysis in various engineering fields. |
| **FLUENT** | Part of the ANSYS suite, it specializes in CFD simulations, including heat transfer. | Finite volume method (FVM), conjugate heat transfer (CHT) | Used for a wide range of heat transfer applications in fluid systems. |

(*Continued*)

**Table 1.12** List of numerical approaches used for heat transfer analysis [62–70]. (*Continued*)

| Software package | Details | Features | Application |
|---|---|---|---|
| **MATLAB** | Numerical computing environment with specialized toolboxes for heat transfer simulations. | Finite difference method (FDM), finite element method (FEM) | Suitable for academic and research purposes, offering flexibility in algorithm development. |
| **SolidWorks Simulation** | Integrated within the solidworks computer aided design (CAD) platform, it provides FEA capabilities for thermal analysis. | Finite element method (FEM), thermal analysis module | Used for thermal studies of components and assemblies in the design phase. |
| **HEC-RAS** | River and channel hydraulics simulation software that includes heat transfer capabilities. | 1D Finite element method (FEM) | Applied to simulate river and reservoir thermal stratification. |
| **Thermal Desktop** | Thermal analysis software for spacecraft, electronics, and other applications. | Finite element method (FEM), heat transfer module | Particularly used in aerospace and electronics industries for detailed thermal analysis. |

## 1.4 Conclusion

In conclusion, the dynamic landscape of heat transfer enhancement reflects a remarkable convergence of diverse technologies and methodologies. From the utilization of cutting-edge materials like nanomaterials and metamaterials to the ingenious design of structured surfaces and the advent of additive manufacturing, the field has witnessed transformative advancements [91–93]. The infusion of artificial intelligence and machine learning further propels heat transfer research into an era of intelligent thermal management, offering adaptability to dynamic conditions and heightened system efficiency. As we continue to integrate novel materials, innovative surface modifications, advanced computational simulations, and smart technologies, the collective impact on heat transfer enhancement is undeniable, fostering a promising trajectory for the optimization of industrial processes, electronic cooling systems, and renewable energy applications [94, 95]. This comprehensive exploration signifies not only the present achievements but also the ongoing evolution that promises a future of heightened efficiency and innovation in the realm of heat transfer.

## References


1. Bergles, A.E., Techniques to Enhance Heat Transfer, in: *Handbook of Heat Transfer*, McGraw-Hill, New York, NY, USA, 1998.
2. Webb, R.L. and Kim, N.H., *Principles of Enhanced Heat Transfer*, 2nd ed., Taylor & Francis, Boca Raton, FL, USA, 2005.
3. Léal, L., Miscevic, M., Lavieille, P., Amokrane, M., Pigache, F., Topin, F., Nogarède, B., Tadrist, L., An overview of heat transfer enhancement methods and new perspectives: Focus on active methods using electroactive materials. *Int. J. Heat Mass Transf.*, *61*, 505–524, 2013.
4. Rashidi, S., Eskandarian, M., Mahian, O., Poncet, S., Combination of nanofluid and inserts for heat transfer enhancement. *J. Therm. Anal. Calorim.*, *135*, 437–460, 2019.
5. Bergles, A.E., Recent developments in enhanced heat transfer. *Heat Mass Transf.*, *47*, 1001–1008, 2011.
6. Gugulothu, R., Reddy, K.V.K., Somanchi, N.S., Adithya, E.L., A review on enhancement of heat transfer techniques. *Mater. Today Proc.*, *4*, 1051–1056, 2017.
7. Maradiya, C., Vadher, J., Agarwal, R., The heat transfer enhancement techniques and their thermal performance factor. *Beni-Suef Univ. J. Basic Appl. Sci.*, *7*, 1–21, 2018.
8. Guo, Z., A review on heat transfer enhancement with nanofluids. *J. Enhanced Heat Transf.*, *27*, 1–70, 2020.
9. Bergles, A.E., Jensen, M.K., Somerscales, E.F.C., Manglik, R.M., *Literature Review of Heat Transfer Enhancement Technology for Heat Exchanges in Gas-Fired Applications*, Technical Report ACRC TR-105; Gas Research Institute, Chicago, IL, USA, 1991.
10. MousaviAjarostaghi, S.S., Zaboli, M., Javadi, H., Badenes, B., Urchueguia, J.F., A Review of Recent Passive Heat Transfer Enhancement Methods. *Energies*, *15*, 986, 2022, https://doi.org/10.3390/en15030986.

11. Alva, G., Lin, Y., Fang, G., An overview of thermal energy storage systems. *Energy*, *144*, 341–378, 2018.
12. Wang, J., Xie, H., Guo, Z., Cai, L., Zhang, K., Using organic phase-change materials for enhanced energy storage in water heaters: An experimental study. *J. Enhanced Heat Transf.*, *26*, 167–178, 2019.
13. Yilmazoglu, M.Z., Gokalp, O., Biyikoglu, A., Heat Removal Improvement in an Enclosure with Electronic Components for Air Conditioning Devices. *J. Enhanced Heat Transf.*, *26*, 1–14, 2019.
14. Salam, B., Biswas, S., Saha, S., Bhuiya, M.M.K., Heat transfer enhancement in a tube using rectangular-cut twisted tape insert. *Procedia Eng.*, *56*, 96–103, 2013.
15. Kumar, R., Nandan, G., Dwivedi, G., Shukla, A.K., Shrivastava, R., Modeling of triangular perforated twisted tape with V-Cuts in double pipe heat exchanger. *Mater. Today Proc.*, *46*, 5389–5395, 2021.
16. Liu, P., Zheng, N., Shan, F., Liu, Z., Liu, W., Heat transfer enhancement for laminar flow in a tube using bidirectional conical strip inserts. *Int. J. Heat Mass Transf.*, *127*, 1064–1076, 2018.
17. Hasanpour, A., Farhadi, M., Sedighi, K., A review study on twisted tape inserts on turbulent flow heat exchangers: The overall enhancement ratio criteria. *Int. Commun. Heat Mass Transf.*, *55*, 53–62, 2014.
18. He, J., Deng, Q., Feng, Z., Heat transfer enhancement of impingement cooling with corrugated target surface. *Int. J. Therm. Sci.*, *171*, 107251, 2022.
19. Rashidi, S., Mahian, O., Languri, E.M., Applications of nanofluids in condensing and evaporating systems. *J. Therm. Anal. Calorim.*, *131*, 2027–2039, 2018.
20. Olabi, A.G., Wilberforce, T., Sayed, E.T., Elsaid, K., Rahman, S.M.A., Abdelkareem, M.A., Geometrical effect coupled with nanofluid on heat transfer enhancement in heat exchangers. *Int. J. Thermofluids*, *10*, 100072, 2021.
21. Zhang, X.-D., Gao, J.Y., Zhang, P.-J., Liu, J., Comparison on enhanced phase change heat transfer of low melting point metal melting using different heating methods. *J. Enhanced Heat Transf.*, *26*, 179–194, 2019.
22. HabibiKhalaj, A. and Halgamuge, S.K., A Review on efficient thermal management of air- and liquid-cooled data centers: From chip to the cooling system. *Appl. Energy*, 205, 1165–1188, 2017, https://doi.org/10.1016/j.apenergy.2017.08.037.
23. Zhang, T., Gao, C., Gao, Q., Wang, G., Liu, M., Guo, Y., Xiao, C., Yan, Y.Y., Status and development of electric vehicle integrated thermal management from BTM to HVAC. *Appl. Therm. Eng.*, 88, 398–409, 2015, https://doi.org/10.1016/j.applthermaleng.2015.02.001.
24. Khanafer, K. and Vafai, K., A critical synthesis of thermophysical characteristics of nanofluids. *Int. J. Heat Mass Transfer*, 54, 4410–4428, 2011.
25. Sheikholeslami, M. and Ganji, D.D., *Applications of Nanofluid for Heat Transfer Enhancement*, Elsevier, 2017.

26. Alami, A.H., Ramadan, M., Tawalbeh, M., Haridy, S., Abdulla, S.A., Aljaghoub, H., Ayoub, M., Alashkar, A., Abdelkareem, M.A., Olabi, A.G., A critical insight on nanofluids for heat transfer enhancement. *Sci. Rep.*, 13, 15303, 2023.
27. Shanthi, R., Anandan, S., Ramalingam, V., Heat transfer enhancement using nanofluids an overview. *Therm. Sci.*, 16, 423–444, 2012.
28. Awasthi, M.K., Uddin, Z., Asthana, R., Temporal instability of a Power-law viscoelastic nanofluid layer. *Eur. Phys. J. Spec. Top.*, 230, 1427–1434, 2021.
29. Awasthi, M.K., Dharamendra, Yadav, D., Temporal instability of nanofluid layer in a circular cylindrical cavity. *Eur. Phys. J. Spec. Top.*, 231, 2773–2779, 2022.
30. Agarwal, S., Awasthi, M.K., Shukla, A.K., Stability analysis of water-alumina nanofluid film at the spherical interface. *Proc. Inst. Mech. Eng., Part E: J. Process Mech. Eng.*, 238, 3, 1195–1205, 2024, doi:10.1177/09544089221150733.
31. Pak, B.C. and Cho, Y.I., Hydrodynamic and heat transfer study of dispersed fluids with submicron metallic oxide particles. *Exp. Heat Transf.*, 11, 151–170, 1999.
32. Ho, C.J., Liu, W.K., Chang, Y.S., Lin, C.C., Natural convection heat transfer of alumina–water nanofluid in vertical square enclosures: an experimental study. *Int. J. Therm. Sci.*, 49, 1345–1353, 2010.
33. Zhou, S.Q. and Ni, R., Measurement of the specific heat capacity of water-based Al2O3 nanofluid. *Appl. Phys. Lett.*, 92, 093123, 2008.
34. Pavlenko, A., Heat and Mass Transfer in Porous Materials. *Mater. (Basel, Switzerland)*, 16, 5591, 2023, https://doi.org/10.3390/ma16165591.
35. Amanifard, N., Borji, M., Haghi, A.K., Heat transfer in porous media. *Braz. J. Chem. Eng.*, 24, 223–232, 2007.
36. Badruddin, I.A., Azeem, Khan, T.M.Y., Baig, M.A.A., Heat Transfer in Porous Media: A Mini Review. *Mater. Today: Proc.*, 24, 1318–1321, 2020.
37. Awasthi, M.K., Shukla, A.K., Singh, A.K., Shukla, P., Interfacial instability analysis of viscous-viscoelastic fluid interface in an annular porous medium. *J. Porous Media*, 26, 1–13, 2023.
38. Awasthi, M.K., Shukla, A.K., Agarwal, S., Singh, A.K., Shukla, P., Stability analysis of rivlin-ericksen fluid filmwith heat transfer through an annular porous medium. *Special Topics & Reviews in Porous Media: An International Journal*, 15, 1, 25–39, 2024, DOI: 10.1615/SpecialTopicsRevPorousMedia.2023046399.
39. Hassan, O.H., Sultan, G.I., Sabry, M.N., Hegazi, A.A., Investigation of heat transfer and pressure drop in a porous media with internal heat generation. *Case Stud. Therm. Eng.*, 32, 101849, 2022.
40. Tarawneh, M., Alshqirate, A., Khasawneh, K., Hammad, M., Experimental study on the effect of porous medium on performance of a single tube heat exchanger: A CO2 case study. *Heat Transfer—Asian Res.*, 42, 6, 473–484, 2013.
41. Diop, S.N., Dieng, B., Senaha, I., A study on heat transfer characteristics by impinging jet with several velocities distribution. *Case Stud. Therm. Eng.*, 26, 101111, 2021.

42. Awasthi, M.K., Study on Kelvin-Helmholtz instability of cylindrical flow with mass and heat transfer through porous media. *J. Porous Media*, 17, 425–437, 2014.
43. Awasthi, M.K., Study on Kelvin-Helmholtz instability with mass transfer through porous media: Effect of irrotational viscous pressure. *J. Hydrodyn. Ser. B*, 26, 624–636, 2014.
44. Awasthi, M.K., Asthana, R., Agrawal, G.S., Viscous correction for the viscous potential flow analysis of Kelvin-Helmholtz instability of cylindrical flow with heat and mass transfer. *Int. J. Heat Mass Transfer*, 78, 251–259, 2014.
45. Zuckerman, N. and Lior, N., Jet impingement heat transfer: physics, correlations, and numerical modeling. *Adv. Heat Transf.*, 39, 565–631, 2006.
46. Kotb, A., Askar, H., Saad, H., On the impingement of heat transfer using swirled air jets. *Front. Mech. Eng.*, 9, 1120985, 2023.
47. Lienhard, J.H., Heat transfer by impingement of circular free-surface liquid jets, in: *18th National and 7th ISHMT-ASME, Heat and Mass Transfer Conference*, Guwahati, India, 2006.
48. Vasiliev, L.L., Heat pipes in modern heat exchangers. *Appl. Therm. Eng.*, 25, 1, 1–19, 2005, https://doi.org/10.1016/j.applthermaleng.2003.12.004.
49. Chan, C.W., Siqueiros, E., Ling-Chin, J., Royapoor, M., Roskilly, A.P., Heat utilisation technologies: A critical review of heat pipes. *Renew. Sustain. Energy Rev.*, 50, 615–627, 2015, ISSN 1364-0321, https://doi.org/10.1016/j.rser.2015.05.028.
50. Tiari, S., Mahdavi, M., Qiu, S., Experimental study of a latent heat thermal energy storage system assisted by a heat pipe network. *Energy Convers. Manage.*, 153, 362–373, 2017, ISSN 0196-8904, https://doi.org/10.1016/j.enconman.2017.10.019.
51. Oyakawa, K., Furukawa, Y., Taira, T., Senaha, I., Nagata, T., Effects of Vortex Generators on Heat Transfer Enhancement, in: *Elsevier Series in Thermal and Fluid Sciences, Experimental Heat Transfer, Fluid Mechanics and Thermodynamics 1993*, A. Duct, M.D. Kelleher, K.R. Sreenivasan, R.K. Shah, Y. Joshi (Eds.), pp. 633–640, Elsevier, 1993, ISBN 9780444816191, https://doi.org/10.1016/B978-0-444-81619-1.50075-7.
52. Karkaba, H., Russeil, S., SimoTala, J.V., Bougeard, D., Boonaert, J., Etienne, L., Pelay, U., Lecoeuche, S., Effect of using multiple vortex generator rows on heat transfer enhancement inside an asymmetrically heated rectangular channel. *Appl. Therm. Eng.*, 227, 120359, 2023, ISSN 1359-4311, https://doi.org/10.1016/j.applthermaleng.2023.120359.
53. Brodnianská, Z. and Kotšmíd, S., Heat transfer enhancement in the novel wavy shaped heat exchanger channel with cylindrical vortex generators. *Appl. Therm. Eng.*, 220, 119720, 2023, ISSN 1359-4311, https://doi.org/10.1016/j.applthermaleng.2022.119720.
54. Tanda, G., Ahmed, E.N., Bottaro, A., Natural convection heat transfer from a ribbed vertical plate: Effect of rib size, pitch, and truncation. *Exp. Therm.*

*Fluid Sci.*, 145, 110898, 2023, ISSN 0894-1777, https://doi.org/10.1016/j.expthermflusci.2023.110898.

55. Gajghate, S.S., Deshpande, O.P., Desai, A.U. *et al.*, Effect of Ribs Configurations on Heat Transfer Enhancement for W-Shaped Ribs in a Square Duct. *Arabian J. Sci. Eng.*, 48, 12141–12160, 2023, https://doi.org/10.1007/s13369-023-07662-7.
56. Tan, J., Lyu, Y., Zhang, J. *et al.*, Experimental study on heat transfer enhancement of square-array jet impingement by using an integrated synthetic jet actuator. *Sci. China Technol. Sci.*, 2023, https://doi.org/10.1007/s11431-022-2384-6.
57. Sarkar, S., Gupta, R., Roy, T., Ganguly, R., Megaridis, C.M., Review of jet impingement cooling of electronic devices: Emerging role of surface engineering. *Int. J. Heat Mass Transf.*, 206, 123888, 2023, ISSN 0017-9310, https://doi.org/10.1016/j.ijheatmasstransfer.2023.123888.
58. Verma, V., Meena, C.S., Thangavel, S., Kumar, A., Choudhary, T., Dwivedi, G., Ground and solar assisted heat pump systems for space heating and cooling applications in the northern region of India – A study on energy and CO2 saving potential. *Sustain. Energy Technol. Assess.*, 59, 103405, 2023, ISSN 2213-1388, https://doi.org/10.1016/j.seta.2023.103405.
59. Awasthi, M.K., Dutt, N., Kumar, A., Kumar, S., Electrohydrodynamic capillary instability of Rivlin–Ericksen viscoelastic fluid film with mass and heat transfer. *Heat Transf.*, 1–19, 2023, doi: 10.1002/htj.22944.
60. Dutt, N., Hedau, A.J., Kumar, A., Awasthi, M.K., Singh, V.P., Dwivedi, G., Thermo-hydraulic performance of solar air heater having discrete D-shaped ribs as artificial roughness. *Environ. Sci. Pollut. Res.*, 2023, https://doi.org/10.1007/s11356-023-28247-9.
61. Kr. Kushwaha, P., Kr. Sharma, N., Kumar, A., Meena, C.S., Recent Advancements in Augmentation of Solar Water Heaters Using Nanocomposites with PCM: Past, Present, and Future. *Buildings*, 13, 79, 2023, https://doi.org/10.3390/buildings13010079.
62. Singh, V.P., Jain, S., Karn, A., Kumar, A., Experimental Assessment of Variation in Open Area Ratio on Thermohydraulic Performance of Parallel Flow Solar Air Heater. *Arabian J. Sci. Eng.*, 2022, https://doi.org/10.1007/s13369-022-07525-7.
63. Meena, C.S., Prajapati, A.N., Kumar, A., Kumar, M., Utilization of Solar Energy for Water Heating Application to Improve Building Energy Efficiency: An Experimental Study. *Buildings*, 12, 2166, 2022, https://doi.org/10.3390/buildings12122166.
64. Singh, V.P., Jain, S., Kumar, A., Mishra, S., Sharma, N.K., Heat transfer and friction factor correlations development for double pass solar air heater artificially roughened with perforated multi-V ribs. *Case Stud. Therm. Eng.*, 39, 102461, 2022, ISSN 2214-157X, https://doi.org/10.1016/j.csite.2022.102461.
65. Singh, V.P., Jain, S., Karn, A., Kumar, A., Dwivedi, G., Meena, C.S., Recent Developments and Advancements in Solar Air Heaters: A Detailed Review. *Sustainability*, 14, 12149, 2022, https://doi.org/10.3390/su141912149.

66. Meena, C.S., Kumar, A., Jain, S., Ur Rehman, A., Mishra, S., Innovation in Green Building Sector for Sustainable Future. *Energies*, 15, 6631, 2022, https://doi.org/10.3390/en15186631.
67. Singh, V.P., Jain, S., Karn, A., Kumar, A., Dwivedi, G., Meena, C.S., Cozzolino, R., Mathematical Modeling of Efficiency Evaluation of Double-Pass Parallel Flow Solar Air Heater. *Sustainability*, 14, 10535, 2022, https://doi.org/10.3390/su141710535.
68. Meena, C.S., Kumar, A., Roy, S., Cannavale, A., Ghosh, A., Review on Boiling Heat Transfer Enhancement Techniques. *Energies*, 15, 5759, 2022, https://doi.org/ 10.3390/en15155759.
69. Singh, V.P., Jain, S., Kumar, A., Establishment of Correlations for the Thermo-Hydraulic Parameters Due to Perforation in a Multi-V Rib Roughened Single Pass Solar Air Heater. *Exp. Heat Transf.*, 2022, doi: DOI: 10.1080/08916152.2022.2064940.
70. Prajapati, Y.K., Gupta, P.K., Kumar, A., Free Surface Undulation and Air Entrainment in a Rectangular Tank. *Int. J. Appl. Eng. Res.*, 6, 4, 409–419, 2011.
71. Pant, G., Meena, C.S., Saxena, A., Kumar, A., Singh, V.P., Dutt, N., Study the Temperature Variation in Alternate Coils of Insulated Condenser Cum Storage Tank: Experimental Study, in: *Advances in Fluid and Thermal Engineering. FLAME 2022.* Lecture Notes in Mechanical Engineering, B.S. Sikarwar, S.K. Sharma, A. Jain, K.M. Singh, (Eds.), Springer, Singapore, 2023, https://doi.org/10.1007/978-981-99-2382-3_52.
72. Saxena, A., Prajapati, A.N., Pant, G., Meena, C.S., Kumar, A., Singh, V.P., Water Consumption Optimization of Hybrid Heat Pump Water Heating System, in: *Recent Advances in Mechanical Engineering. FLAME 2022.* Lecture Notes in Mechanical Engineering, A.K. Shukla, B.P. Sharma, A. Arabkoohsar, P. Kumar, (Eds.), Springer, Singapore, 2023, https://doi.org/10.1007/978-981-99-1894-2_61.
73. Kundu, A., Kumar, A., Dutt, N., Meena, C.S., Singh, V.P., Introduction to Thermal Energy Resources and their Smart Applications, in: *Thermal Energy Systems: Design, Computational Techniques and Applications*, A. Kumar, V.P. Singh, C.S. Meena, N. Dutt, (Eds.), pp. 1–15, CRC Press, Boca Raton, FL, USA, 2023, Chapter 1; http://dx.doi.org/10.1201/9781003395768-1.
74. Dewangan, A.K., Moinuddin, S.Q., Cheepu, M., Sajjan, S.K., Kumar, A., Thermal Energy Storage: Opportunities, Challenges and Future Scope, in: *Thermal Energy Systems: Design, Computational Techniques and Applications*, A. Kumar, V.P. Singh, C.S. Meena, N. Dutt, (Eds.), pp. 17–28, CRC Press, Boca Raton, FL, USA, 2023, Chapter 2; http://dx.doi.org/10.1201/9781003395768-2.
75. Kundu, A., Kumar, A., Dutt, N., Singh, V.P., Meena, C.S., Modelling and Simulation of Thermal Energy System for Design Optimization, in: *Thermal Energy Systems: Design, Computational Techniques and Applications*, A. Kumar, V.P. Singh, C.S. Meena, N. Dutt, (Eds.), pp. 103–140, CRC Press, Boca Raton, FL, USA, 2023, Chapter 7; http://dx.doi.org/10.1201/9781003395768-7.

76. Bansal, A., Bhardwaj, H.K., Sharma, V.K., Kumar, A., Static and dynamic behavior analysis of Al-6063 alloy using modified Hopkinson bar, in: *Additive Manufacturing in Industry 4.0: Methods, Techniques, Modeling and Nano Aspects*, V.K. Sharma, A. Kumar, M. Gupta, V. Kumar, D.K. Sharma, S.K. Sharma, (Eds.), pp. 107–124, CRC Press, Boca Raton, FL, USA, 2022, Chapter 6; DOI: 10.1201/9781003360001-6.
77. Sharma, V.K., Kumar, V., Joshi, R.S., Kumar, A., Effect of REOs on tribological behavior of aluminum hybrid composites using ANN, in: *Additive Manufacturing in Industry 4.0: Methods, Techniques, Modeling and Nano Aspects*, V.K. Sharma, A. Kumar, M. Gupta, V. Kumar, D.K. Sharma, S.K. Sharma, (Eds.), pp. 153–168, CRC Press, Boca Raton, FL, USA, 2022, Chapter 9; DOI: 10.1201/9781003360001-9.
78. Sharma, V.K., Kumar, A., Gupta, M., Kumar, V., Sharma, D.K., Sharma, S., *Additive Manufacturing in Industry 4.0: Methods, Techniques, Modeling, and Nano Aspects*, Taylor & Francis (CRC Press), 2022, https://doi.org/10.1201/9781003360001.
79. Kumar, A., Singh, V.P., Meena, C.S., Dutt, N., *Thermal Energy Systems: Design, Computational Techniques and Applications*, Taylor & Francis (CRC Press), 2023, https://doi.org/10.1201/9781003395768.
80. Kumar, A., Kumar, A., Kumar, A., *Laser Based Technologies for Sustainable Manufacturing*, Taylor & Francis (CRC Press), 2023, ISBN: 9781003402398, https://doi.org/10.1201/9781003402398.
81. Kumar, R., Prasad, A., Kumar, A., *Sustainable Smart Manufacturing Processes in Industry 4.0*, Taylor & Francis (CRC Press), 2023, ISBN: 9781003436072, https://doi.org/10.1201/9781003436072 (Annexure 14).
82. Chakraborty, G., Pandey, V., Prasad, A., Kumar, A., Introduction to Sustainable Manufacturing for Industries 4.0, in: *Sustainable Smart Manufacturing Processes in Industry 4.0*, R. Kumar, A. Prasad, A. Kumar (Eds.), pp. 01–17, CRC Press, Boca Raton, FL, USA, 2023, Chapter 01; https://doi.org/10.1201/9781003436072-1.
83. Francis Luther King, M., Kumar, A., Kumar, A., Kumar, A., Introduction to Optics and Laser-based Manufacturing Technologies, in: *Laser-based Technologies for Sustainable Manufacturing*, A. Kumar, A. Kumar, A. Kumar (Eds.), pp. 1–43, CRC Press, Boca Raton, FL USA, 2023, Chapter 01; DOI: 10.1201/9781003402398-1.
84. Patil, P.P. and Kumar, A., Dynamic Structural and Thermal Characteristics Analysis of Oil Lubricated Multi Speed Transmission Gearbox: Variation of Load, Rotational Speed and Convection Heat Transfer. *Iran. J. Sci. Technol. Trans. Mech. Eng. (ISTM)*, 41, 281–291, 2017, https://doi.org/10.1007/s40997-016-0063-z.
85. Patil, P., Gori, Y., Kumar, A., Tyagi, M.R., Experimental Analysis of Tribological Properties of Polyisobutylene Thickened Oil in Lubricated Contacts. *Tribol. Int.*, 159, 1–7, 106983, 2021, https://doi.org/10.1016/j.triboint.2021.106983.
86. Kumar, A., Pathak, A., Kumar, A., Kumar, A., Physics of Laser-Matter Interaction in Laser-Based Manufacturing, in: *Laser-based Technologies for Sustainable*

*Manufacturing*, A. Kumar, A. Kumar, A. Kumar (Eds.), pp. 45–54, CRC Press, Boca Raton, FL, USA, 2023, Chapter 02; DOI: 10.1201/9781003402398-2.

87. Kumar, R., Kumar, A., Kant, L., Prasad, A., Bhoi, S., Meena, C.S., Singh, V.P., Ghosh, A., Experimental and RSM-based Process-Parameters Optimisation for Turning Operation of EN36B Steel. *Materials*, 16, 339, 2023, https://doi.org/10.3390/ma16010339.
88. Bhoi, S., Kumar, A., Prasad, A., Meena, C.S., Sarkar, R.B., Mahto, B., Ghosh, A., Performance Evaluation of Different Coating Materials in Delamination for Micro- Milling Applications on High-Speed Steel Substrate. *Micromachines*, 13, 1277, 2022, https://doi.org/10.3390/mi13081277.
89. Pathak, A., Kumar, A., Kumar, A., Kumar, A., Francis Luther King, M., Application of Laser Technology in the Mechanical and Machine Manufacturing Industry, in: *Laser-based Technologies for Sustainable Manufacturing*, A. Kumar, A. Kumar, A. Kumar (Eds.), pp. 107–155, CRC Press, Boca Raton, FL, USA, 2023, Chapter 06; DOI: 10.1201/9781003402398-6.
90. Dutt, N., Hedau, A., Awasthi, M.K., Kumar, A., Meena, C.S., Thermo-hydraulic performance investigation of solar air heater duct having staggered D-shaped ribs: Numerical approach. *Heat Transf.*, 53, 3, 1501–1531, 2024, https://doi.org/10.1002/htj.22998.
91. Verma, V., Thangavel, S., Dutt, N., Kumar, A., Weerasinghe, R., Recent Development of Thermal Energy Storage: Solar, Geothermal and Hydrogen Energy, in: *Highly Efficient Thermal Renewable Energy Systems: Design, Design, Optimization and Applications*, V. Verma, S. Thangavel, N. Dutt, A. Kumar, R. Weerasinghe (Eds.), CRC Press, Boca Raton, FL, USA, 2024, Chapter 01; ISBN 9781032595641.
92. Ahmadizadeh, M., Heidari, M., Thangavel, S., Naamani, E.A., Khashehchi, M., Verma, V., Kumar, A., Technological Advancements in Sustainable and Renewable Solar Energy Systems, in: *Highly Efficient Thermal Renewable Energy Systems: Design, Design, Optimization and Applications*, V. Verma, S. Thangavel, N. Dutt, A. Kumar, R. Weerasinghe (Eds.), CRC Press, Boca Raton, FL, USA, 2024, Chapter 02; ISBN 9781032595641.
93. Chitt, M., Thangavel, S., Verma, V., Kumar, A., Green Hydrogen Productions: Methods, Designs and Smart Applications, in: *Highly Efficient Thermal Renewable Energy Systems: Design, Design, Optimization and Applications*, V. Verma, S. Thangavel, N. Dutt, A. Kumar, R. Weerasinghe (Eds.), CRC Press, Boca Raton, FL, USA, 2024, Chapter 16; ISBN 9781032595641.
94. Khashehchi, M., Thangavel, S., Rahmanivahid, P., Heidari, M., Moazzeni, T., Verma, V., Kumar, A., Solar Desalination Techniques: Challenges and Opportunities, in: *Highly Efficient Thermal Renewable Energy Systems: Design, Design, Optimization and Applications*, V. Verma, S. Thangavel, N. Dutt, A. Kumar, R. Weerasinghe (Eds.), CRC Press, Boca Raton, FL, USA, 2024, Chapter 19; ISBN 9781032595641.
95. *Highly Efficient Thermal Renewable Energy Systems: Design, Design, Optimization and Applications*, V. Verma, S. Thangavel, N. Dutt, A. Kumar, R. Weerasinghe (Eds.), CRC Press, Boca Raton, FL, USA, 2024, ISBN 9781032595641.

2

# Renewable Thermal Energy Systems: Sustainable, Modern and Reliable Energy

**Bipasa B. Patra[1]* and Pratik Sharad Chirmade[2]**

[1]*Department of Electrical Engineering, Padmashri Dr. V B Kolte College of Engineering Malkapur, Maharashtra, India*
[2]*G H Raisoni Institute of Engineering & Business Management, Jalgaon, Maharashtra, India*

***Abstract***
The undesirable side effects of fossil fuel utilization include air pollution, acid rain, water pollution, and habitat destruction. Additionally, the extraction and transportation of fossil fuels can have adverse social and environmental impacts, such as habitat destruction, water contamination, and displacement of Indigenous communities. The increasing global awareness of these detrimental effects has led to a growing demand for cleaner and more sustainable energy alternatives. Renewable energy sources, such as solar, wind, hydro, and bioenergy, have gained attention due to their potential to mitigate climate change and reduce environmental impacts. Governments, organizations, and communities worldwide are working towards implementing renewable energy policies, investing in research and development, and promoting the adoption of clean energy technologies. These efforts aim to ensure a transition to a more sustainable energy future while addressing energy poverty, reducing environmental impacts, and fostering economic growth.

***Keywords*:** Renewable energy, solar, geothermal, reliable energy, thermal energy systems

## 2.1 Introduction

Energy indeed plays a vital role in the development of any nation. Throughout history, renewable energy sources, such as biomass, hydropower, wind, and

**Corresponding author*: akasvora@gmail.com

Ashwani Kumar, Nitesh Dutt and Mukesh Kumar Awasthi (eds.) Heat Transfer Enhancement Techniques: Thermal Performance, Optimization and Applications, (39–62) © 2025 Scrivener Publishing LLC

solar energy, have been the primary means of energy generation. These sources were harnessed for various purposes, including heating, cooking, and mechanical power. However, with the advent of the industrial revolution, there was a significant shift towards the use of non-renewable energy sources. Hard coal became increasingly important as a fuel for steam engines, driving the industrialization process. Later on, the discovery and utilization of crude oil further transformed the energy landscape, offering advantages such as easy transportation and versatile applications as a raw material. Crude oil quickly became a prime energy carrier, powering various sectors such as transportation, industry, and electricity generation. Similarly, natural gas gained prominence as a convenient and abundant energy source for space heating, power generation, and transportation. As fossil fuels gained dominance, the utilization of renewable energy sources declined both in absolute and relative terms, particularly in industrialized countries. The focus shifted towards the convenience and high energy density provided by fossil fuels, leading to a decrease in investments and research in renewable energy technologies [1–3]. Here are a few significant advancements in this field:

- Solar thermal energy: recent years have witnessed notable advancements in solar thermal systems. For instance, concentrated solar power (CSP) technology employs mirrors or lenses to focus sunlight onto a tiny area, creating high temperatures that can be utilized for industrial processes or the production of energy. Solar thermal energy is now more dependable thanks to developments in heat storage devices, which enable energy to be stored and used at night or during cloudy conditions [4–6].
- Geothermal energy: in order to increase the use of geothermal energy outside of areas where geothermal reservoirs naturally occur, enhanced geothermal systems (EGS) are being created. In order to construct or improve geothermal reservoirs and increase the availability of geothermal energy (EGS) entails digging deep wells.
- Biomass thermal energy: wood, agricultural waste, and other organic materials are examples of biomass that can be utilized to provide thermal energy. Technological developments in biomass combustion and gasification have increased the efficiency and sustainability of this energy source. In power plants, co-firing biomass and coal is one way to cut greenhouse gas emissions.

- Waste heat recovery: the technologies used to collect and use waste heat produced by power plants or industrial operations have advanced. Cogeneration, or combined heat and power (CHP) systems, increase overall efficiency by enabling the simultaneous production of useable heat and electricity from a single energy source [7–9].
- Energy-efficient buildings: advances in building design and construction have led to more energy-efficient structures. Passive solar design, better insulation materials, and advanced HVAC (heating, ventilation, and air conditioning) systems help reduce the energy required for heating and cooling.
- Heat pump systems: the efficiency and efficacy of heat pumps—which can provide both heating and cooling—have improved. Geothermal heat pumps, also known as ground-source heat pumps, use the Earth's constant temperature to produce heating and cooling while using less energy [10–13].
- Thermal energy storage: the intermittent nature of various renewable energy sources necessitates the use of efficient thermal energy storage technologies. Excess thermal energy may be saved for later use thanks to advancements in thermal energy storage technology, such as phase-change materials and molten salt [14–16].
- District heating and cooling: the use of district energy systems, which provide thermal energy to several buildings, is growing in popularity. They lessen the impact of heating and cooling on the environment by enabling the efficient distribution and centralized generation of renewable thermal energy.
- Combined technologies: hybrid systems are the result of merging many technologies in many breakthroughs. For instance, heat and electricity can be produced simultaneously by integrating solar thermal collectors with photovoltaic (PV) solar panels [17–21].

These technical developments help to create a more robust and sustainable energy infrastructure in addition to lowering greenhouse gas emissions and reliance on fossil fuels. Continued research and development in renewable thermal energy systems will be essential to accomplishing global energy transition goals as we move towards greener and more sustainable energy sources.

## 2.2 Sustainable Development Goals (SDG)

Total 17 sustainable goals can be summarized with the requirement of technological advancements in sustainable and renewable thermal energy systems:

- To put an end everywhere to all the forms of poverty: implementation of national appreciation social protection system and the measures to all including the same by agenda of 2023 archive includes sustained coverage of the affordability. One of the goals by 2030 is for men and women to have equal rights to economic resources, equal access to basic services and ownership, equal control over land and other forms of property and naturals resources, and equal opportunities to learn new technology and avail of financial services, including small financing. Also included in the SDG of the 2030 Agenda is the objective to build the resilience of the poor and those in vulnerable conditions. It also endeavors to reduce their exposure and vulnerability to extreme natural events, disasters, and socioeconomic challenges. It is crucial to significantly mobilize resources from various sources including fortifying the development collaboration among nations to provide adequate and predictable means for developing countries. Countries' formulated policy is a framework at the national level, divisional level, and international level based on the affordability for the people and gender sensitivity among developed stages to the support required to accelerate investment in poverty conditions [22–27].
- Ensure that food security is achieved and hunger is put to end and nutrition is improved while promoting sustainable agriculture: the complete elimination of all types of malnutrition is necessary to achieve the Sustainable Development Goal (SDG) 2030. Meeting the globally set goals for lowering in children under five by 2025 is part of this. In addition, it calls for increasing the agricultural output and incomes of small-scale food producers, particularly women and marginalized populations, in order to fulfill the nutritional needs of particular demographics, including girls, pregnant and nursing mothers, and the elderly. This means giving them access to

financial services and information. Additionally, there is a commitment to developing resilient agricultural practices and sustainable food production systems by 2030. These methods are intended to increase output while protecting ecosystems and strengthening the ability of the community to adapt to calamities like floods, droughts, and harsh weather. Additionally, this procedure will prioritize gradually enhancing the quality of the land and soil. Preserving genetic variety in domesticated animals, farmed plants, seeds, and allied wild species is equally essential. In compliance with international agreements, this entails the responsible administration of a variety of seed and plant banks at the national, regional, and global levels as well as the promotion of fair access and benefit-sharing from the utilization of genetic resources and related traditional knowledge. Furthermore, initiatives are underway to remove trade barriers and market inefficiencies in international agriculture. In conclusion, steps are being taken to guarantee the efficient operation of food commodity markets and its derivatives, as well as to offer prompt access to market data, such as information about food reserves [28–33].

- To ensure healthy lives and thus promoting at all ages a healthy well-being: the SDG of the 2030 Agenda also aims to reduce the global maternal mortality ratio to <70 per 100,000 live births. Moreover, part of the goal is to end preventable deaths of newborns and children <5 years old, reducing neonatal mortality to at least as low as 12, and for children under-5, to at least as low as 25 per 1000 live births. By the agenda 2030, we need to put to an end to the epidemics of AIDS, tuberculosis, malaria and neglected tropical diseases along-with combat hepatitis, diseases that are water-borne and various diseases that are communicable. By the agenda 2030, increase to one-third of premature mortality from diseases that are non-communicable over and done with prevention and mental health treatment. Strength of the avoidance and cure of the substance mishandling comprising of narcotics drugs and curb the dangerous use of alcohol. Another goal of the 2030 Agenda is to have a reduction in the number of deaths around the globe, and injuries from accidents due to road traffic accidents. By 2030, we need to ensure universal access to sexual and reproductive healthcare amenities, family planning, including education and information, and the incorporation of reproductive well-being into nationwide approaches and programs. Thus, the goal is to achieve wide coverage to healthcare that is universal, comprising economical risk security, and access to superior and crucial healthcare [34–38].

- To ensure quality education that is equitable and inclusive, hence, to promote opportunities for all for lifelong learning: the sustainable development goal 4 is to confirm comprehensive and equability, superiority in education and promoted to the lifetime learning opportunity for all in one. The indication of this target is the best example of attained charges at to primary schools and also conclusion rate of the primary schools educations and the school participated in the tertiary education and so on. In each case, the parity indices are to be looked at on to the same to make sure that the students primarily are not at all missed out. Therefore, male and female, rural and urban, top and bottom, education is precious. There is a need to curb the disability of the status and individual purposes as there are the indication around the school that it is the facilities that have to gear up like the electricity, internet, computers, water of drinking, toilet and washrooms [39–42].

The sustainable development goal 4 measurable indicators and target output are:

1. Free of cost primary and secondary educations
2. Same access to the quality of the first primary education
3. Offerings relating to technical education
4. Offerings relating to vocational education
5. And the offerings of higher education

Eliminate all discrimination in education, universal details and literacy, and education for the sustainable development and citizen development as a whole required for the development of the Nation as a whole.

There are the mentioned means of the same to implement a target that builds and looks into the inclusive upgrade and safety of the school, and expanding higher education scholarships for the development of the country, and also, to increase the continuation of the education of teachers in the development of the country.

- Gender equality to be achieved and women empowerment: the goal for sustainable development SDG 5 is to achieve gender equality and empower all women and girls; also, it indicates the inclusion for the examples of the sustainable legal framework and the representations by the feminine gender in the national parliament or in local bodies. The number of forced marriage must be curbed totally [43–45]. These are the sections related to the sustainable development goal number 5:
  - ➢ End all forms of discrimination against all women and girls everywhere
  - ➢ End violence and exploitation of women and girls
  - ➢ Elimination of harmful practices such as child, early, and forced marriage for women.
  - ➢ Increase the value of owing care and promotes mutuality in the domestic responsibility.
  - ➢ The full contribution of women (ladies) in leadership and decision-making power bodies.
  - ➢ Access to global reproductive authority and health.

There are three resources of implementation of the target:

1. The same authority to financial resources, ownership of property, and economical services.
2. Promote effects of the environment of ladies/women and girls through technologies.
3. Adoption and strengthening policies for gender equality and supporting the same.

- Ensure all sustainable management and availability of water and sanitation: the SDG 6 for sustainable development is to ensure availability and sustainable management of water and sanitization, with combined monitoring programs of the World Health Organization (WHO) and UNICEF being totally accountable for the program monitoring and progress of programs to achieve the essential objectives in this goal. Mainly, the very important indications for this goal are the percentage of human population that use safety measures for managing the drinking water and it has been made accessible for the safety of those managing the sanitizing facilities and sanitization. The other indication deals with the share of domestics and industrialized wastage water that is safely treated. The sustainable development goals outcomes is to provide safe and affordable water for use as drinking water, end open deflection, and provide access to sanitization and hygiene, hence, an increase in the water quality. Therefore, the wastewater treatment and safe re-use, improvement of the efficiency of water use, and surety of freshwater supply—the implementation of these goals will expand the water sanitization support to developing countries and to support the local engagement in water and sanitization management [46–49].

- Ensure for all the affordable access to sustainable, modern, and reliable energy: SDG number 7 is to ensure for all the affordable access to sustainable, modern, and reliable energy. The percentage of the population with access to electricity is one of the markers for this goal. Beyond the signs, consider the percentage of renewable energy sources in energy efficiency. Several nations, most notably India, Bangladesh, and Kenya, have made progress in improving access to power. The objective has a few targets to be achieved by 2030. Progress towards the goals is measured by indication of outcomes and goals—worldwide access to the present energy; an increase in global percentile of the renewable energy source, doubled improvement in energy efficiency. The targets and goals for implementation are made up of the remaining goals. Encouraging access to research and technology, as well as investment in clean energy, is crucial. Expanding and then improving energy sources is also important for developing nations' amenities. Stated differently, these objectives and aims include raising the proportion of renewable energy in the world's energy mix while facilitating access to inexpensive, dependable energy. They also have the focus on improvising energy efficacy global establishment investing in clean energy framework [50–53].
- Promote for all, sustainable economic growth, inclusive and sustained productive employment, and work that is decent: SDG number 8 is promote for all, economic and sustainable growth, inclusive and sustained, as well as employment that

is productive. Complete and creative engagement in work is essential for all. The main essential indication of the target includes economic growth concerning developing nations and the frequency of factual GDP per capita. Also, the instances are rate of youth that is lacking job and work-related insecurities or the count of the ladies involved in the task force as relative to the men. For SDG number 8, the total target is to be achieved by 2030, and while some targets are for 2020, others are for 2030. The first ten outcome goals are sustainable economics growths, diversity, innovation and up-gradation for economic productive, promoted policy enterprise, support the job creation and growth of enterprises, increase the resource efficiency in the consumption and productivity, full employment, and good work with the equal pay. Promotion of employment of the youth, training and education, trafficking and child laboring, protection of labor rights and justice, and safe employment promoting the environment, promote advantageous and sustainable tourisms, and access to banking being universal [54–57].

- Building of resilient infrastructure, promoting sustainable and inclusive industrialization, and innovation to be fostered: the SDG target 9 is related to the building of resilient infrastructure, promoting sustainable and inclusive industrialization, and fostering innovation. These include an example of the proportion of people who are engaged in the manufacturing activities and programs or who are residing in the areas sheltered by mobile networks, or those who ensure access to the internet, and the indication that is connected to the environmental challenges that carbon dioxide emission per unit of value is monitored and looked upon successfully.

  The target refers to the realization that the ability of humanity to connect and communicate effectively, move people and things efficiently, and create new industries, skills, and techniques will be necessary to overcome the many interconnected socio-economic and environmental concerns of the twenty-first century.

  To support the growth economically for the well-being of mankind, create trustworthy, resilient, and infrastructure that is superior in quality, including infrastructure both transnational and regional, with an emphasis on justifiable and affordable for everyone to access. Develop a sustainable Industrial Revolution, industrial development, and by 2030, fourfold the jobs at industry, the share of GDP along with the increase in it by a significant amount in line with national circumstances [58–59].

- Inequality to be reduced among and within the countries: SDG 10 is one of the significant goals out of the 17 SGDs set by the United Nations in 2015. It concentrates on lowering inequality. "Reduce inequality within and among countries" is the complete title. By 2030, the 10 goals following the target must be accomplished. Indicators will trail how fine we are doing as far as our goals are concerned. The first objectives to be served include:
  - ➢ Reduce disparities related to income;
  - ➢ Encourage inclusion at global, social, economic, and political aspects;
  - ➢ Put an end to discrimination and guarantee opportunities that are equal;
  - ➢ To promote equality, adopt social and fiscal policies;
  - ➢ Improve financial market and organization regulations globally;
  - ➢ Representation must be increased of nations that are developing in financial structure;
  - ➢ Ethics to be practiced responsibly.

  There are many distinct types of inequality, including racial, social, economic, sex, and disability discrimination. Inequality is frequently inextricably linked to problems with pollution, health, and environmental justice. These problems are occasionally linked to Indigenous and Aboriginal populations, ethnic minorities, and communities with poor socio-economic status.
- Make inclusive, safe, resilient, and sustainable cities and human settlements: sustainable development goal 11 is regarding the making of inclusive, safe, resilient, and above all, sustainable cities and human settlements as communities. The most important indication for this goal or target is the population count of people living in the urban slums areas. The urban population proportion who has access to convenient public transportation, including the extent of constructed transportation for public area per person [60–62].

The official mission or target and focus of SDG 11 is to make cities inclusively safe, resilient, and sustainable.

Therefore, SDG 11 is achieved by the following:

1. Housing that is safe and affordable.
2. System of transport that is affordable.
3. Mainly, urbanization to be sustainable.
4. The cultural heritage of the world must be protected and safeguarded.
5. Cope with the adverse effects of natural disasters.
6. The natural impacts of cities must be looked upon.
7. Provide safe and inclusive access to green and public places.
8. Strong national and divisional development plans.
9. The sustainable and resilient building in developed countries must be supported.
10. Implement policies for resource efficiency, inclusion, and reduction of risk of disasters.

- Ensure sustainable production with consumption patterns: "ensure sustainable consumption and production patterns," states SDG 12. The amount of tools of national policy that are used to support justifiable patterns of consuming and producing the same. Subsidies for fossil fuels worldwide are an additional one. Reducing reliance on global plastic waste trade and increasing domestic recycling are two other actions that might help reach the goal.

  The sustainable development goals consist of twelve objectives. The first outcome and targets include implementing the sustainable development goals and attaining sustainable and efficient management and use of resources that are natural. Out of the 17 sustainable development goals, goal 12 was established by the United Nations in 2015. It is also known as SDG 12 or Global Goal 12. Sustainable consumption and production patterns are required, together with responsible consumption and production, according to SDG 12. Sustainable development goal 12 aims to provide access to basic services, encourage the efficient use of resources, improve the efficiency of energy and infrastructure, create decent and environmentally friendly jobs, and guarantee a higher standard of living for all. Sustainable development goal 12 consists of objectives that, at the latest, must be accomplished by 2030 [63–65].

- Climate change and the following impacts must be dealt upon urgently: SDG 13 is to limit and adapt to climate changes. One of the 17 sustainable development targets is taking urgent action to combat climate changes and the impacts of the sustainable development target 13 and sustainable development on clean energy are closely related and complementary.
  Target 13 for sustainable development consists of five goals that must be completed by 2030. These goals address a variety of issues related to climate action; the first three are outcome goals, such as resilience and increasing understanding and preparedness for climate change; creating resilience to climate-related calamities; and integrating climate change mitigation strategies into planning and policy. The latter two objectives are implementation means objectives. Low-carbon energy sources must be employed to generate power in place of fossil fuels in order to cut emissions.
- Use of the oceans, seas and marine resources must be sustainable and conserved for sustainable development - target 14 of the sustainable development agenda is the sustainable use and conservation of the oceans, seas, and marine resources for the ecological development sustainably. The existing state of ocean resource protection does not adequately safeguard marine life, the environment, small-scale fishermen, or the seas themselves. The rise in the warmth of water and reduced oxygen loss work in tandem with ocean acidification to form the lethal trifecta of climate change stresses. The decreasing alkalinity of the ocean is known as acidification of the ocean. This does not imply that saltwater has an acidic pH already. Among the marine species most at risk are calcifying corals and mollusks. A drop in ocean pH might have an impact on a variety of marine organisms. Among these are decreased calcification, slowed metabolic rates, weakened immune responses, and less energy for vital functions including reproduction. The acidification of the ocean has an adverse result on marine environments, which provide a significant portion of humanity's requirements for food, livelihood, and other ecosystem services. A billion people rely wholly or partially on coral reefs for their requirements in terms of tourism, fishing, and waterway management. This means that food systems that are linked to the seas are at risk from ocean acidification [66–68].

- Combat desertification, and halt and reverse land degradation and halt biodiversity loss, sustainable use of terrestrial eco-system to protect, restore and promote, manage forests sustainably - SDG 15 is to combat desertification, and halt and reverse land degradation and halt biodiversity loss, sustainable use of terrestrial eco-system to protect, restore and promote, manage forests sustainably. One of the targets established by the United Nations in the year of 2015 is to protect, restore, and promote sustainable uses of the terrestrial ecosystem and sustainably manage the forest, combat desertification, and halt reserving land degradation. The outcome focuses on the following:

1. Preserve equitable benefit sharing and access to genetic resources;
2. Put an end to the poaching and trafficking of endangered animals. Stop foreign invasive organisms from entering aquatic and terrestrial habitats;
3. Put a stop to deforestation and repair damaged forests;
4. Make sure that mountain habitats are conserved;
5. Safeguard natural environments and biodiversity; and
6. Protect access to genetic resources.

By 2020, with an emphasis on forests, wetlands, mountains, and dry lands, ensure the preservation, restoration, and sustainable use of terrestrial and inland freshwater ecosystems and their services in accordance with obligations arising from international agreements.

- Promote and provide access to justice for all to build effective peaceful and inclusive societies for sustainable development - one of the 17 SDGs established by the United Nations in 2015, Sustainable Development Goal 16 (SDG 16 or Global Goal 16) is officially stated as follows: "Promote peaceful and inclusive societies for sustainable development, provide access to justice for all, and build effective, accountable, and inclusive institutions at all levels." Honor-based violence (HBV) is referred to as acts of violence or murder that are motivated by religious principles and are often carried out by male family members. According to estimates from the UN, HBV causes the deaths of about 5000 women every year. However, because these crimes are poorly handled, the true number is much greater. Authorities are unwilling to release data that would jeopardize their standing abroad or are bought off to keep the family's reputation intact. Target 16.5 suggests that it reinforces the idea of bribing. One goal is to put a stop to child abuse, sex trafficking, forced labor, and all other forms of brutality and torture. However, because there is a lack of information on some crimes, such as the severe punishment of children, it is challenging to keep track of this aim. There is a lot of abuse committed by adults against children. The difficulties with child protection and mental health services have increased as a result of the worldwide COVID-19 pandemic. Due to the current global situation, no nation is on track to fulfil the 2030 Agenda for Sustainable Development's commitments to eradicating all forms of violence and promoting mental health and well-being.

- Revitalize the global partnership and strengthen the means of implementation for sustainable development - SDG 17 (also known as Global Goal 17) is one of the 17 sustainable development goals that the United Nations set forth in 2015. It is about "partnerships for the goals." In order to achieve all of the objectives by 2030, non-hegemonic, egalitarian cross-sector and cross-national cooperation are essential, according to SDG 17. By promoting more equitable and expanded trade as well as coordinated investment activities, SDG 17 seeks to advance sustainable development on a global scale.

It seeks to enhance and simplify international interactions between developed and developing nations by using the SDGs as a framework and a common objective to steer the path towards cooperation. It seeks to promote equitable trade practices as well as international trade. The 17 goals that need to be completed by 2030 in order to reach the goal are divided into five categories: trade, systemic challenges, strengthening capacities, finance, and technology. There will be 25 indicators to track the aims' development. These objectives are all regarded as means-of-implementation objectives [69].

The correlation of the relation between SDGs and the technological advances in the field of sustainable and renewable thermal energy systems is provided in Table 2.1.

**Table 2.1** Renewable thermal energy systems and SDG 17.

| Sr. no. | Feature | Key points |
|---|---|---|
| 1 | Renewable energy source | Utilizes sources like solar, geothermal, and biomass to generate energy sustainably |
| 2 | Sustainability | Demonstrates environmental friendliness, reducing greenhouse gas emissions and minimizing reliance on non-renewable resources |
| 3 | Modern technology integration | Incorporates advanced technologies, including smart controls, automation, and monitoring systems for optimized performance. |
| 4 | Reliability and availability | Provides consistent energy production with minimal downtime, ensuring high availability for a stable energy supply. |
| 5 | Efficiency | Maximizes energy output while minimizing waste through advanced heat capture and conversion technologies |
| 6 | Carbon footprint | Significantly lowers the carbon footprint compared to traditional fossil fuels, contributing to the fight against climate change. |
| 7 | Energy storage integration | Integrates energy storage solutions for improved grid management, load balancing, and uninterrupted power supply |
| 8 | Scalability | Adaptable to various scales, suitable for residential, commercial, and large-scale industrial applications |
| 9 | Government incentives | Eligible for incentives, subsidies, and tax credits, promoting widespread adoption and investment in renewable thermal energy |

(*Continued*)

**Table 2.1** Renewable thermal energy systems and SDG 17. (*Continued*)

| Sr. no. | Feature | Key points |
|---|---|---|
| 10 | Cost competitiveness | Competitive with traditional energy sources, with decreasing costs over time due to technological advancements and economies of scale |
| 11 | Job creation | Supports the growth of a skilled workforce in the renewable energy sector, fostering economic development and job opportunities |
| 12 | Environmental impact | Minimizes environmental impact through responsible sourcing, construction, and operation, preserving ecosystems and biodiversity |
| 13 | Community benefits | Provides decentralized energy solutions, empowering local communities and enhancing energy resilience |
| 14 | Technological innovations | Ongoing research and development lead to innovations that enhance system efficiency, durability, and overall performance |
| 15 | Global energy transition | Plays a crucial role in the transition to a sustainable, low-carbon energy future, contributing to global efforts to address climate change and reduce reliance on fossil fuels. |

## 2.3 Discussion

Renewable thermal systems have made a number of significant advancements. These systems employ solar energy to create heat, which is useful for a variety of tasks like heating, producing electricity, and running industrial operations. Recent significant advancements are:

- Efficiency of concentrated solar power (CSP): the goal of CSP technological advancements was to maximize the conversion of sunlight into electrical power. This required refining the heat transfer fluid utilized in the system, raising the precision of solar tracking devices, and improving the architecture of solar collectors.

- Thermal energy storage: one important area of CSP improvement was the development of thermal energy storage (TES) devices. High-temperature molten salt storage systems were becoming more efficient and economical. These systems store heat produced during the day for usage at night or during cloudy periods.
- Next-generation collector designs: the main goals of these designs were to increase the concentration of sunlight and decrease heat loss. Sophisticated materials were being used to improve collector performance, including specialized mirrors and selective coatings.
- Hybrid systems: research was being done on hybrid systems that integrated CSP with other energy sources, such as biomass or natural gas. Even during times of low sunshine, electricity generation was able to continue because to these hybrid systems.
- High-temperature CSP: in order to boost efficiency, researchers were attempting to create CSP systems that could operate at higher temperatures. Supercritical carbon dioxide (sCO2) power cycles and hydrogen synthesis are two potential uses for high-temperature CSP systems.
- Modular CSP: due to their ease of installation and scalability, modular CSP systems were becoming more and more popular. A greater spectrum of consumers might have access to CSP thanks to the more flexible and cost-effective deployment of these smaller-scale systems.
- Heat from industrial processes and desalination: applications for solar thermal systems extended beyond the production of energy. Improvements were being made in the desalination of water by solar energy and the provision of high-temperature heat for industrial operations, including those in the textile, chemical, and food sectors.
- Better sun tracking: the precision of CSP systems was improved by advancements in solar tracking technology. As a result, solar collectors were able to gather energy more efficiently by precisely tracking the sun's path.
- Advanced control systems: sophisticated control and optimization algorithms have made it possible to effectively manage

CSP facilities and guarantee their maximum dependability and efficiency.

- Global deployment: due to abundant sunshine, CSP was becoming more and more popular around the world, especially in the Middle East, North Africa, and the Southwest region of the United States. To lessen dependency on fossil fuels and provide energy to the grid, large-scale CSP units were being constructed.

To increase the usage of renewable energy sources and lower greenhouse gas emissions, researchers and engineers are always trying to enhance the efficacy, dependability, and affordability of these systems.

## References


1. United Nations, Resolution adopted by the General Assembly on 6 July 2017, Work of the Statistical Commission pertaining to the 2030 Agenda for Sustainable Development, 2017.
2. United Nations official website on SDGs: https://sustainabledevelopment.un.org/sdg17.
3. Goal 17: Partnerships for the goals, United Nations Development Programme (UNDP), Retrieved 24 October 2023.
4. Envision2030Goal17: Partnerships for the goals, United Nations Department of Economic and Social Affairs (UNDESA), Retrieved 24 October 2020.
5. Bartram, J., Brocklehurst, C., Bradley, D., Muller, M., Evans, B., Policy review of the means of implementation targets and indicators for the sustainable development goal for water and sanitation. *Clean Water*, 1, 1, 3, December 2018, doi: DOI:10.1038/s41545-018-0003-0. S2CID 169226066.
6. United, Nations, Resolution adopted by the General Assembly on 6 July 2017: Work of the statistical commission pertaining to the 2030 Agenda for sustainable development (PDF), United Nations General Assembly, pp. 1–25, 10 July 2017.
7. Verma, V., Meena, C.S., Thangavel, S., Kumar, A., Choudhary, T., Dwivedi, G., Ground and solar assisted heat pump systems for space heating and cooling applications in the northern region of India – A study on energy and CO2 saving potential. *Sustain. Energy Technol. Assess.*, 59, 103405, 2023, ISSN 2213-1388, https://doi.org/10.1016/j.seta.2023.103405.
8. Awasthi, M.K., Dutt, N., Kumar, A., Kumar, S., Electrohydrodynamic capillary instability of Rivlin–Ericksen viscoelastic fluid film with mass and heat transfer. *Heat Transf.*, 1–19, 2023, doi: 10.1002/htj.22944.

9. Dutt, N., Hedau, A.J., Kumar, A., Awasthi, M.K., Singh, V.P., Dwivedi, G., Thermo-hydraulic performance of solar air heater having discrete D-shaped ribs as artificial roughness. *Environ. Sci. Pollut. Res.*, 2023, https://doi.org/10.1007/s11356-023-28247-9.
10. Kr. Kushwaha, P., Kr. Sharma, N., Kumar, A., Meena, C.S., Recent Advancements in Augmentation of Solar Water Heaters Using Nanocomposites with PCM: Past, Present, and Future. *Buildings*, 13, 79, 2023, https://doi.org/10.3390/buildings13010079.
11. Singh, V.P., Jain, S., Karn, A., Kumar, A., Experimental Assessment of Variation in Open Area Ratio on Thermohydraulic Performance of Parallel Flow Solar Air Heater. *Arabian J. Sci. Eng.*, 2022, https://doi.org/10.1007/s13369-022-07525-7.
12. Smith, J., Advancements in Solar Thermal Energy Systems. *Sol. Energy J.*, 45, 3, 123–136, May 2022.
13. Kreider, J.F. and Kreith, F., *Solar Energy Handbook*, McGraw-Hill, 1981.
14. Sali, G.P., Deshmukh, M.J., Wankhede, M.S., Patra, B.B., Smart IOT Automation for Advanced Home Security. *Int. J. Eng. Res. Electr. Electron. Eng. (IJEREEE)*, 6, 4, 1–6, IFERP, April 2020, ISSN (Online) - 2395-2717, DOI: 01.1617/vol7/iss4/pid4582.
15. Fuso-Nerini, F., Tomei, J., To, L.S., Bisaga, I., Parikh, P., Black, M., Mulugetta, Y., Mapping synergies and trade-os between energy and the Sustainable Development Goals. *Nat. Energy*, 3, 10–15, 2017.
16. Meena, C.S., Prajapati, A.N., Kumar, A., Kumar, M., Utilization of Solar Energy for Water Heating Application to Improve Building Energy Efficiency: An Experimental Study. *Buildings*, 12, 2166, 2022, https://doi.org/10.3390/buildings12122166.
17. Singh, V.P., Jain, S., Kumar, A., Mishra, S., Sharma, N.K., Heat transfer and friction factor correlations development for double pass solar air heater artificially roughened with perforated multi-V ribs. *Case Stud. Therm. Eng.*, 39, 102461, 2022, ISSN 2214-157X, https://doi.org/10.1016/j.csite.2022.102461.
18. Singh, V.P., Jain, S., Karn, A., Kumar, A., Dwivedi, G., Meena, C.S., Recent Developments and Advancements in Solar Air Heaters: A Detailed Review. *Sustainability*, 14, 12149, 2022, https://doi.org/10.3390/su141912149.
19. Meena, C.S., Kumar, A., Jain, S., Ur Rehman, A., Mishra, S., Innovation in Green Building Sector for Sustainable Future. *Energies*, 15, 6631, 2022, https://doi.org/10.3390/en15186631.
20. Singh, V.P., Jain, S., Karn, A., Kumar, A., Dwivedi, G., Meena, C.S., Cozzolino, R., Mathematical Modeling of Efficiency Evaluation of Double-Pass Parallel Flow Solar Air Heater. *Sustainability*, 14, 10535, 2022, https://doi.org/10.3390/su141710535.
21. Patra, B. *et al.*, Optimization of solar energy using MPPT techniques and industry 4.0 Modelling. *Sustain. Oper. Comput.*, 4, 22–28, 2023.

22. Patra, B. and Nema, P., Analysis of Solar Integrated Multilevel Inverter for Smart Grid Power Filters. *Proceedings of ICAECT 2021*, 06 April 2021, IEEE Xplore, Bhilai, and Chhattisgarh, Doi: 1109/ICAECT49130.2021.9392527.
23. Meena, C.S., Kumar, A., Roy, S., Cannavale, A., Ghosh, A., Review on Boiling Heat Transfer Enhancement Techniques. *Energies*, 15, 5759, 2022, https://doi.org/ 10.3390/en15155759.
24. Singh, V.P., Jain, S., Kumar, A., Establishment of Correlations for the Thermo-Hydraulic Parameters Due to Perforation in a Multi-V Rib Roughened Single Pass Solar Air Heater. *Exp. Heat Transf.*, 2022, doi: DOI: 10.1080/08916152.2022.2064940.
25. Prajapati, Y.K., Gupta, P.K., Kumar, A., Free Surface Undulation and Air Entrainment in a Rectangular Tank. *Int. J. Appl. Eng. Res.*, 6, 4, 409–419, 2011.
26. Pant, G., Meena, C.S., Saxena, A., Kumar, A., Singh, V.P., Dutt, N., Study the Temperature Variation in Alternate Coils of Insulated Condenser Cum Storage Tank: Experimental Study, in: *Advances in Fluid and Thermal Engineering. FLAME 2022.* Lecture Notes in Mechanical Engineering, B.S. Sikarwar, S.K. Sharma, A. Jain, K.M. Singh, (Eds.), Springer, Singapore, 2023, https://doi.org/10.1007/978-981-99-2382-3_52.
27. Saxena, A., Prajapati, A.N., Pant, G., Meena, C.S., Kumar, A., Singh, V.P., Water Consumption Optimization of Hybrid Heat Pump Water Heating System, in: *Recent Advances in Mechanical Engineering. FLAME 2022.* Lecture Notes in Mechanical Engineering, A.K. Shukla, B.P. Sharma, A. Arabkoohsar, P. Kumar (Eds.), Springer, Singapore, 2023, https://doi.org/10.1007/978-981-99-1894-2_61.
28. Kundu, A., Kumar, A., Dutt, N., Meena, C.S., Singh, V.P., Introduction to Thermal Energy Resources and their Smart Applications, in: *Thermal Energy Systems: Design, Computational Techniques and Applications*, A. Kumar, V.P. Singh, C.S. Meena, N. Dutt, (Eds.), pp. 1–15, CRC Press, Boca Raton, FL, USA, 2023, Chapter 1; http://dx.doi.org/10.1201/9781003395768-1.
29. Dewangan, A.K., Moinuddin, S.Q., Cheepu, M., Sajjan, S.K., Kumar, A., Thermal Energy Storage: Opportunities, Challenges and Future Scope, in: *Thermal Energy Systems: Design, Computational Techniques and Applications*, A. Kumar, V.P. Singh, C.S. Meena, N. Dutt, (Eds.), pp. 17–28, CRC Press, Boca Raton, FL, USA, 2023, Chapter 2; http://dx.doi.org/10.1201/9781003395768-2.
30. Kundu, A., Kumar, A., Dutt, N., Singh, V.P., Meena, C.S., Modelling and Simulation of Thermal Energy System for Design Optimization, in: *Thermal Energy Systems: Design, Computational Techniques and Applications*, A. Kumar, V.P. Singh, C.S. Meena, N. Dutt, (Eds.), pp. 103–140, CRC Press, Boca Raton, FL, USA, 2023, Chapter 7; http://dx.doi.org/10.1201/9781003395768-7.
31. Kumar, A., Singh, V.P., Meena, C.S., Dutt, N., *Thermal Energy Systems: Design, Computational Techniques and Applications*, Taylor & Francis (CRC Press), 2023, https://doi.org/10.1201/9781003395768.

32. Dutt, N., Hedau, A., Awasthi, M.K., Kumar, A., Meena, C.S., Thermo-hydraulic performance investigation of solar air heater duct having staggered D-shaped ribs: Numerical approach. *Heat Transf.*, 53, 3, 1501–1531, 2024, https://doi.org/10.1002/htj.22998.
33. Pakkan, S., Sudhakar, C., Tripathi, S. *et al.*, A correlation study of sustainable development goal (SDG) interactions. *Qual. Quant.*, 57, 1937–1956, 2023, https://doi.org/10.1007/s11135-022-01443-4.
34. Patra, B., Nema, P. *et al.*, Eleven-Level Inverter Topology with Photovoltaic Interface. *Proceedings of ICEPES 2021 MANIT*, Bhopal, 04 Feb 2022, IEEE Xplore, pp. 1–5, Doi: 10.1109/ICEPES52894.2021.9699805.
35. Patra, B. and Nema, P., Comprehensive topological analysis of solar integrated multilevel inverter for power filters. *Smart Technologies for Energy, Environment and Sustainable Development, Vol 1: Select Proceedings of ICSTEESD 2020*, Springer Nature Singapore, Singapore, pp. 475–485, 2022.
36. Patra, B.B., Necessity for Future Smarter Nation with a Sustainable Trend-Smart Grid. *Bus. Technol. (IJSSBT)*, 6, 2, 35, September 2018, ISSN (Print) 2277-7261.
37. Bansal, A., Bhardwaj, H.K., Sharma, V.K., Kumar, A., Static and dynamic behavior analysis of Al-6063 alloy using modified Hopkinson bar, in: *Additive Manufacturing in Industry 4.0: Methods, Techniques, Modeling and Nano Aspects*, V.K. Sharma, A. Kumar, M. Gupta, V. Kumar, D.K. Sharma, S.K. Sharma, (Eds.), pp. 107–124, CRC Press, Boca Raton, FL, USA, 2022, Chapter 6; DOI: 10.1201/9781003360001-6.
38. Sharma, V.K., Kumar, V., Joshi, R.S., Kumar, A., Effect of REOs on tribological behavior of aluminum hybrid composites using ANN, in: *Additive Manufacturing in Industry 4.0: Methods, Techniques, Modeling and Nano Aspects*, V.K. Sharma, A. Kumar, M. Gupta, V. Kumar, D.K. Sharma, S.K. Sharma, (Eds.), pp. 153–168, CRC Press, Boca Raton, FL, USA, 2022, Chapter 9; DOI: 10.1201/9781003360001-9.
39. Sharma, V.K., Kumar, A., Gupta, M., Kumar, V., Sharma, D.K., Sharma, S., *Additive Manufacturing in Industry 4.0: Methods, Techniques, Modeling, and Nano Aspects*, Taylor & Francis (CRC Press), 2022, https://doi.org/10.1201/9781003360001.
40. Kumar, A., Kumar, A., Kumar, A., *Laser Based Technologies for Sustainable Manufacturing*, Taylor & Francis (CRC Press), 2023, ISBN: 9781003402398, https://doi.org/10.1201/9781003402398.
41. Kumar, R., Prasad, A., Kumar, A., *Sustainable Smart Manufacturing Processes in Industry 4.0*, Taylor & Francis (CRC Press), 2023, ISBN: 9781003436072, https://doi.org/10.1201/9781003436072.
42. Chakraborty, G., Pandey, V., Prasad, A., Kumar, A., Introduction to Sustainable Manufacturing for Industries 4.0, in: *Sustainable Smart Manufacturing Processes in Industry 4.0*, R. Kumar, A. Prasad, A. Kumar (Eds.), pp. 01–17, CRC Press, Boca Raton, FL, USA, 2023, Chapter 01; https://doi.org/10.1201/9781003436072-1.

43. *Sustainable Development Goals: Actors and Implementation: A Report from the International Conference*, J.M. Vilalta, A. Betts, V. Gómez V. (Eds.), pp. 10–14, GUNI, Barcelona, Spain, 2018, Available online: http://www.acup.cat/sites/default/files/2018-06/Higher%20Education%27s%20Role.pdf (Accessed on 25November 2023).
44. Bipasa, P. and Nema, P., Comparative Design Analysis of Modified Solar Based 15-Level Multi Level Inverter for Power Quality Improvement. *ESCI-J. Algebr. Stat.*, 13, 3, 1084–1095, 2022.
45. Francis Luther King, M., Kumar, A., Kumar, A., Kumar, A., Introduction to Optics and Laser-based Manufacturing Technologies, in: *Laser-based Technologies for Sustainable Manufacturing*, A. Kumar, A. Kumar, A. Kumar (Eds.), pp. 1–43, CRC Press, Boca Raton, FL USA, 2023, Chapter 01; DOI: 10.1201/9781003402398-1.
46. Patil, P.P. and Kumar, A., Dynamic Structural and Thermal Characteristics Analysis of Oil Lubricated Multi Speed Transmission Gearbox: Variation of Load, Rotational Speed and Convection Heat Transfer. *Iran. J. Sci. Technol. Trans. Mech. Eng. (ISTM)*, 41, 281–291, 2017, https://doi.org/10.1007/s40997-016-0063-z.
47. Patil, P., Gori, Y., Kumar, A., Tyagi, M.R., Experimental Analysis of Tribological Properties of Polyisobutylene Thickened Oil in Lubricated Contacts. *Tribol. Int.*, 159, 106983, 1–7, 2021, https://doi.org/10.1016/j.triboint.2021.106983.
48. Kumar, A., Pathak, A., Kumar, A., Kumar, A., Physics of Laser-Matter Interaction in Laser-Based Manufacturing, in: *Laser-based Technologies for Sustainable Manufacturing*, A. Kumar, A. Kumar, A. Kumar (Eds.), pp. 45–54, CRC Press, Boca Raton, FL, USA, 2023, Chapter 02; DOI: 10.1201/9781003402398-2.
49. Kumar, R., Kumar, A., Kant, L., Prasad, A., Bhoi, S., Meena, C.S., Singh, V.P., Ghosh, A., Experimental and RSM-based Process-Parameters Optimisation for Turning Operation of EN36B Steel. *Materials*, 16, 339, 2023, https://doi.org/10.3390/ma16010339.
50. Bhoi, S., Kumar, A., Prasad, A., Meena, C.S., Sarkar, R.B., Mahto, B., Ghosh, A., Performance Evaluation of Different Coating Materials in Delamination for Micro- Milling Applications on High-Speed Steel Substrate. *Micromachines*, 13, 1277, 2022, https://doi.org/10.3390/mi13081277.
51. Pathak, A., Kumar, A., Kumar, A., Kumar, A., Francis Luther King, M., Application of Laser Technology in the Mechanical and Machine Manufacturing Industry, in: *Laser-based Technologies for Sustainable Manufacturing*, A. Kumar, A. Kumar, A. Kumar (Eds.), pp. 107–155, CRC Press, Boca Raton, FL, USA, 2023, Chapter 06; DOI: 10.1201/9781003402398-6.
52. Patra, B. and Nema, P., Advancements in Solar Thermal and Photovoltaic System, in: *Thermal Energy Systems: Design, Computational Techniques, and applications*, CRC press (Taylor & Francis), USA, 2023.

53. Patra, B., Introduction and Fundamentals of Solar Energy Collectors, in: *Thermal Energy Systems: Design, Computational Techniques, and applications*, CRC press (Taylor & Francis), USA, 2023.
54. Awasthi, M.K., Dutt, N., Kumar, A., Hedau, A., Introduction to Mathematical and Computational Methods, in: *Computational Fluid Flow and Heat Transfer: Advances, Design, Control and Applications*, M.K. Awasthi, A. Kumar, N. Dutt, S. Singh (Eds.), CRC Press, Boca Raton, FL, USA, 2024, Chapter 01; ISBN 9781032603186.
55. Awasthi, M.K., Kumar, A., Dutt, N., Modeling Rayleigh-Taylor Instability in Nanofluid Layers, in: *Computational Fluid Flow and Heat Transfer: Advances, Design, Control and Applications*, M.K. Awasthi, A. Kumar, N. Dutt, S. Singh (Eds.), CRC Press, Boca Raton, FL, USA, 2024, Chapter 13; ISBN 9781032603186.
56. *Computational Fluid Flow and Heat Transfer: Advances, Design, Control and Applications*, M.K. Awasthi, A. Kumar, N. Dutt, S. Singh (Eds.), CRC Press, Boca Raton, FL, USA, 2024, ISBN 9781032603186.
57. Nilsson, M., Griggs, D., Visbeck, M., Policy: map the interactions between sustainable development goals. *Nature*, 534, 320–322, 2016.
58. Bogers, M., Biermann, F., Kalfagianni, A., Kim, R.E., Sustainable Development Goals fail to advance policy integration: A large-n text analysis of 159 international organizations. *Environ. Sci. Policy*, 138, 134–145, 2022, ISSN 1462-9011, https://doi.org/10.1016/j.envsci.2022.10.002.
59. Verma, V., Thangavel, S., Dutt, N., Kumar, A., Weerasinghe, R., Recent Development of Thermal Energy Storage: Solar, Geothermal and Hydrogen Energy, in: *Highly Efficient Thermal Renewable Energy Systems: Design, Design, Optimization and Applications*, V. Verma, S. Thangavel, N. Dutt, A. Kumar, R. Weerasinghe (Eds.), CRC Press, Boca Raton, FL, USA, 2024, Chapter 01; ISBN 9781032595641.
60. Ahmadizadeh, M., Heidari, M., Thangavel, S., Naamani, E.A., Khashehchi, M., Verma, V., Kumar, A., Technological Advancements in Sustainable and Renewable Solar Energy Systems, in: *Highly Efficient Thermal Renewable Energy Systems: Design, Design, Optimization and Applications*, V. Verma, S. Thangavel, N. Dutt, A. Kumar, R. Weerasinghe (Eds.), CRC Press, Boca Raton, FL, USA, 2024, Chapter 02; ISBN 9781032595641.
61. Chitt, M., Thangavel, S., Verma, V., Kumar, A., Green Hydrogen Productions: Methods, Designs and Smart Applications, in: *Highly Efficient Thermal Renewable Energy Systems: Design, Design, Optimization and Applications*, V. Verma, S. Thangavel, N. Dutt, A. Kumar, R. Weerasinghe (Eds.), CRC Press, Boca Raton, FL, USA, 2024, Chapter 16; ISBN 9781032595641.
62. Khashehchi, M., Thangavel, S., Rahmanivahid, P., Heidari, M., Moazzeni, T., Verma, V., Kumar, A., Solar Desalination Techniques: Challenges and Opportunities, in: *Highly Efficient Thermal Renewable Energy Systems: Design, Design, Optimization and Applications*, V. Verma, S. Thangavel, N.

Dutt, A. Kumar, R. Weerasinghe (Eds.), CRC Press, Boca Raton, FL, USA, 2024, Chapter 19; ISBN 9781032595641.

63. *Highly Efficient Thermal Renewable Energy Systems: Design, Design, Optimization and Applications*, V. Verma, S. Thangavel, N. Dutt, A. Kumar, R. Weerasinghe (Eds.), CRC Press, Boca Raton, FL, USA, 2024, ISBN 9781032595641.
64. Kumar, A., Singla, Y., Namboodri, T., Globalization and International Issues in Sustainable Manufacturing, in: *Sustainability in Smart Manufacturing: Trends, Scope, and Challenges*, S. Shah, H. Nautiyal, G. Gugliani, A. Kumar, T. Namboodri, Y.K. Singla (Eds.), CRC Press, Boca Raton, FL, USA, 2024, Chapter 01; ISBN 9781032 740713.
65. *Sustainability in Smart Manufacturing: Trends, Scope, and Challenges*, S. Shah, H. Nautiyal, G. Gugliani, A. Kumar, T. Namboodri, Y.K. Singla (Eds.), CRC Press, Boca Raton, FL, USA, 2024, ISBN 9781032 740713.
66. Singh, V.P., Kumar, A., Meena, C.S., Dwivedi, G., Introduction to Energy Efficient Vehicles for Sustainable Transportation: Transitions and Challenges, in: *Energy Efficient Vehicles: Technologies and Challenges*, V.P. Singh, A. Kumar, C.S. Meena, G. Dwivedi (Eds.), CRC Press, Boca Raton, FL, USA, 2024, Chapter 01; ISBN 9781032548111.
67. Upadhyay, R.K., Singh, V.P., Kumar, A., Sustainable Transportation: Policy, Planning and Implementation, in: *Energy Efficient Vehicles: Technologies and Challenges*, V.P. Singh, A. Kumar, C.S. Meena, G. Dwivedi (Eds.), CRC Press, Boca Raton, FL, USA, 2024, Chapter 05; ISBN 9781032548111.
68. Singh, V.P. and Kumar, A., Techno-Economic and Future Aspects of the HEV-EV-FCV: Decarbonisation, Digitalization and Sustainability, in: *Energy Efficient Vehicles: Technologies and Challenges*, V.P. Singh, A. Kumar, C.S. Meena, G. Dwivedi (Eds.), CRC Press, Boca Raton, FL, USA, 2024, Chapter 11; ISBN 9781032548111.
69. *Energy Efficient Vehicles: Technologies and Challenges*, V.P. Singh, A. Kumar, C.S. Meena, G. Dwivedi (Eds.), CRC Press, Boca Raton, FL, USA, 2024, ISBN 9781032548111.

# 3

# HVAC System Efficiency Improvement Through Heat Transfer Enhancement Techniques

**Md Naim Hossain and Arijit Kundu***

*Department of Mechanical Engineering, Jalpaiguri Government Engineering College, Jalpaiguri, West Bengal, India*

***Abstract***

The heating, ventilation, and air conditioning (HVAC) systems have an important role in maintaining indoor environments for comfort with energy efficiency. The enhancement of the efficiency of HVAC systems through heat transfer enhancement techniques is of paramount importance in reducing energy consumption and operating costs. The HVAC system is equipped with many sub-systems that commonly exhibit two-phase internal flow; a phenomenon widely embraced for its efficient heat transfer capabilities. Recent strides in sustainability-driven innovations have propelled efforts to enhance heat transfer between tubes and two-phase working fluids. The techniques are systematically classified as passive or active. Passive techniques are classified as geometrical and surface modifications of heat transfer surfaces. The technique includes the use of fins and inserts, explores the effects of altering geometric parameters, etc. On the other hand, the active techniques include the electro-hydrodynamic approaches, the use of surface vibration, etc. This chapter describes a concise overview of the fundamental methods that are employed to optimize the HVAC system performance through heat transfer enhancement. The enhancement of HVAC system efficiency through heat transfer enhancement techniques may offer multiple feasible design solutions that can be used by suitable trade-offs to meet specific building requirements and local climate conditions.

*Corresponding author*: arijit.kundu@me.jgec.ac.in; ORCID: 0000-0003-0719-5907
Md Naim Hossain: ORCID: 0000-0002-7547-832X

Ashwani Kumar, Nitesh Dutt and Mukesh Kumar Awasthi (eds.) Heat Transfer Enhancement Techniques: Thermal Performance, Optimization and Applications, (63–86) © 2025 Scrivener Publishing LLC

***Keywords***: HVAC efficiency, heat transfer enhancement, two-phase flow, passive techniques, active techniques, sustainable innovations, building requirements, local climate conditions

## 3.1 Introduction

Reducing the energy usage of heating, ventilation, and air conditioning (HVAC) systems is a crucial concern, given their substantial environmental impact, high energy costs, and a notable surge in market demand. As the global industrial output continues to surge, sustainability and energy efficiency have become focal points of concern. In recent years, the effectiveness of thermal management approaches has emerged as a critical factor directly or indirectly influencing the sustained growth and productivity of industries worldwide. The increasing emphasis on enhancing the efficiency of HVAC systems has spurred advancements in heat exchanger technology through the implementation of techniques that augment heat transfer. These techniques range from fundamental adjustments in fluid velocity within the unit to intricate innovations involving the design of new surface geometries and inserts, especially in tubular configurations. Heat transfer enhancement techniques are broadly classified into two groups: active and passive [1, 2]. Active methods demand external power, such as fluid suction or injection, and surface fluid vibrations [2]. In contrast, passive methods involve modifying the heat transfer surface of the system [1]. Due to their effectiveness, cost-efficiency, easy installation and removal for cleaning, versatility in construction materials, and reliability, passive methods have emerged as pivotal heat transfer enhancement techniques in recent years. This chapter delves into the diverse approaches to heat transfer enhancement, exploring the development of new exchanger technology, whether in the form of compact designs or the incorporation of internal inserts into conventional tubular technology.

The HVAC&R system includes air cooling, air heating (heat pump), refrigeration, and dehumidification. All of these devices handle condenser and evaporator tubes. The two-phase transition from liquid to vapor (boiling) and vapor to liquid (condensation) is a common phenomenon in HVAC systems. The use of different enhanced geometries will boost the boiling and condensation phenomena that further result in a reduction in equipment size, refrigerant charge, enhanced process control, improved coefficient of performance (COP), and reduced environmental impact. On the other hand, enhanced efficiency of the heat exchanging surfaces enables the design of more compact and cost-effective heat exchangers that require reduced raw materials which is too high in the recent scenario (93 billion tons) [3]. As a consequence, the associated carbon emissions from transportation and raw material production can be mitigated. Concurrently, the decrease in material consumption results in a reduction in waste production, as the need for fewer materials during production correlates with diminished waste at the end of the device's lifecycle [3]. Hence the design operation and modification of HVAC heat exchanging devices with enhanced techniques will promote sustainable development and green building construction globally.

Present literature indicates that a significant portion of research efforts was dedicated to optimizing the heat transfer in heat exchangers, specifically in the context of two-phase flow. The primary approach involved enhancing the thermal performance of heat transfer surfaces, such as tubes and channels. Typically, the key strategies centered around disrupting fluid flow to induce turbulence, improving mixing, and maximizing the contact area between the fluid and tube walls [4–6].

Heat transfer enhancement techniques are categorized into active and passive methods [7], respectively. Passive techniques typically rely on modifications to the geometry or the insertion of elements into an existing tube or channel to induce flow disturbance [2]. In contrast, active techniques depend on the external force application like the mechanical or electrical power to generate oscillations or disturbances. While passive techniques are generally easier and more cost-effective to implement, they are more commonly used in the industry [1, 2]. Regardless of being active or passive, it is crucial to account for the associated increase in pressure drop resulting from heat transfer enhancement. All techniques discussed in this paper involve disrupting fluid flow, leading to an additional pressure drop along the tube. The heightened pressure drop necessitates the use of a larger and more expensive pump to compensate for the losses. If the heat transfer enhancement from a solution causes an excessive pressure drop, the cost of employing a larger pump may outweigh the savings achieved by a more power-dense heat exchanger.

To facilitate a comprehensive comparison and offer design guidelines, various enhancement factors have been developed and are outlined in this chapter. However, it is important to use enhancement factors cautiously, as some may not account for changes in pressure drop. This review not only provides a thorough summary of historical enhancement techniques for two-phase flows but also formulates design guidelines for the advancement of sophisticated heat exchangers, driven by the contemporary emphasis on the implementation of renewable and sustainable energy systems [76–79].

## 3.2 Passive Heat Transfer Enhancement Techniques

The passive heat transfer enhancement techniques associated with two-phase heat transfer along pipe are described in this section. In both boiling and condensation processes, enhancement of the heat transfer coefficient is achievable by enlarging the surface area conducive to heat exchange [8].

For vertical flow boiling, various flow regimes can manifest in vertical heated riser tubes. The flow inside the riser tube may exhibit subcooled conditions, followed by bubbly, slug, annular, mist flow, and single-phase vapor. Alternatively, it may display subcooled conditions followed by bubbly, inverted annular mist flow, and single-phase vapor [9–14].

The flow within the heated riser tube could encounter the critical heat flux (CHF) or burnout phenomenon, leading to a sharp reduction in the heat transfer coefficient [9]. This occurrence may result from departure from nucleate boiling (DNB) after the bubbly flow phase, caused by a high heat flux and low quality. Alternatively, it may arise from dry-out inception following the annular flow condition in the high-quality region due to suppressed nucleation [9]. The heat transfer mechanisms in bubbly, slug, and annular flows primarily involve a combination of convective and nucleate boiling [15–18]. If the CHF phenomenon occurs due to dry-out, the flow characteristics transition to 'mist flow,' characterized by vapor flow with dispersed liquid droplets [9–11]. Heat transfer in mist flow involves single-phase vapor convection and the interaction of dispersed bubbles with the heated tube wall—a phenomenon known as dispersed flow film boiling (DFFB) [10, 19, 20]. Conversely, if the CHF incident results from DNB, inverted annular flow film boiling heat transfer (IAFB) occurs [10, 21, 22], followed by dispersed single-phase vapor convective heat transfer [10, 17]. The flow regimes along the heated vertical pipe are shown in Figure 3.1.

The flow regimes along heated horizontal pipe are subcooled flow flowed by bubbly flow, slug flow, wavy flow, annular flow, and wavy annular flow. Finally, the wall dry-out inception happens and the flow is shifted to single-phase vapor. The flow regimes are presented in Figure 3.2 [15].

However, in the context of flow boiling, the thermal resistance to vapor heat transfer significantly surpasses that of liquid, emphasizing the preference for a larger wetted area within the tube walls [9–12]. The desirable flow regimes during the safe and efficient operation of two-phase flow along heated pipes should not meet the tube dry-out or departure from the nucleate boiling scenario. Consequently, various techniques aim to enhance the presence of annular flow within the tube. Alternatively, some methods adopt a distinct approach, focusing on surface modification (roughness) to amplify nucleate boiling. It is to be mentioned that the augmentation in heat transfer surfaces

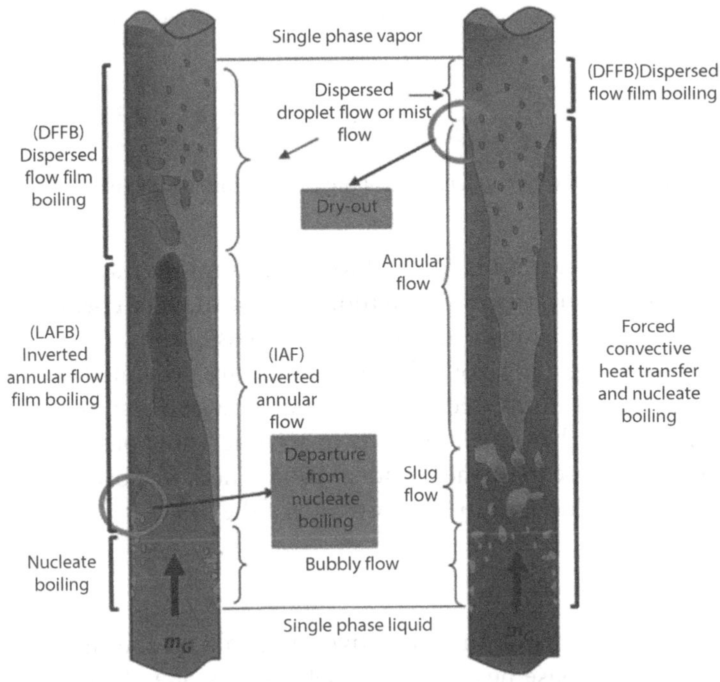


**Figure 3.1** Flow regimes along the heated vertical pipe (plain tube).

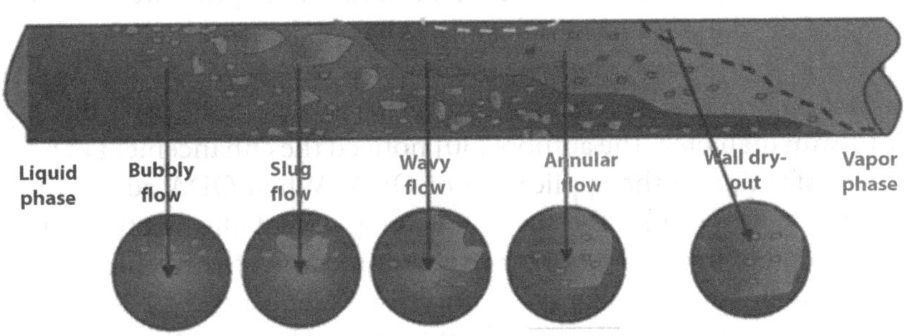


**Figure 3.2** Flow regimes along the heated horizontal pipe (plain tube).

significantly changes the flow regimes. That further may enhance or decrease the two-phase heat transfer coefficient based on the new flow regimes as per the geometrical configurations of the new augmented surfaces [23, 25].

The primary obstacle in improving condensation heat transfer lies in effectively removing liquid condensate from the condensing surface as soon as it forms [55]. As liquid accumulates on the surface and the thickness of the liquid film grows [56], the thermal resistance to heat transfer from the vapor to the surface also increases [57], impeding efficient heat

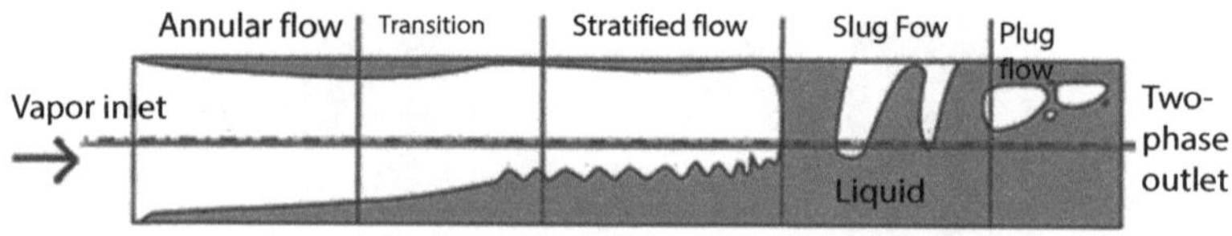


**Figure 3.3** Flow regimes for condensation along the horizontal pipe (plain tube).

transfer [58]. This section is devoted to examining studies that investigate the impact of enhancements to the tube on heat transfer coefficient in the context of two-phase flow condensation. The possible flow regimes for flow condensation along a horizontally cooled tube are presented in Figure 3.3 [9, 24]. The flow is converted to an annular flow as soon as it enters the cooled tube. After that, the flow shifts to stratified flow passing through a transition region due to further condensation. Then the flow shifted to slug flow followed by plug flow in the upstream of the flow.

### 3.2.1 Surfactants

The use of surfactants or surface-active materials on the tube surface to enhance the two-phase heat transfer coefficient is one of the oldest techniques of passive heat transfer enhancement. Surfactant substances are added in small quantities to the coolant fluid [26–29] to lower the surface tension of the liquids and significantly enhance the pool boiling.

Qiu *et al.* [30] investigated the impact of octadecylamine (ODA) on the flow boiling heat transfer of water in a vertical copper tube of 0.6-m length, and 12-mm diameter. The authors [30] noticed the enhancement of boiling heat transfer due to the application of ODA. When ODA adheres to the tube surface, it modifies the surface heating, potentially creating artificial sites for bubble initiation and growth. Additionally, the bubble departure diameter decreases, leading to an increased frequency of bubble departure and consequently enhancing heat transfer. However, there is a critical mass fraction (3×10-5) after which the application of ODA reduces the heat transfer coefficient significantly [30]. At the critical mass flux, the surfactants saturate the liquid-vapor interface, and micelles form in the fluid bulk due to the hydrophobic attraction between the tails of amphiphilic molecules (surfactants). Beyond the critical mass flux, where surface tension stabilizes, the impact of higher surfactant concentration becomes negligible. However, beyond the critical mass flux, the heat transfer coefficient was still higher than the plain coolant.

Hetsroni *et al.* [31] investigated the heat transfer and pressure drop of water and a water-surfactant solution (Habon G) at concentrations of 530 and 1060 ppm. The authors [31] carried out their investigation for fully developed laminar flow. They mentioned the range of Reynolds number within 10 < Red < 450 for the laminar flow.

Where

$$\mathrm{Re}_d = \frac{Ud_i}{\nu} \tag{3.1}$$

Where $U$ is the average fluid speed, $d$ is the internal tube diameter, and $\nu$ is the liquid kinematic viscosity.

The authors investigated that in a capillary tube with $di = 1.07$ mm inner diameter, the addition of surfactant increased pressure drop in both adiabatic and diabatic flows. Finally, they observed the increase in heat transfer coefficient due to the application of surfactants.

### 3.2.2 Straight Microfins and Helical Microfins Tubes

Microfinned tubes, initially pioneered by Hitachi Cable [32], find widespread application in evaporator or condenser tubes. The microfin tubes can enhance the in-tube heat transfer. However, they increase the tube pressure drop. Hence, an optimal use of fin geometry to enhance the heat transfer while incurring an acceptable range of pressure drop will facilitate the HVAC system to increase efficiency. Various factors influence the performance of microfinned tubes, including the outer diameter ($d_o$), root diameter ($d_r$), tip diameter ($d_t$), fin height (H), helix angle ($\beta$), and apex angle ($\theta$). On the other hand, the microfins may adopt different shapes such as rectangular or trapezoidal. Depending on the specific application, the number of fins, fin height, and helix angle of microfin tubes typically fall within the ranges of 50–70 fins, 0.075–0.4 mm, and 10°–35°, respectively [33, 34].

By augmenting the heat transfer area, the microfinned tubes promote wetting inside the tube, leading to an increased heat transfer rate. Numerous studies have explored the impact of different fluids within microfinned tubing during flow boiling. There are two governing factors to compare the effectiveness of the microfin tube compared to the plain tube. These are the heat transfer enhancement factor or enhancement factor (EF) and pressure drop factor (PF), respectively. The factors are presented by the following expressions.

$$EF = \frac{h_{tp}\big|_{enhanced\ tube}}{h_{tp}\big|_{plain\ tube}} \tag{3.2}$$

$$PF = \frac{h_{tp}\big|_{enhanced\ tube}}{h_{tp}\big|_{plain\ tube}} \tag{3.3}$$

There are many research in open literature describing the usefulness of the microfin tube. These are presented here. Schlager *et al.* [35] examined the influence of 12.7-mm spiraled microfinned copper tubes on heat transfer and pressure drop with R22 as the working fluid. The heat transfer enhancement factor (Equation 3.2) for evaporation ranged from 1.6 to 2.2. Furthermore, with increasing heat transfer coefficient, the mass flux exhibited an increase, with the heat transfer coefficient at 300 kg/(m2·s) being 400% higher than that observed at 75 kg/(m2·s). The pressure drop factor (Equation 3.3) was lower than the EF, reaching a maximum of 1.4.

Notably, the enhancement in heat transfer provided by fins decreased with increasing mass flux, attributable to the analysis of flow regimes within the tube. At lower mass flux, the flow pattern in the smooth tube is stratified-wavy, whereas in the microfin tube, the flow regime is primarily annular. Consequently, significant heat transfer enhancement occurs in the low mass flux region. However, at higher mass flux, annular flow patterns are observed in both tubes, and the heat transfer enhancement at higher mass fluxes mainly arises from the increase in area. Various studies have delved into the impact of mass flux on heat transfer coefficient and pressure drop.

Utilizing R417A, Zhang *et al.* [36] conducted flow boiling experiments within internally grooved tubes, yielding measured EF and PF in the ranges of 1.26–2.8 and 1.8–2.6, respectively. Jiang *et al.* [37] explored flow boiling heat transfer and pressure drop for R22, R134a, R407C, and R410A in a 9.52-mm diameter tube with both smooth and microfin configurations (Figure 3.4). The EFs for R22, R134a, R407C, and R410A were 1.86, 1.80, 1.69, and 1.78, while the PFs were 1.42, 1.30, 1.45, and 1.40, respectively.

The helix angle (β) emerged as a crucial factor, influencing heat transfer coefficient (HTC) and pressure drop. Schlager *et al.* [38] found that, for R22 in horizontal microfin tubes, the optimal angle for peak performance was $\beta = 25°$. Oh, and Bergles [39] reported that the optimal helix angle depended on mass flux for a 9.52-mm OD microfinned tube using R134a. Kimura and Ito [40] and Yang and Hrnjak [41] further explored the impact of β on evaporative heat transfer in microfin tubes, with varying results based on flow regimes and mass flux.

While much of the discussion focused on helical microfin tubes, axial or straight microfin tubes ($\beta = 0°$) emerged as an economical alternative for heat transfer enhancement, particularly in aluminum tubes. Chiang [42]

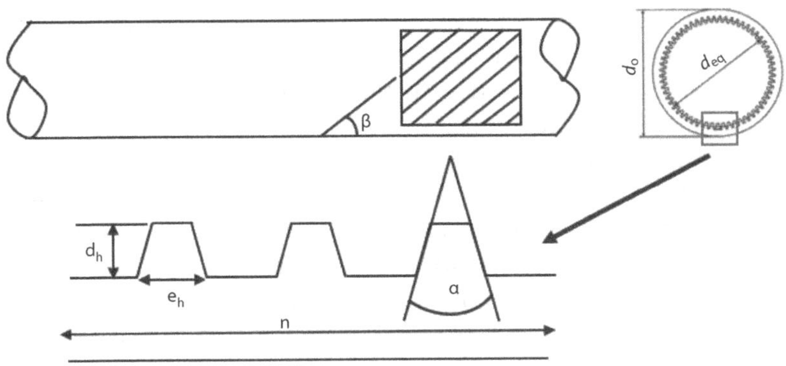


**Figure 3.4** Helical microfin tube geometry.

reported that the mass flux influenced heat transfer performance in axial microfin tubes, outperforming helical microfin tubes at higher fluxes but losing efficiency at lower fluxes. Wu *et al.* [43] highlighted the potential of axial microfins for enhancing evaporation heat transfer, especially for R32. Yang and Hrnjak [44] and [45] investigated axial microfin tubes, considering tube expansion effects on heat transfer performance, reporting average EF and PF of 1.34 and 1.23, respectively. Additionally, they noted that liquid refrigerant during stratified flow increased heat transfer inside the tube due to lower thermal resistance.

Apart from boiling, microfins contribute to the enhancement of condensation heat transfer coefficient by increasing the surface area and promoting the formation of thin-film condensate on their tips. In a study by Jung *et al.* [59], the impact of microfin tubes on flow condensation heat transfer coefficient was investigated across a range of mass and heat fluxes, utilizing R22, R134a, R407C, and R410A as working fluids. The results demonstrated a two- to threefold increase in heat transfer coefficient when using microfin tubes compared to smooth tubes, with the enhancement diminishing at higher mass fluxes. Similar findings were corroborated by Kwon *et al.* [60], who studied condensation in a microfin tube using R22 and R410A. The study revealed an enhancement ratio increase of 1.5 to 2.2 times based on the nominal heat transfer area, showing a negative correlation with mass flux.

Eckels and Pate [61] conducted an investigation using a tube with 60 fins, each with a height of 0.2 mm and a 17° spiral angle. Employing R134a and CFC-12 as the flowing fluids, the study covered various mass fluxes and wall temperatures. The fins positively impacted heat transfer, with the heat transfer coefficient during condensation increasing by 1.75–2.5 times for HFC-134a and 1.7–2.3 times for R12. Despite an associated increase in pressure drop, the enhanced heat transfer coefficient outweighed the pressure drop in almost all cases, establishing microfins as an effective heat transfer enhancement technique.

Eckels *et al.* [62] investigated the use of 150 saybolt universal seconds (SUS) kinematic viscosity pentaerythritol ester lubricants in a microfin tube with 60 fins and an 18° spiral angle. Using R134a, they reported a 50–200% increase in condensation HTC and a 10–50% increase in pressure drops compared to smooth tubes. Higher lubricant concentration adversely affected heat transfer but had no impact on pressure drop. In a similar study, Cho and Tae [63] used polyol-ester (POE) and mineral oils as alternatives to the ester lubricant, finding that oil concentrations negatively affected heat transfer augmentation due to increased fluid viscosity. However, in U-bend sections, the heat transfer enhancement reached its maximum value, attributed to the promotion of secondary flows and flow disturbances by the U-shape bend. Additionally, the straight section following the bend exhibited a higher average HTC than the straight section preceding the bend due to the extension of secondary flow and disturbances. The pressure drop was reported to increase with oil concentration.

### 3.2.3 Herringbone Tube

One alternative arrangement for extended surfaces involves the utilization of a herringbone pattern. The tube featured a V-groove-shaped pattern (shown in Figure 3.5) guiding the fluid in two distinct flow directions, causing non-uniform film layer distribution around the circumference. Prior research has established that the structural orientation of the pattern influences the liquid distribution along the walls and consequently affects the heat transfer coefficient [46]. The study, conducted under low mass flux conditions using R22, R410A, and R32 as working fluids, revealed minimal heat transfer enhancement in the herringbone tube. It was observed that the EF was considerably lower than the surface area ratio, indicating that not all inner wall areas contributed to heat transfer enhancement. This discrepancy arose from sections of the tube being exposed to vapor and thus ineffective in heat transfer. Additionally, as the saturation temperature was lower than the condensation temperature, liquid dry-out occurred, further diminishing heat transfer enhancement.

Olivier *et al.* [47] compared a herringbone tube with exploring a microfin copper tube and a plain copper tube. The authors [47] experimented with a mass flux range of 400 to 800 kg/m$^2$s using R22, R407C, and R134a as working fluids. They observed that there were 80% higher-pressure

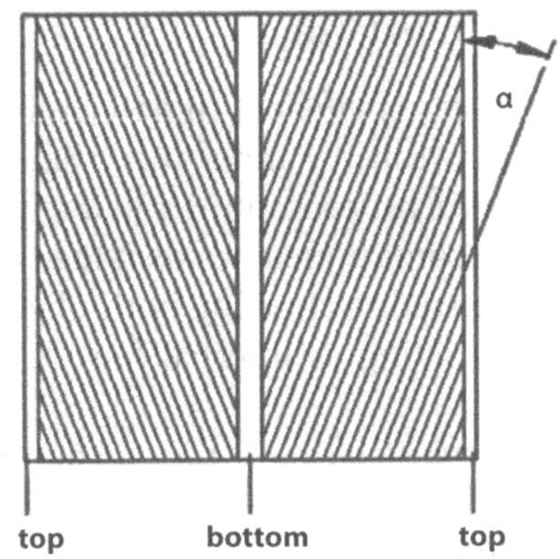


**Figure 3.5** Schematic diagram of V-shaped herringbone surface.

drop-in herringbone tubes compared to plain copper tubes and 27% compared to microfin tubes. This higher-pressure drop was attributed to the extension of surfaces from the tube walls prolonging the annular flow regime, subsequently increasing vapor velocity and friction. However, the herringbone tube exhibited enhanced heat transfer, and it was 70% higher than plain tubes and 40% higher than microfin tubes.

Apart from boiling, herringbone tube has a great impact on flow condensation. In their study, Miyara *et al.* [64] investigated the impact of herringbone patterns on the condensation of R410A and R22, comparing the results with conventional helical microfin tubes. The findings revealed that, at high fluid velocities, the herringbone pattern outperformed the helical microfin tube, but at low velocities, the opposite was observed. Herringbone tubes demonstrated an ability to collect liquid at the top and bottom sections, resulting in thinner films on the sides of the tubes and consequently enhancing heat transfer. Additionally, similar to conventional helical fins, herringbone tubes promoted turbulence and mixing, further enhancing heat transfer. Regardless of velocity, the herringbone microfin tube consistently exhibited a higher pressure drop than the helical fin tubes.

Continuing their investigation, Miyara *et al.* [65] explored the effect of fin shape on the herringbone-patterned extended surface, using R410A. The results indicated that longer fins led to increased heat transfer, but this enhancement plateaued when fins reached a length of 0.18 mm. Furthermore, the study reported that the heat transfer coefficient increased with higher helix angles, attributed to the increased circumferential force affecting film thickness. Although altering the fin shape augmented heat transfer, the pressure drop increased with fin height or helical angle.

### 3.2.4 Twisted Tapes Insert Tube

To enhance the heat transfer coefficient in flow boiling scenarios, optimization of heat transfer between the inner tube walls and the fluid is crucial. This optimization involves promoting the intermixing of the flowing fluid and maximizing the contact area between the tube wall and the fluid. Twisted tapes prove effective in achieving both objectives, inducing swirl flow inside the tube and acting as internal fins. A twisted tape-inserted tube is shown in Figure 3.6.

In a study by Sarmadian *et al.* [48], the impact of twisted tapes with twist ratios of 4, 10, and 15 on heat transfer coefficient and pressure drop was investigated using R600a as the working fluid. The experiment covered various vapor qualities and mass fluxes, revealing an increase in both pressure drop and heat transfer for tubes with inserts compared to plain tubes. However, they observed that for most mass flux ranges studied, the rise in pressure drop outweighed the increase in heat transfer coefficient, rendering the tape insertion impractical. Moreover, it was noted that smaller twist ratios correlated with higher increases in heat transfer coefficient and pressure drop.

Shishkin *et al.* [49] conducted a comparable investigation using R134a, exploring various twist ratios, tube diameters, mass fluxes, saturation temperatures, vapor qualities, and heat fluxes. The stainless-steel twisted inserts, covered with thermostable powder paint to electrically insulate them from tube walls, induced an annular flow regime at lower-than-usual vapor qualities. It was noted that varying mass flux had negligible effects on the heat transfer coefficient for mass and heat fluxes surpassing 400 $kg/m^2s$ and 100 $kW/m^2$, respectively, as turbulence dominated at higher fluxes, minimizing the impact of flow velocity fluctuations.

Kanazawa *et al.* [50] compared aluminum twisted tapes with twist ratios of 3, 4, 9, and 14 in tubes using R134a as the working fluid. Over a range of heat and mass fluxes, the authors reported up to a 45% improvement in heat transfer coefficient in tubes with tape inserts. Although this enhancement was accompanied by a substantial increase in frictional pressure drop

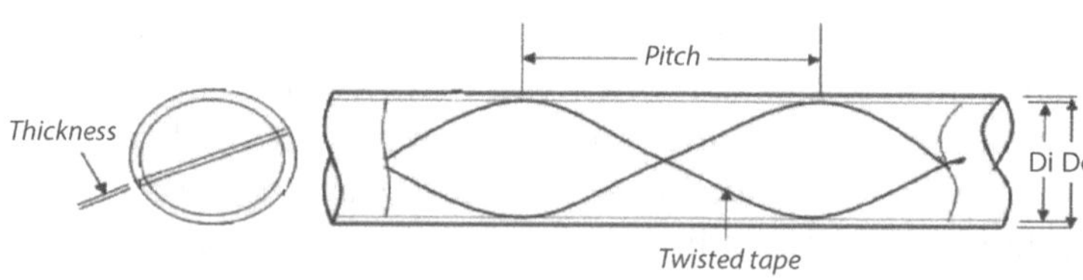


**Figure 3.6** Twisted tape inserted tube.

(up to 30 times), the overall heat transfer augmentation exceeded the rise in pressure drop in most operating conditions, endorsing the use of twisted tapes. The study further emphasized the significant influence of vapor quality on tube performance. Intermittent and annular flow conditions corresponding to certain vapor qualities witnessed improved heat transfer, while higher vapor qualities experienced a degradation in heat transfer. Similar trends were observed for pressure drops, with higher vapor qualities corresponding to increased friction due to the heightened fluid velocity during the liquid-vapor phase change.

As emphasized, the utilization of a twisted tape insert can enhance heat transfer through multiple mechanisms. Initially, the twisted tape constrains the fluid to follow its convoluted path, thereby increasing the overall flow length. Additionally, acting as a fin, the twisted tape augments the contact area between the tube and the flowing fluid. Moreover, the presence of a twisted tape induces swirl mixing, creating turbulence and vortices [66]. The resulting increase in axial velocity, along with the generation of vortices by the tapes, facilitates heightened vapor shear and condensate entrainment, contributing to further enhancement in the heat transfer coefficient [67]. However, it is important to note that employing twisted tape inserts for heat transfer enhancement introduces flow obstruction, leading to a significant pressure drop.

In a study utilizing R134a as the working fluid, Hejazi *et al.* [68] investigated the impact of a twisted tape on in-tube heat transfer during condensation. The findings indicated that while twisted tape inserts enhanced heat transfer coefficient, they also increased pressure drop. Under the most challenging test conditions, the pressure drop experienced was 2.4 times greater than that in a smooth tube. Nonetheless, under optimal test conditions, the condensation heat transfer coefficient increased by up to 40%. Overall, it was established that the additional pressure drop associated with the twisted insert was substantial enough to question its justification.

Agrawal *et al.* [69] investigated the heat transfer augmentation provided by segmented and full-length twisted tape inserts, using R22 as the working fluid and varying mass fluxes. The enhancement in heat transfer during condensation reached up to 25% for a full-length twisted tape insert. While segmented twisted tapes also enhanced the heat transfer coefficient, the improvement was relatively modest. This enhancement increased with the length of the segmented tape and was maximized for full-length tape inserts. The authors further noted that all tape inserts were generally more effective at low mass fluxes, where the additional enhancement in the flow field became more significant. Overall, it is evident that the heat transfer coefficient increases with the length of the tape segments and is highest for full-length tape inserts, which remained effective even at higher mass flow rates.

### 3.2.5 Wired Coils

Much like twisted tapes, wired coils contribute to flow agitation, enhancing fluid intermixing. However, this agitation is more concentrated on the boundary layer near the tube wall.

Behabadi *et al.* [51] investigated the impact of wired coils, with varying diameters and pitches, inserted into a 1.2-m long copper tube with an inner diameter of 7.5 mm, using R134a as the working fluid. The study revealed that the insertion of a coil could lead to a pressure drop increase of up to 1000% compared to plain tubes without inserts. The enhancement of heat transfer was found to be dependent on vapor quality, with the maximum increase in heat transfer coefficient reaching 98%. Additionally, it was observed that the heat transfer coefficient increased with decreasing coil pitch, up to a minimum of 8 mm; further reduction in pitch resulted in decreased heat transfer coefficient as the insert resisted heat transfer like a solid shell. Considering that the pressure drop caused by the inserts far exceeded the increase in heat transfer coefficient, the study concluded that wired coils are not a favorable heat transfer enhancement method.

Niezgoda-Żelasko [52] explored wired coils with different pitches and diameters in vertical tubes, using R507, R410, and R407C as working fluids. The heat transfer coefficient for azeotropic fluids increased by 1.1–1.7 times, and for zeotropic agents, it increased by 1.1–1.3 times, accompanied by a pressure drop increase of 1.8–4.5 times compared to plain tubes.

Alimardani *et al.* [53] conducted a similar study with R600a as the working fluid in horizontally oriented tubes, observing a 90–958% increase in pressure drop with wired coil insertion. However, this pressure drop decreased as the wire diameter decreased and the coil pitch increased.

Another prevalent type of inserts widely employed is wired coils. Functioning as additional roughness, wired coils disrupt the boundary layer of the fluid near the tube walls, resulting in improved heat transfer alongside an increase in pressure loss. In a study by Agrawal *et al.* [70], the impact of a coiled wire on R22 condensation inside a horizontal tube was examined, considering various wire diameters and coil pitches. The findings indicated that the presence of a wired coil insert could elevate the condensing heat transfer coefficient by up to 100%. Furthermore, the heat transfer coefficient enhancement showed an increase in wire diameter for high vapor qualities and a decrease for low vapor qualities. At high vapor quality, the reduction in the liquid film thickness on the inner walls of the tube contributed to a decrease in thermal resistance, a trend similarly observed in tubes with microfins and twisted tape inserts [71]. Additionally, for a given wire diameter and low vapor quality, heat transfer enhancement increased as the coil pitch decreased.

### 3.2.6 Dimpled Tubes

The use of dimpled tubes as shown in Figure 3.7 also facilitates better heat transfer for flow boiling and condensation. Shafaee *et al.* [54] explored a passive approach to heat transfer enhancement, focusing on the impact of dimples on a copper tube. The experiments were conducted under specific conditions, including a mass flux (G) range of 155–470 kg/m$^2$s, vapor quality below 0.8, heat flux at 15.8 kW/m$^2$, a saturation temperature of 56.5°C, and a tube with a 9.5-mm outer diameter and 1-m length. R600a, chosen for its lower ozone depletion potential, served as the working fluid. Dimples on the tubes were arranged with a pitch ratio of 1.21 and were of two types: shallow and deep. Shallow dimples had diameters of 1 mm and depths of 0.5 mm, while deep dimples had diameters of 2 mm and depths of 1 mm.

Results for both dimple types revealed two distinct regions based on vapor quality. In the low-vapor-quality region, an increase in vapor quality led to a decrease in the heat transfer coefficient, reaching a minimum between $0.15 < x < 0.45$. In the high-vapor-quality region, the heat transfer coefficient increased with vapor quality. Analysis of the nucleate boiling mechanism explained behavior in the low-vapor-quality regions. Nucleate boiling dominated at lower vapor qualities, but as quality increased and the annular flow regime initiated, liquid in contact with tube walls thinned (film effect). With high vapor velocity carrying heat away, wall superheats dropped below the threshold for bubble nucleation, reducing nucleate boiling and subsequently decreasing the heat transfer coefficient. In dimpled tubes, the vapor quality range where nucleate boiling dominated was

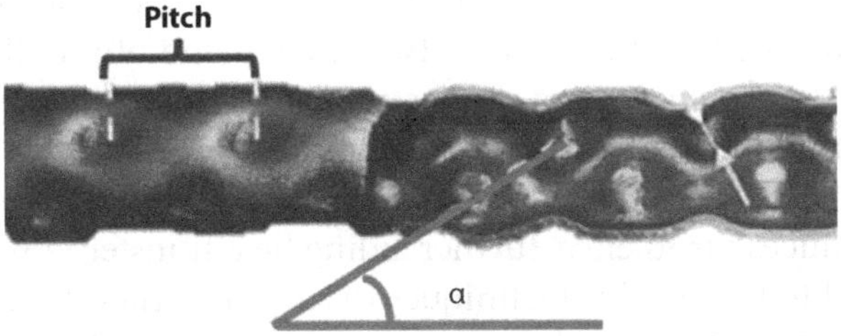


**Figure 3.7** Dimpled tube.

smaller than in plain tubes. Enhanced heat transfer during low vapor quality resulted from protrusions increasing superheated regions and creating more nucleation sites. In the high-vapor-quality region, R600a refrigerant vaporization caused a decrease in density, increasing flow velocity and promoting liquid convection boiling dominance. For dimpled tubes, this high-vapor-quality region was wider, and the heat transfer coefficient increase was more pronounced than in plain tubes.

Experiments demonstrated that the heat transfer coefficient of dimpled tubes was 1.29–2 times greater than plain tubes. Dimpled tubes promoted swirling near the wall, disrupting the boundary layer, increasing agitation, and fostering mixing and turbulence. The increased tube surface area also facilitated more nucleation sites. Mass flux played a crucial role in heat transfer, with increased mass flux leading to higher refrigerant velocity, liquid convection, and heat transfer coefficient. However, this influence became marginal when vapor quality was low.

## 3.3 Electro-Passive Heat Transfer Enhancement Techniques

The ultimate enhancement approach examined in two-phase flow boiling involves an active technique. In this approach, a dielectric fluid undergoes the influence of an electric field to induce directed motions. Similar to dimples and extended surfaces, electrohydrodynamics (EHD) has the potential to amplify heat transfer in both single-phase flow and two-phase flow boiling.

In the context of two-phase boiling, EHD contributes to heat transfer enhancement in multiple ways. Initially, at low heat flux values, heat is transferred through forced or natural convection. As the heat flux or superheat intensifies, the onset of nucleate boiling (ONB) [9] takes place, activating nucleation sites. The application of EHD has been demonstrated to initiate ONB at an earlier stage, reducing the hysteresis portion before ONB [72, 73] and elevating the critical heat flux (CHF) limit. Electrohydrodynamics also eliminates the heat flux drop between CHF and film boiling [72, 73]. Additionally, the application of an electric field reduces the surface tension of the dielectric fluid, enhancing wettability on the surface [74], analogous to the action of surfactants. Moreover, the increased frequency of bubble departure induces turbulence, further aiding heat transfer [73].

Compared to the passive techniques discussed earlier, EHD has a clear advantage in its ability to enhance every part of the boiling curve, unlike passive techniques focusing solely on nucleate boiling [72]. However, active methods are less commonly employed due to their inherent complexity.

During condensation, less volatile substances condense initially. Over time, the vapor near the free surface experiences an increase in the concentration of volatile liquids, leading to the development of a concentration gradient between this vapor region and the bulk flow. Consequently, diffusion ensues, impeding the flow of condensing vapor and resulting in a decrease in heat transfer due to the heightened interfacial resistance. Electrohydrodynamics (EHD) is employed to mitigate this resistance and enhance heat transfer [75].

## 3.4 Conclusion

This chapter provides an extensive overview of heat transfer enhancement techniques for two-phase flows, crucial for the advancement of next-generation heat exchangers associated with HVAC systems. The first method examined involved incorporating surfactants into the two-phase fluid, a non-intrusive and cost-effective approach requiring no alterations to existing designs. However, surfactants are most effective at lower heat fluxes and with higher surface tension fluids, limiting their applicability. Increased heat flux may cause surfactant detachment from the tube surface, rendering the technique less effective, and potential concerns arise in industries, such as food, where surfactants could be deemed contaminants. The second set of techniques involved inserts like coils or tapes, offering a cost-effective means to enhance in-tube heat transfer. While easy to install, these inserts may require occasional replacement due to wear or fouling. Other techniques, such as extended surfaces, dimples, and ribs, directly modify tube wall geometry. Dimpled tubes or tubes with ribs incur less maintenance cost and have minimal risk of blockage, but their implementation in existing systems is challenging. Microfins on tubes, despite being relatively challenging to manufacture, offer substantial heat transfer enhancement, particularly when compared to pressure drop increase. For evaporative two-phase flows, microfins are most beneficial at lower mass fluxes. Fins also facilitate self-drainage in condensation applications, reducing liquid film resistance. Herringbone tubes, akin to microfin tubes but with a higher heat transfer enhancement, may result in larger pressure drops and increased manufacturing complexity. Active techniques, specifically EHD, offer substantial heat transfer augmentation and controllable enhancement levels. However, their integration into pre-existing systems is challenging and may pose safety hazards, with the added limitation of operating solely with dielectric fluids. The choice of the most suitable technique depends on specific circumstances. Electrohydrodynamics offers significant enhancement but is costly and complex to implement, making it less prevalent in the industry. If ease of implementation is paramount, altering fluid behavior with surfactants proves economical but may not be suitable in certain applications. Microfins strike a balance, being widely adopted in industries due to their advantages and acceptable limitations. The optimal microfin geometry varies and is application-dependent, with commercially available vendors assisting in decision-making.

## References

1. Sheikholeslami, M., Gorji-Bandpy, M., Ganji, D.D., Review of Heat Transfer Enhancement Methods: Focus on Passive Methods Using Swirl Flow Devices. *Renew. Sustain. Energy Rev.*, 49, 444–469, 2015.
2. Sheikholeslami, M. and Bhatti, M., Active Method for Nanofluid Heat Transfer Enhancement by Means of EHD. *Int. J. Heat Mass Transf.*, 109, 115–122, 2017.
3. Nations, U., Special edition: progress towards the Sustainable Development Goals, 2019.
4. Sheikholeslami, M., Abohamzeh, E., Jafaryar, M., Shafee, A., Babazadeh, H., CuO nanomaterial two-phase simulation within a tube with enhanced turbulator. *Powder Technol.*, 373, 1–13, 2020, https://doi.org/10.1016/j.powtec.2020.06.017.
5. He, W., Toghraie, D., Lotfipour, A., Pourfattah, F., Karimipour, A., Afrand, M., Effect of twisted- tape inserts and nanofluid on flow field and heat transfer characteristics in a tube. *Int. Commun. Heat Mass Transf.*, 110, 104440, 2020, https://doi.org/10.1016/j.icheatmasstransfer.2019.104440.
6. Eiamsa-ard, S., Wongcharee, K., Eiamsa-ard, P., Thianpong, C., Heat transfer enhancement in a tube using delta-winglet twisted tape inserts. *Appl. Therm. Eng.*, 30, 310–8, 2010, https://doi.org/10.1016/j.applthermaleng.2009.09.006.
7. Habibi Khalaj, A. and Halgamuge SK, A., Review on efficient thermal management of air- and liquid-cooled data centers: From chip to the cooling system. *Appl. Energy*, 205, 1165–88, 2017, https://doi.org/10.1016/j.apenergy.2017.08.037.
8. Li, J., Kang, D., Fazle Rabbi, K., Fu, W., Yan, X., Fang, X. *et al.*, Liquid film–induced critical heat flux enhancement on structured surfaces. *Sci. Adv.*, 7, eabg4537, 2021, https://doi.org/10.1126/sciadv.abg4537.
9. Collier, J.G. and Thome, J.R., *Convective Boiling and Condensation*, Third Edition, Oxford Science Publication, Clarendon Press, Oxford, 1994.
10. Ishii, M. and Hibiki, T., *Thermo-Fluid Dynamics of Two-Phase Flow*, Second Edition, Springer, New York, ISBN: 978-1-4419-7984-1, 2011.

11. Callizo, C.M., Flow Boiling Heat Transfer in Single Vertical Channels of Small Diameter, Doctoral Thesis, Division of Applied Thermodynamics and Refrigeration, Department of Energy Technology, Royal Institute of Technology, Stockholm, Sweden, ISBN: 978-91-7415-780-2, 2010.
13. Chen, Y. and Chen, H., Dispersed Flow Film Boiling Heat Transfer of Flowing Water in Vertical Tubes — CIAE Steady State Data and Prediction Methods, China Institute of Atomic Energy Report No- XA0055010.
14. Chen, Y., Heat Transfer in Film Boiling of Flowing Water. *Heat Transfer - Theoretical Analysis", Experimental Investigations and Industrial Systems*, In Tech, pp. 235–260, 2011, ISBN: 978-953-307-226-5.
15. Kandlikar, S.G., A General Correlation for Saturated Two-Phase Flow Boiling Heat Transfer Inside Horizontal and Vertical Tubes. *J. Heat Transfer*, 112, 119–228, 1990.
16. Gungor, K.E. and Winterton, R.H.S., A General Correlation for Flow Boiling in Tubes and Annuli. *Int. J. Heat Mass Transf.*, 29, 3, 351–358, 1986.
17. Kim, S.M. and Mudawar, I., Universal Approach to Predicting Saturated Flow Boiling Heat Transfer in Mini/Micro Channels-Part-II. Two-Phase Heat Transfer Coefficient. *Int. J. Heat Mass Transf.*, 64, 1239–1256, 2013.
18. Kim, S.M. and Mudawar, I., Universal Approach to Predicting Saturated Flow Boiling Heat Transfer in Mini/Micro Channel-Part-I. Dryout Incipience Quality. *Int. J. Heat Mass Transfer*, 64, 1226–1238, 2013.
19. Shah, M.M. and Siddiqui, M.A., A General Correlation for Heat Transfer During Dispersed-Flow Film Boiling in Tubes. *Heat Transf. Eng.*, 21, 18–32, 2000.
20. Groeneveld, D.C., Leung, L.K.H., Vasic, A.Z., Guo, Y.J., Cheng, S.C., A Look-Up Table for Fully Developed Film-Boiling Heat Transfer. *Nucl. Eng. Des.*, 225, 83–97, 2003.
21. Hammouda, N., Groeneveld, D.C., Cheng, S.C., Two-Fluid Modeling of Inverted Annular Film Boiling. *Int. J. Heat Mass Transf.*, 40, 11, 2655–2670, 1997.
22. Agafonova, N.D. and Paramonova, I.L., Heat Transfer in Inverted Annular Mode of Steam–Water Flow. *Therm. Eng.*, 60, 3, 176–181, 2013.
23. Hossain, M.N. and Ghosh, K., A comparative study of plain and enhanced helical microfin evaporator tube with drift flux-based entropy generation model for R32 and R410A refrigerants. *Int. J. Heat Mass Transf.*, 215, 2023, https://doi.org/10.1016/j.ijheatmasstransfer.2023.124441.
24. Wallis, G.B., *One-dimensional two-phase flow*, McGraw Hill Publication, 1969.
25. Hossain, M.N., Ghosh, K., Manna, N.K., Integrated thermal modeling, analysis, and sequential design of heat exchanger surfaces of a natural circulation RDF steam generator including evaporator tubes. *Appl. Therm. Eng.*, 211, 118455, 2022, https://doi.org/10.1016/j.applthermaleng.2022.118455.
26. Tzan, Y.L. and Yang, Y.M., Experimental Study of Surfactant Effects on Pool Boiling Heat Transfer. *J. Heat Transf.*, 112, 207, 1990, https://doi.org/10.1115/1.2910346.

27. Yang, Y.M. and Maa, J.R., Pool Boiling of Dilute Surfactant Solutions. *J. Heat Transfer*, 105, 190, 1983.
28. Yang, Y.M., Dynamic surface effect on boiling of aqueous surfactant solutions. *Int. Commun. Heat Mass Transf.*, 17, 711–27, 1990, https://doi.org/10.1016/0735- 1933(90)90018-F.
29. Wu, W.-T., Yang, Y.-M., Maa, J.-R., Enhancement of Nucleate Boiling Heat Transfer and Depression of Surface Tension by Surfactant Additives. *J. Heat Transf.*, 117, 526, 1995, https://doi.org/10.1115/1.2822558.
30. Qiu, Y., Chen, W., Si, Q., Enhancement of flow boiling heat transfer with surfactant. *J. Cent. South Univ. Technol.*, 7, 219–22, 2007, https://doi.org/10.1007/s11771-000-0058-0.
31. Hetsroni, G., Gurevich, M., Mosyak, A., Rozenblit, R., Drag reduction and heat transfer of surfactants flowing in a capillary tube. *Int. J. Heat Mass Transf.*, 47, 3797–809, 2004, https://doi.org/10.1016/j.ijheatmasstransfer.2004.03.009.
32. K. Fujie, M. Itoh, T. Innami, H. Kimura, W. Nakayama, T. Yanugidi, . US patent No. 4044797 A, 1977.
33. Newell, T.A. and Shah, R.K., An Assessment of Refrigerant Heat Transfer, Pressure Drop, and Void Fraction Effects In Microfin Tubes. *HVAC&R Res.*, 7, 125–53, 2001, https://doi.org/10.1080/10789669.2001.10391267.
34. Hossain, Md.N. and Ghosh, K., A comparative study of plain and enhanced helical microfin evaporator tube with drift flux-based entropy generation model for R32 and R410A refrigerants. *Int. J. Heat Mass Transf.*, 215, 124441, 2023, https://doi.org/10.1016/j.ijheatmasstransfer.2023.124441.
35. Schlager, L.M., Pate, M.B., Bergles, A.E., Evaporation and Condensation Heat Transfer and Pressure Drop in Horizontal, 12.7-mm Microfin Tubes With Refrigerant 22. *J. Heat Transf.*, 112, 1041, 1990, https://doi.org/10.1115/1.2910476.
36. Zhang, X., Zhang, J., Ji, H., Zhao, D., Heat transfer enhancement and pressure drop performance for R417A flow boiling in internally grooved tubes. *Energy*, 86, 446–54, 2015, https://doi.org/10.1016/j.energy.2015.04.054.
37. Jiang, G.B., Tan, J.T., Nian, Q.X., Tang, S.C., Tao, W.Q., Experimental study of boiling heat transfer in smooth/micro-fin tubes of four refrigerants. *Int. J. Heat Mass Transf.*, 98, 631–42, 2016, https://doi.org/10.1016/j.ijheatmasstransfer.2016.03.024.
38. Schlager, L., Pate, M., Bergles, A., Heat transfer and pressure drop during evaporation and condensation of R22 in horizontal micro-fin tubes. *Int. J. Refrig.*, 12, 6–14, 1989, https://doi.org/10.1016/0140-7007(89)90006-6.
39. Oh, S.-Y. and Bergles, A.E., Microfin Tube Technology — the Effects of Spiral Angle on Evaporative Heat Transfer Enhancement, in: *Heat Transf. Enhanc. Heat Exch*, pp. 67–78, Springer Netherlands, Dordrecht, 1999, https://doi.org/10.1007/978-94-015-9159-1_5.
40. H.K. and M.I., Evaporating Heat Transfer in Horizontal Internal Spiral-grooved Tubes in the Region of Low Flow Rates. *JSME Int. J.*, 24, 1602–7, 1981.

41. Yang, C.-M. and Hrnjak, P., Effect of helical micro-fins on two-phase flow behavior of R410A evaporating in horizontal round tubes obtained through visualization. *Int. J. Heat Mass Transf.*, 144, 118654, 2019, https://doi.org/10.1016/j.ijheatmasstransfer.2019.118654.
42. Chiang, R., Heat transfer and pressure drop during evaporation and condensation of refrigerant-22 in 7.5 mm and 10mm diameter axial and helical grooved tubes. *AlChE J.*, 89, 205–10, 1993.
43. Wu, X., Zhu, Y., Huang, X., Influence of 0° helix angle micro fins on flow and heat transfer of R32 evaporating in a horizontal mini multichannel flat tube. *Exp. Therm Fluid Sci.*, 68, 669–80, 2015, https://doi.org/10.1016/j.expthermflusci.2015.07.010.
44. Yang, C.-M. and Hrnjak, P., Effect of straight micro fins on heat transfer and pressure drop of R410A during evaporation in round tubes. *Int. J. Heat Mass Transf.*, 117, 924–39, 2018, https://doi.org/10.1016/j.ijheatmasstransfer.2017.10.064.
45. Yang, C.-M. and Hrnjak, P., Visualization of two-phase flow of R410A in horizontal smooth and axial micro-finned tubes. *Int. J. Heat Mass Transf.*, 138, 49–58, 2019, https://doi.org/10.1016/j.ijheatmasstransfer.2019.03.174.
46. Wellsandt, S. and Vamling, L., Prediction method for flow boiling heat transfer in a herringbone microfin tube. *Int. J. Refrig.*, 28, 912–20, 2005, https://doi.org/10.1016/j.ijrefrig.2005.01.009.
47. Olivier, J.A., Liebenberg, L., Thome, J.R., Meyer, J.P., Heat transfer, pressure drop, and flow pattern recognition during condensation inside smooth, helical micro-fin, and herringbone tubes. *Int. J. Refrig.*, 30, 609–23, 2007, https://doi.org/10.1016/j.ijrefrig.2006.11.003.
48. Sarmadian, A., Moghaddam, H.A., Asnaashari, A., Joushani, H.A.N., Moosavi, M., Islam, M.S. *et al.*, Flow boiling heat transfer and pressure drop characteristics of Isobutane in horizontal channels with twisted tapes. *Int. J. Heat Mass Transf.*, 162, 120345, 2020, https://doi.org/10.1016/j.ijheatmasstransfer.2020.120345.
49. Shishkin, A., Kanizawa, F.T., Ribatski, G., Tarasevich, S., Yakovlev, A., Experimental investigation of the heat transfer coefficient during convective boiling of R134a in tubes with twisted tape insert. *Int. J. Refrig.*, 92, 196–207, 2018, https://doi.org/10.1016/j.ijrefrig.2018.06.002.
50. Kanizawa, F.T., Mogaji, T.S., Ribatski, G., Evaluation of the heat transfer enhancement and pressure drop penalty during flow boiling inside tubes containing twisted tape insert. *Appl. Therm. Eng.*, 70, 328–40, 2014, https://doi.org/10.1016/j.applthermaleng.2014.05.029.
51. Akhavan-Behabadi, M.A., Mohseni, S.G., Najafi, H., Ramazanzadeh, H., Heat transfer and pressure drop characteristics of forced convective evaporation in horizontal tubes with coiled wire inserts. *Int. Commun. Heat Mass Transf.*, 36, 1089–95, 2009, https://doi.org/10.1016/j.icheatmasstransfer.2009.07.009.
52. Niezgoda-Żelasko, B., Heat Transfer and Pressure Drop Measurements During Boiling in Vertical Tubes with Coiled-Wire Inserts. *Heat Transf. Eng.*, 40, 1633–48, 2019, https://doi.org/10.1080/01457632.2018.1480880.

53. Alimardani, F., Moghaddam, H.A., Sarmadian, A., Shafaee, M., Pressure loss and performance assessment of horizontal spiral coil inserted pipes during forced convective evaporation of R-600a. *Int. J. Refrig.*, 107, 20–30, 2019, https://doi.org/10.1016/j.ijrefrig.2019.08.016.
54. Shafaee, M., Mashouf, H., Sarmadian, A., Mohseni, S.G., Evaporation heat transfer and pressure drop characteristics of R-600a in horizontal smooth and helically dimpled tubes. *Appl. Therm. Eng.*, 107, 28–36, 2016, https://doi.org/10.1016/j.applthermaleng.2016.06.148.
55. Weisensee, P.B., Wang, Y., Qian, H., Schultz, D., King, W.P., Miljkovic, N., Condensate droplet size distribution on lubricant-infused surfaces. *Int. J. Heat Mass Transf.*, 109, 187–99, 2017, https://doi.org/10.1016/j.ijheatmasstransfer.2017.01.119.
56. Miljkovic, N. and Wang, E.N., Condensation heat transfer on superhydrophobic surfaces. *MRS Bull.*, 38, 397–406, 2013, https://doi.org/10.1557/mrs.2013.103.
57. Cho, H.J., Preston, D.J., Zhu, Y., Wang, E.N., Nanoengineered materials for liquid–vapour phase-change heat transfer. *Nat. Rev. Mater.*, 2, 16092, 2017, https://doi.org/10.1038/natrevmats.2016.92.
58. Shahriari, A., Birbarah, P., Oh, J., Miljkovic, N., Bahadur, V., Electric Field–Based Control and Enhancement of Boiling and Condensation. *Nanoscale Microscale Thermophys. Eng.*, 21, 102–21, 2017, https://doi.org/10.1080/15567265.2016.1253630.
59. Jung, D., Cho, Y., Park, K., Flow condensation heat transfer coefficients of R22, R134a, R407C, and R410A inside plain and microfin tubes. *Int. J. Refrig.*, 27, 25–32, 2004, https://doi.org/10.1016/S0140-7007(03)00122-1.
60. Kwon, J.-T., Park, S.K., Kim, M.H., Enhanced Effect of a Horizontal Micro-fin Tube for Condensation Heat Transfer with R22 and R410A. *J. Enhanced Heat Transf.*, 7, 97–107, 2000, https://doi.org/10.1615/JEnhHeatTransf.v7.i2.30.
61. Eckels, S.J. and Pate, M.B., Evaporation And Condensation Of Hfc-134a And Cfc-12 In A Smooth Tube And A Micro-Fin Tube. *ASHRAE Trans.*, 97, 68–78, 1389.
62. Eckels, S.J., Doerr, T.M., Pate, M.B., Heat transfer coefficients and pressure drops for R-134a and an ester lubricant mixture in a smooth tube and a micro-fin tube. *ASHRAE Trans.*, 1998.
63. Cho, K. and Tae, S.-J., Condensation heat transfer for R-22 and R-407C refrigerant–oil mixtures in a microfin tube with a U-bend. *Int. J. Heat Mass Transf.*, 44, 2043–51, 2001, https://doi.org/10.1016/S0017-9310(00)00253-2.
64. Miyara, A., Nonaka, K., Taniguchi, M., Condensation heat transfer and flow pattern inside a herringbone-type micro-fin tube. *Int. J. Refrig.*, 23, 141–52, 2000, https://doi.org/10.1016/S0140-7007(99)00037-7.
65. Miyara, A., Otsubo, Y., Ohtsuka, S., Mizuta, Y., Effects of fin shape on condensation in herringbone microfin tubes. *Int. J. Refrig.*, 26, 417–24, 2003, https://doi.org/10.1016/S0140- 7007(02)00154-8.

66. Manglik, R.M., Heat transfer enhancement of intube flows in process heat exchangers by means of twisted-tape inserts, Rensselaer Polytechnic Institute, 1991.
67. Kreith, F. and Margolis, D., Heat transfer and friction in turbulent vortex flow. *Appl. Sci. Res.*, 8, 457–73, 1959, https://doi.org/10.1007/BF00411769.
68. Hejazi, V., Akhavan-Behabadi, M.A., Afshari, A., Experimental investigation of twisted tape inserts performance on condensation heat transfer enhancement and pressure drop. *Int. Commun. Heat Mass Transf.*, 37, 1376–87, 2010, https://doi.org/10.1016/j.icheatmasstransfer.2010.07.021.
69. Agrawal, K.N., Kumar, R., Lal, S.N., Varma, H.K., Heat Transfer Augmentation by Segmented Tape Inserts During Condensation of R-22 Inside a Horizontal Tube. *ASHRAE Trans.*, 110, 143–50, 2004.
70. Agrawal, K.N., Kumar, A., Akhavan Behabadi, M.A., Varma, H.K., Heat transfer augmentation by coiled wire inserts during forced convection condensation of R-22 inside horizontal tubes. *Int. J. Multiphase Flow*, 24, 635–50, 1998, https://doi.org/10.1016/S0301-9322(97)00061- X.
71. Said, S.A. and Azer, N.Z., Heat transfer and pressure drop during condensation inside horizontal tubes with twisted tape inserts. *ASHRAE Trans.*, 89, 96–113, 1983.
72. Allen, P.H.G. and Karayiannis, T.G., Electrohydrodynamic enhancement of heat transfer and fluid flow. *Heat Recov. Syst. CHP*, 15, 389–423, 1995, https://doi.org/10.1016/0890- 4332(95)90050-0.
73. Karayiannis, T.G., Collins, M.W., Allen, P.H.G., Electrohydrodynamic Enhancement Of Nucleate Boiling Heat Transfer In Heat Exchangers. *Chem. Eng. Commun.*, 81, 15–24, 1989, https://doi.org/10.1080/00986448908940528.
74. Eames, I.W. and Sabir, H.M., Potential benefits of electrohydrodynamic enhancement of two-phase heat transfer in the design of refrigeration systems. *Appl. Therm. Eng.*, 17, 79–92, 1997, https://doi.org/10.1016/1359-4311(96)00012-9.
75. Gidwani, A., Molki, M., Ohadi, M.M., EHD-Enhanced Condensation of Alternative Refrigerants in Smooth and Corrugated Tubes. *HVAC&R Res.*, 8, 219–37, 2002, https://doi.org/10.1080/10789669.2002.10391439.
76. Kundu, A., Kumar, A., Dutt, N., Meena, C.S., Singh, V.P., Introduction to Thermal Energy Resources and their Smart Applications, in: *Thermal Energy Systems: Design, Computational Techniques and Applications*, A. Kumar, V.P. Singh, C.S. Meena, N. Dutt, (Eds.), pp. 1–15, CRC Press, Boca Raton, FL, USA, 2023, Chapter 1; http://dx.doi.org/10.1201/9781003395768-1.
77. Kundu, A., Kumar, A., Dutt, N., Singh, V.P., Meena, C.S., Modelling and Simulation of Thermal Energy System for Design Optimization, in: *Thermal Energy Systems: Design, Computational Techniques and Applications*, A. Kumar, V.P. Singh, C.S. Meena, N. Dutt, (Eds.), pp. 103–140, CRC Press, Boca Raton, FL, USA, 2023, Chapter 7; http://dx.doi.org/10.1201/9781003395768-7.

78. Kumar, A., Singh, V.P., Meena, C.S., Dutt, N., *Thermal Energy Systems: Design, Computational Techniques and Applications*, Taylor & Francis (CRC Press), 2023, https://doi.org/10.1201/9781003395768.
79. Verma, V., Thangavel, S., Dutt, N., Kumar, A., Weerasinghe, R., (Eds.), *Highly Efficient Thermal Renewable Energy Systems: Design, Optimization and Applications,* CRC Press, Boca Raton, FL USA, 2024, https://doi.org/10.1201/9781003472629.

# 4

# Indoor Thermal Performance Enhancement of Sustainable Buildings

**D.B. Jani**

*Government Engineering College, Dahod, Gujarat Technological University (GTU), Gujarat, India*

***Abstract***

The traditional vapo ur compression and absorption cooling systems can be replaced successfully by desiccant-assisted dehumidification and cooling systems, which are heat-driven cooling devices. A rotating dehumidifier, also known as a desiccant wheel, is used to dehumidify moist process air as a main part of a desiccant-assisted dehumidification and cooling system. After that, the dry air is cooled down a little bit in a rotary regenerator (heat wheel), a sensible heat exchanger, and then cooled down even more with a sensible cooler. The system can be run either in ventilation or recirculation mode in an open cycle, or, more frequently, in a closed cycle. For the desiccant material used in rotary dehumidifier to renew, a heat source is required. For regeneration, low-grade heat at a temperature of around 60–80°C is suitable, therefore renewable energy such as solar energy or industrial waste heat can be used in a cost-effective, low energy, and continuously repeating cycle.

***Keywords*:** Desiccant dehumidifier, energy saving, regeneration, sustainable buildings

## List of Nomenclature

| | |
|---|---|
| COP | Coefficient of performance |
| E | Electrical energy consumption (kW) |
| EV | Expansion valve |

*E-mail*: dbjani@rediffmail.com; ORCID: https://orcid.org/0000-0002-0221-6572

Ashwani Kumar, Nitesh Dutt and Mukesh Kumar Awasthi (eds.) Heat Transfer Enhancement Techniques: Thermal Performance, Optimization and Applications, (87–104) © 2025 Scrivener Publishing LLC

| | |
|---|---|
| h | Enthalpy of humid air (kJ/kg dry air) |
| IEQ | Indoor air quality |
| $Q_{cc}$ | Cooling capacity of cooling coil (kW) |
| T | Dry bulb temperature (°C) |
| V | Flow rate of air stream ($m^3$/hr) |
| VCR | Vapor compression refrigeration |

*Greek letters*

| | |
|---|---|
| $\varepsilon$ | Effectiveness for dehumidifier |
| $\omega$ | Humidity ratio |

*Subscripts*

| | |
|---|---|
| 1,2, etc. | State numbers |
| CC | Cooling coil |
| DW | Desiccant wheel |
| HRW | Heat recovery wheel |
| HT | Heater |
| p | Process air |
| r | Regeneration air |

## 4.1 Introduction

There is a great demand for the development and circulation of electrical energy, thus the available energy must be used wisely. Every family uses between 40% and 45% of its electric power for HVAC (heating, ventilation, and air conditioning) systems. A huge amount of heat is rejected as waste by all of these systems to the environment can raise their load further. Therefore, the coefficient of performance of this vapor compression refrigeration (VCR)-based conventional cooling systems and their slight negative impact on the environment might both be increased by properly reclaiming this wasted heat. A VCR-integrated desiccant dehumidification and cooling hybrid systems can effectively remove sensible and latent load in hot and humid areas. Additionally, it lowers the energy consumption of a VCR system and raises COP. The temperature at which desiccant regenerates will be lowered from 60–80°C to 45–55°C when a polymer-based advanced desiccant dehumidification and cooling is integrated with a traditional VCR system. Thus, the condenser's heat rejection can aid in the dehumidifier's desiccant's renewal. These additional advantages have been demonstrated by desiccant-based air conditioners than by traditional dehumidification and cooling systems. As a result, their application is also sporadic for tertiary and residential buildings, primarily in situations when desiccant renewal can be obtained by employing available wasted heat. The need for air conditioning is growing due to increased occupant comfort needs, while environmental preservation and energy conservation are becoming increasingly important as the world's energy and environmental crises worsen. It is therefore more important than ever to develop innovative air conditioning systems that are both environmentally friendly and energy efficient. These demands may be well met by the desiccant-assisted dehumidification and cooling systems coupled with VCS powered by low-grade heat sources; in the meantime, they offer a perfect location for the use of waste heat released from nearby companies, which causes less environmental harm [1–3].

A desiccant is a material or chemical that draws moisture from the humid air and absorbs it, leaving an area around it dry (adsorption). This chemical is used usually in industry to draw moisture from moist air, remove it, and keep the items dry, which stops corrosion. In order to guard against moisture damage, these chemicals are put in product containers during transportation. Additionally, drips stop moisture from condensing between window panes. Desiccants are widely utilized in numerous sectors, including manufacturing, electronics, packing, and manufacturing. They are available in a variety of solid or liquid forms [4, 5]. Desiccants may be divided into two main groups, each with unique chemical characteristics and functional needs. These categories consist of both liquid and solid desiccants. Rotating desiccant wheels, processed air, sensible cooling systems, and hot air for desiccant regeneration are the different components of desiccant cooling systems. Depending on the kind of desiccant material used, desiccant wheel air conditioning systems can be classified as follows: There are two types of desiccant air conditioning systems: (1) liquid and (2) solid.

By lowering the moisture content in the appropriate areas, desiccant-assisted dehumidification and cooling systems can prevent needless repairs from costing hundreds of rupees. Additionally, desiccant-assisted dehumidification and cooling systems can improve the quality of indoor air, increase air ventilation rate, and eliminate odors and inborn air pollutants. Similarly, desiccants cooling systems can be used in restaurants, schools, hospitals, and other commercial spaces to eliminate pollutants, odors, and moisture from the air for a more hygienic and productive atmosphere. Additionally, it can help to improve increased shelf life for chilled and frozen goods in cold supply chain [6–9].

An economic analysis and comparative study between a desiccant-assisted dehumidification and air conditioning system and a conventional

VCR system reveals that using hot ambient air during summer cooling season and freely available renewable solar thermal energy for desiccant regeneration may improve the performance of the hybrid desiccant system further [10–12].

## 4.2 Background of the Present Study

Air conditioning systems are necessary for human indoor thermal comfort in areas with hot and humid climates, such as those close to the coast, in tropical regions, and in green rain forests. Air conditioning systems also effectively control humidity while controlling indoor air quality (IAQ) it effectively meeting the building's cooling requirements in terms of temperature drop and effective moisture control is a challenge for air conditioning systems, particularly in extreme hot and humid weather [13]. The proper design of air conditioning systems is essential because hot and humid countries can experience extreme temperatures and moisture control for high latent heat load due to high occupancy or use of ventilated air in great proportion. Improper air conditioning system design frequently results in lacking the fresh air supply because of inadequate ventilation resulted to serious health risks, decreased thermal comfort, and increased energy usage as well as high $CO_2$ concentrations [14–17].

Vapor compression type conventionally used refrigeration systems based on mechanical compression cooling are well-liked by HVAC engineers as well as designers and builders. This is a result of the system's simplicity, widespread use among designers, and ease of commercial availability. The temperature for conditioned air is the only parameter that can be handled properly by VCR system design, which treats both the sensible and latent loads of a given space by the same air handling unit. Humidity is handled separately as a byproduct of the temperature control. Even in excess moist environment, where outdoor humidity levels are very high, there is no direct control over the humidity. The system design's intrinsic limitation allows for the achievement of the demanded conditions by indoor temperature requirement, but the relative humidity was not effectively handled due to moist air outdoor environment. Moreover, the overcooling that causes condensate to evaporate from the unit and into the conditioned space is another disadvantage of traditional air conditioners. This can lead to higher moisture concentration in indoor space. An evaporative cooler and a rotating desiccant dehumidifier are the two main parts of a desiccant cooling system. Freely available solar heat and surplus industrial waste heat from cogeneration system can be used to regenerate the desiccant dehumidifier. Due to the system's ability to handle temperature and humidity independently, it can function on a broad range of sensible humidity ratios [18–21].

Therefore, desiccant dryer-assisted air dehumidification and air cooling systems for reducing humidity and temperature will become popular in the future due to the use of renewable-based sustainable energy sources such as renewable energy and energy products. Evaporative cooling systems are effective cooling devices used for cooling purposes in hot climates. However, in low humidity conditions such as the 72–78% range, many researchers have proposed combined desiccant dehumidifier and evaporative cooler technologies (also known as desiccant-assisted evaporative cooling systems). Previous studies have also reported that all-season dryer-assisted evaporative cooling systems provide energy savings of up to 54% in hot and humid seasons [22].

Although successful today, desiccant-enhanced evaporative cooling systems are still under development and further research and collaboration is required to develop these systems suited to especially hot and humid climatic conditions. This study entails dryer-based evaporative cooling applications and energy recovery from desiccant-assisted hybrid cooling systems, focusing its application especially for hot and humid environmental areas. In commercial and large-scale applications, desiccant dehumidification is the most suitable way to remove water vapor from moist air while carrying comfort air conditioning for required indoor conditions [23–25].

## 4.3 System Operation

After being brought the incoming moist room air by passing through process air flow section of dehumidifier, desiccant material in the rotary desiccant wheel's process segment is then contacted with humid air. Because of the more humid air state, processed air's vapour pressure is higher than the desiccant's dry surface vapour pressure when it streams over the desiccant exterior in the adsorption segment, creating a noticeable difference in vapour pressure. As a result, the adsorption sector removes water vapour from damp air. This water vapour condenses above the desiccant after being adsorbed on its exterior. Thus, adsorption takes place. After that, the air is sensibly cooled to the design temperature by passing over the cooling coil of the sensible cooler. Desiccant surface heating occurs in the regeneration segment when restoration air passes over the desiccant exterior. Its increased vapour pressure causes the vapour pressure difference between

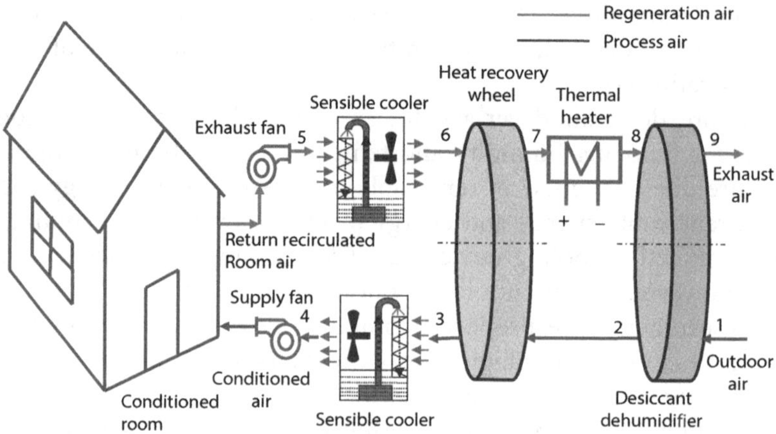


**Figure 4.1** Desiccant dehumidification-based sustainable cooling system (ventilation mode).

the desiccant surface and regeneration air to reverse. As a result, moisture in a desiccant layer is removed by restoration air. Thus, there is regeneration of desiccant material stored in dehumidifier [26–28].

The dehumidification-integrated cooling system consists of a rotary dehumidifier, air-to-air heating heat recovery wheel, sensible cooling coil and regenerative heat source, as shown in Figure 4.1 for the ventilation mode and Figure 4.2 shows recirculation mode. In ventilation mode, the fresh ambient air is used in the ventilation system for dehumidification. The main dehumidifying wheel dehumidifies the process air chamber but increases its temperature slightly (case points 1–2). The initial cooling of this heating and drying process occurs in the air-to-air heating reverse wheel (state points 2–3), where the humidity is constant throughout the cooling period. Before being supplied to the conditioned room indoor space, the dehumidifier process air is further cooled by passing over the sensible cooling coil (state point 3–4) where the air humidity almost remains constant but kept well below the comfort requirement. As the supply air enters the conditioned room indoor space, its temperature and humidity both maintained according to the room supply design conditions. The conditioned room air leaving the indoor room space is directed to the exhaust where it undergoes several psychrometric processes starting with cooling in an evaporative cooler (state point 5–6). This brings the air close to the saturation line with or without humidification, depending on the evaporative cooling process (direct or indirect). The air leaving the sensible cooler is then directed to the opposite in the regeneration side, first in the heat recovery wheel, so the air is preheated (state point 6–7) before it passes through the heater (state point 7–8) where its

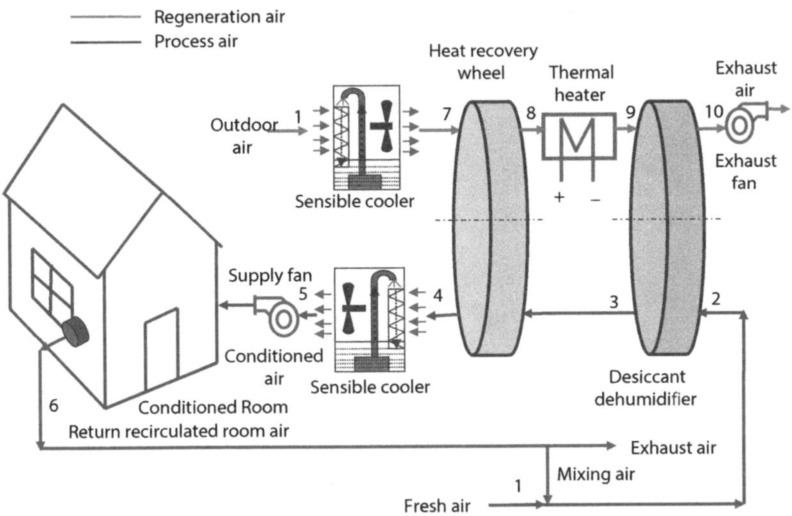


**Figure 4.2** Desiccant dehumidification-based sustainable cooling system (recirculation mode).

temperature is further raised so that it will be able to desorb the water from the desiccant in the regeneration side of the desiccant wheel (state point 8–9). Regenerative heat sources can be electricity, solar thermal systems or waste. The rotor type can be selected with different treatments for the air (R/P) ratio, *e.g.*, 25:75, 50:50, 75:25, etc. It can be divided between dehumidification and regeneration areas. In case of recirculation mode room return room air is feed to the dehumidifier for effective moisture control [29, 30].

A comparison of ventilation (blue line) and recirculation (red line) mode is shown in the psychrometric diagram in Figure 4.3. From the diagram, it can be seen that switching the recirculation to operating mode when the ambient humidity is high will save the required energy to reboot. At the same time, especially in hot and humid climates, the return air for home dehumidification can be used at lower temperature and humidity than the outside air, resulting in more cooling in the internal system. Therefore, less energy is needed for cooling, while reactivation energy is saved due to better heat exchange between the two air streams. Ultimately leading to higher performance of system operation [31–34].

Figure 4.4 illustrates how heat recovery from waste heat rejected by the VCR (vapor compression refrigeration) condenser can also be used to regenerate the desiccant used in rotary dehumidifiers. As a result, the condenser cools more effectively and the solid desiccant-based hybrid comfort space cooling system performs better overall. Therefore, VCR condenser

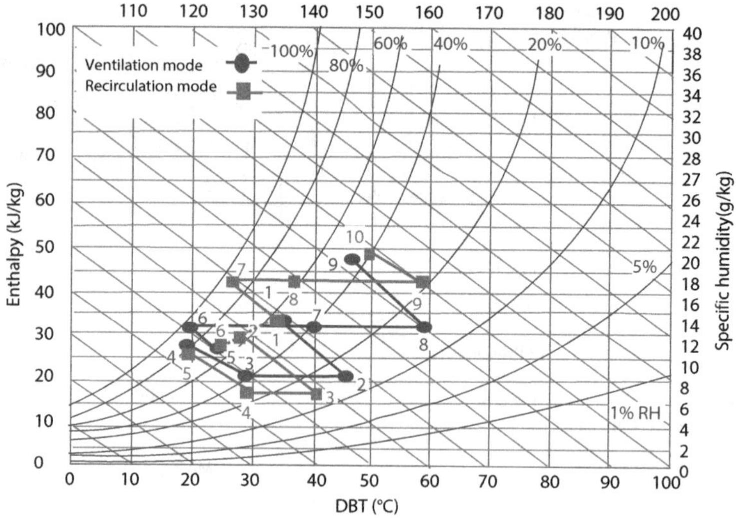


**Figure 4.3** Comparison among ventilation and recirculation working modes of desiccant-based sustainable cooling.

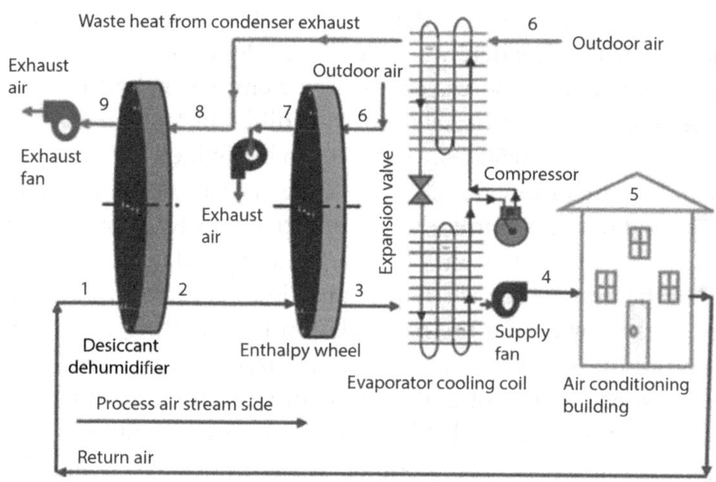


**Figure 4.4** Schematic layout of desiccant-assisted dehumidification and cooling system.

exhaust heat can also be used to meet the dehumidifier's reactivation heat requirement in solid desiccant-based vapor compression-integrated hybrid cooling; otherwise, the heat is expelled into the open atmosphere. This demonstrates that using waste heat for dehumidifier reactivation not only results in higher energy savings but also in a slight decrease in system operating costs, making the system economically feasible [35–38]. A number of studies have been carried out recently with the primary goal of improving the overall performance of desiccant-assisted dehumidification and cooling systems through the development of novel desiccant materials, novel system configurations, improved system designs and controls, and the integration of various hybrid energy sub-systems technologies [39–41].

As mentioned above, the main disadvantage of desiccant-assisted air conditioning is that it requires a large amount of renewable energy sources for the desorption of saturated desiccant during part of the exchange process. If this heat can be provided free of charge from renewable energy

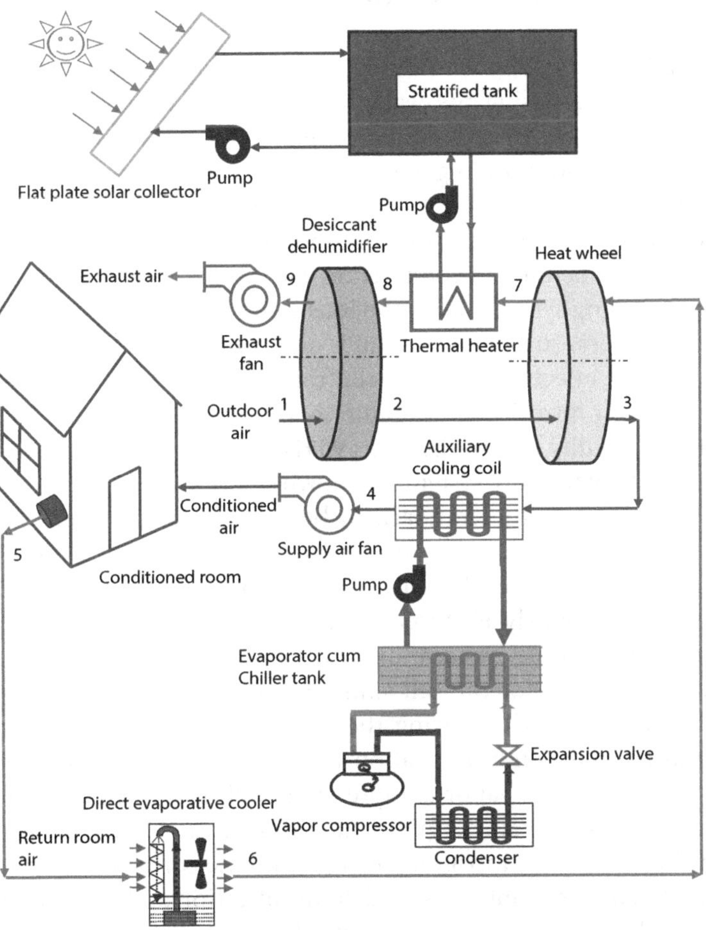


**Figure 4.5** Regenerating the desiccant wheel by using renewable solar heat in vapor compression desiccant-integrated hybrid cooling system.

sources such as solar energy, the system will be efficient. In addition, the increase in cooling need also coincides with solar radiation. Different solar products can regenerate the dryer by regenerating the heat generated by solar heat, as shown in Figure 4.5. The heating system used in winter or seasonal weather causes solar radiation to be insufficient to produce the necessary heat. Using solar energy to regenerate the dryer wheel is an interesting option to meet the needs of cold storages because solar energy has been found to be highest at the same time [42].

## 4.4 Comparison of Desiccant Cooling with Traditional VCR Cooling

It is possible to regulate the moisture or water vapor content of conditioned air by either condensing the water vapor in VCR cooling or by utilizing appropriate absorbents in desiccant cooling. A desiccant system only dehumidifies the air before cooling it, in contrast to conventional air conditioners that cool and dehumidify the air at the same time. Furthermore, to keep the moisture content and temperature of conditioned room air constant, a desiccant system can be used in conjunction with an evaporative cooling system. Desiccant cooling systems were previously employed in the agricultural and industrial sectors, such as textile mills and post-harvest crop storage units for drying and effective moisture control. However, desiccant cooling systems have been introduced as an air conditioning can be an efficient way to control humidity due to the energy crisis and the need to develop more environmentally friendly systems as shown in Table 4.1 [43–45] .

Figure 4.6 shows the comparison between a desiccant-assisted dehumidification and cooling systems and traditional vapor compression air conditioning-based the electrical energy consumption. This shows the comparison for hourly electricity consumption of a conventional vapor compression cooling system and a desiccant-assisted dehumidification cooling system for the full summer cooling season in a tropical hot and humid climate. Comparing the cooling energy requirement for the desiccant-assisted dehumidification and cooling system as compared to that of the traditional vapor compression cooling system, it is found that the significant electrical energy consumption reduction could be achieved. The findings thus offer intriguing insight into the considerable energy-saving potential of desiccant-assisted dehumidification and cooling systems in tropical and subtropical humid climates [46].

**Table 4.1** Summary of comparison among VCR-integrated conventional cooling and desiccant-assisted dehumidification and cooling systems.

| Parameter for comparison | Traditional vapor compression cooling | Desiccant dehumidification-based innovative cooling |
|---|---|---|
| Dehumidification capacity | Average | High |
| Energy for running the system | Electrical power to run the compressor and fan | Renewable free energy like solar heat and industrial waste heat |
| Operating/running cost of system | High | About 42–48 % power savings |
| IEQ-indoor air quality | Poor | Large fresh ventilated air flow improves IEQ |
| Impact on environment | Harmful as prone to global warming | Green cooling technology |
| Control over temperature and humidity | Combined control with single cooling coil | Separate control by use of dehumidifier and individual cooling coil |
| Integration with evaporative cooling | No | Judiciously coupled to evaporative cooling for mild climate ambient |

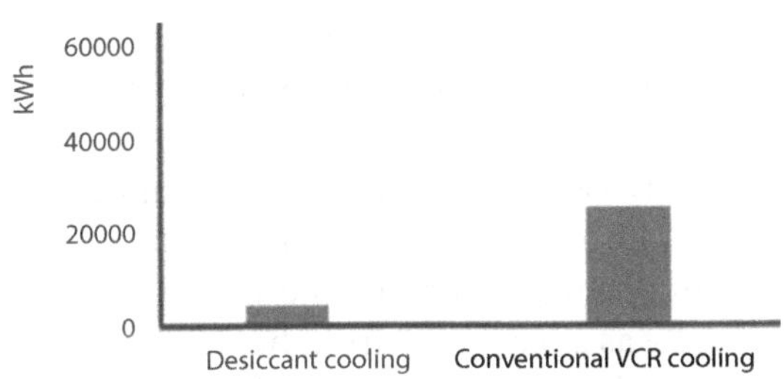


**Figure 4.6** Summary of electrical energy requirement for the cooling in desiccant-assisted dehumidification and conventional vapor compression cooling systems.

The conventional vapor compression cooling cycle system significantly contributes to major environmental effects like global warming, ozone layer depletion, and the greenhouse effect. Desiccant cooling systems must be used in place of conventional cooling systems in order to lessen these effects. While vapor compression air conditioners use some refrigerant to cool the air, desiccant-assisted dehumidification and cooling produces fresh air by cooling and dehumidifying. Unlike VC systems, desiccant-assisted dehumidification and cooling systems do not use refrigerant, therefore they do not have any negative effects. Because the working fluid in desiccant wheel air conditioning systems is natural, they have zero potential to cause global warming or ozone depletion. As a result, an eco-friendly air conditioning system can be represented by a desiccant-assisted dehumidification and cooling system. It is seen that the desiccant-assisted dehumidification and cooling systems can be regarded as affordable and environmentally friendly HVAC cooling technology that lowers significant environmental problems and ensures better air quality for cooling, both of which ensure better user health. There is a 13% reduction in carbon dioxide as compared to the conventional VCR cooling by the use of desiccant-assisted dehumidification and cooling system. The desiccant cooling method's primary benefit is its heat-driven cycle [47, 48]. As a result, it permits the use of various renewable energy sources, including freely available renewable solar energy and low-grade industrial waste heat energy. The running cost of the system is lowered or saved up to 87% when waste heat is impregnated into a desiccant-assisted dehumidification and cooling system.

The vapor compression cycle used to run a traditional HVAC unit can direct process air only as much as necessary. In order to cool, the temperature of the air layer must be well below the dew point, so that water vapor is freed from condensation in the cooling air. Cooling below the dew point can result in decreased humidity and overcooling of indoor air products. After the temperature of the process air is reduced, the air is reheated to provide the desired temperature in the cooling area. This process is depicted schematically as a black line from hot state to hot state in the psychrometric diagram in Figure 4.7. Excessively humid cooling air requires more energy to dry than the cooling air. Desiccant-assisted air conditioning systems take advantage of this concept by dehumidifying the process air before it is sent to the HVAC system where it is cooled. Desiccant systems effectively reduce air conditioning requirements by reducing cooling

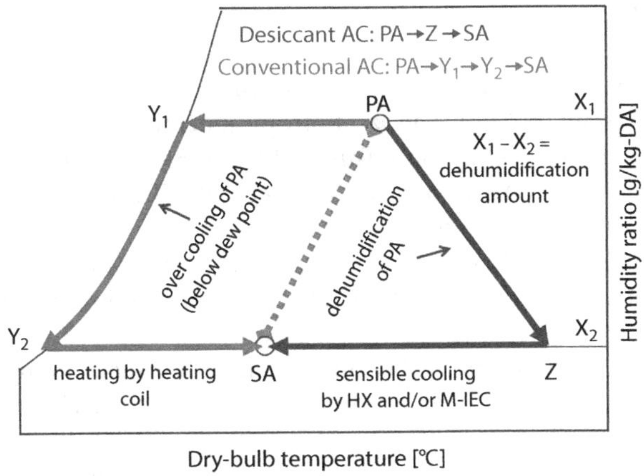


**Figure 4.7** Comparison between VCR cooling (black line) and desiccant cooling (red line) on psychrometric chart.

capacity [49, 50]. Desiccant-driven dehumidification and cooling room cooling process air as shown by the red line in the psychrometric diagram in Figure 4.7. The dehumidification process of desiccant cooling is represented by the proximity of the red line. When the air is dehumidified, its temperature rises slightly before it enters the air conditioner [51]. The dehumidified process air is cooled to a suitable temperature by the air conditioner to reach the desired temperature of the cooling area.

## 4.5 Conclusions

An inventive strategy for sustainable cooling can be achieved by the use of desiccant-assisted dehumidification and cooling systems in place of contemporary air conditioning systems to save energy and employ green cooling techniques for building comfort. By using waste heat to reactivate the desiccant material used in the system's rotary dehumidifiers, desiccant-based innovative dehumidification and cooling systems can achieve significant energy savings and increase the system's cost-effectiveness. By using waste energy harvesting, it can also assist in reducing the peak electrical power demand made by the traditional vapor compression air conditioning system.

## References

1. Bejan, A., Second law analysis in heat transfer. *Energy*, 5, 8-9, 720–732, 1980.
2. San, J.Y., Worek, W.M., Lavan, Z., Entropy generation in combined heat and mass transfer. *Int. J. Heat Mass Transf.*, 30, 7, 1359–1369, 1987.
3. Van den Bulck, E.S.A.K., Klein, S., Mitchell, J.W., Second law analysis of solid desiccant rotary dehumidifiers, 1988.
4. Ahmed, C.K., Gandhidasan, P., Zubair, S.M., Al-Farayedhi, A.A., Exergy analysis of a liquid-desiccant-based, hybrid air-conditioning system. *Energy*, 23, 1, 51–59, 1998.
5. Camargo, J.R., Ebinuma, C.D., Silveira, J.L., Thermoeconomic analysis of an evaporative desiccant air conditioning system. *Appl. Therm. Eng.*, 23, 12, 1537–1549, 2003.
6. Dincer, I. and Sahin, A.Z., A new model for thermodynamic analysis of a drying process. *Int. J. Heat Mass Transf.*, 47, 4, 645–652, 2004.
7. Rosen, M.A. and Dincer, I., Effect of varying dead-state properties on energy and exergy analyses of thermal systems. *Int. J. Therm. Sci.*, 43, 2, 121–133, 2004.
8. Caliskan, H., Dincer, I., Hepbasli, A., Exergetic and sustainability performance comparison of novel and conventional air cooling systems for building applications. *Energy Build.*, 43, 6, 1461–1472, 2011.
9. Hürdoğan, E., Buyükalaca, O., Hepbasli, A., Yılmaz, T., Exergetic modeling and experimental performance assessment of a novel desiccant cooling system. *Energy Build.*, 43, 6, 1489–1498, 2011.
10. Koronaki, I.P., Rogdakis, E., Kakatsiou, T., Thermodynamic analysis and neural network model of open cycle desiccant cooling systems with silica gel, 2011.
11. Parameshwaran, R., Kalaiselvam, S., Harikrishnan, S., Elayaperumal, A., Sustainable thermal energy storage technologies for buildings: A review. *Renew. Sustain. Energy Rev.*, 16, 5, 2394–2433, 2012.
12. Enteria, N., Yoshino, H., Takaki, R., Yonekura, H., Satake, A., Mochida, A., First and second law analyses of the developed solar-desiccant air-conditioning system (SDACS) operation during the summer day. *Energy Build.*, 60, 239–251, 2013.
13. Koronaki, I.P., Rogdakis, E., Kakatsiou, T., Experimental assessment and thermodynamic analysis of a solar desiccant cooling system. *Int. J. Sustain. Energy*, 32, 2, 121–136, 2013.
14. Kwong, Q.J., Adam, N.M., Sahari, B.B., Thermal comfort assessment and potential for energy efficiency enhancement in modern tropical buildings: A review. *Energy Build.*, 68, 547–557, 2014.

15. Rafique, M.M., Gandhidasan, P., Al-Hadhrami, L.M., Rehman, S., Energy, exergy and anergy analysis of a solar desiccant cooling system. *J. Clean Energy Technol.*, 4, 1, 78–83, 2016.
16. Jani, D.B., Mishra, M., Sahoo, P.K., Solid desiccant air conditioning–A state of the art review. *Renew. Sustain. Energy Rev.*, 60, 1451–1469, 2016.
17. Uçkan, İ., Yılmaz, T., Hürdoğan, E., Büyükalaca, O., Exergy analysis of a novel configuration of desiccant based evaporative air conditioning system. *Energy Convers. Manage.*, 84, 524–532, 2014.
18. Jani, D.B., Mishra, M., Sahoo, P.K., Experimental investigations on hybrid solid desiccant–vapor compression air-conditioning system for Indian climate, in: *The proceedings of the 24th IIR international congress of refrigeration*, Yokohama, Japan, pp. 16–22, 2015.
19. Zendehboudi, A., Implementation of GA-LSSVM modelling approach for estimating the performance of solid desiccant wheels. *Energy Convers. Manage.*, 127, 245–255, 2016.
20. Jani, D.B., Mishra, M., Sahoo, P.K., Application of artificial neural network for predicting performance of solid desiccant cooling systems–A review. *Renew. Sustain. Energy Rev.*, 80, 352–366, 2017.
21. Dadi, M. and Jani, D.B., TRNSYS simulation of an evacuated tube solar collector and parabolic trough solar collector for hot climate of Ahmedabad, 2019, Available at SSRN 3367102.
22. Jani, D.B., An overview on desiccant assisted evaporative cooling in hot and humid climates. *Alger. J. Eng. Technol.*, 1, 1–7, 2019.
23. Peng, D., Zhou, J., Luo, D., Exergy analysis of a liquid desiccant evaporative cooling system. *Int. J. Refrig.*, 82, 495–508, 2017.
24. Soares, N., Bastos, J., Pereira, L.D., Soares, A., Amaral, A.R., Asadi, E., Rodrigues, E., Lamas, F.B., Monteiro, H., Lopes, M.A.R., Gaspar, A.R., A review on current advances in the energy and environmental performance of buildings towards a more sustainable built environment. *Renew. Sustain. Energy Rev.*, 77, 845–860, 2017.
25. Jani, D.B., Mishra, M., Sahoo, P.K., Exergy analysis of solid desiccant-vapour compression hybrid air conditioning system. *Int. J. Exergy*, 20, 4, 517–535, 2016.
26. Bachrun, A.S., Ming, T.Z., Cinthya, A., Building Envelope Component to Control Thermal Indoor Environment in Sustainable Building: A Review. *Sinergi*, 23, 2, 79–98, 2019.
27. Guan, B., Liu, X., Zhang, T., Analytical solutions for the optimal cooling and heating source temperatures in liquid desiccant air-conditioning system based on exergy analysis. *Energy*, 203, 117860, 2020.
28. Jani, D.B., Bhabhor, K., Dadi, M., Doshi, S., Jotaniya, P.V., Ravat, H., Bhatt, K., A review on use of TRNSYS as simulation tool in performance prediction of desiccant cooling cycle. *J. Therm. Anal. Calorim.*, 140, 5, 2011–2031, 2020.

29. Jani, D.B., An Overview on Recent Development in Desiccant Materials. *Int. J. Adv. Mater. Res.*, 5, 2, 31–37, 2019.
30. Jani, D.B., Mishra, M., Sahoo, P.K., Performance analysis of a solid desiccant assisted hybrid space cooling system using TRNSYS. *J. Build. Eng.*, 19, 26–35, 2018.
31. Jani, D.B., Mishra, M., Sahoo, P.K., Solar assisted solid desiccant—vapor compression hybrid air-conditioning system, in: *Applications of Solar Energy*, pp. 233–250, 2018.
32. Bhabhor, K.K. and Jani, D.B., Progressive development in solid desiccant cooling: A review. *Int. J. Ambient Energy*, 1–24, 2019.
33. Jani, D.B., Mishra, M., Sahoo, P.K., A critical review on solid desiccant-based hybrid cooling systems. *Int. J. Air-Cond. Refrig.*, 25, 03, 1730002, 2017.
34. Ge, F. and Wang, C., Exergy analysis of dehumidification systems: A comparison between the condensing dehumidification and the desiccant wheel dehumidification. *Energy Convers. Manage.*, 224, 113343, 2020.
35. Jani, D.B., Mishra, M., Sahoo, P.K., Investigations on effect of operational conditions on performance of solid desiccant based hybrid cooling system in hot and humid climate. *Therm. Sci. Eng. Prog.*, 7, 76–86, 2018.
36. Vyas, V. and Jani, D.B., An overview on application of solar thermal power generation. *Int. J. Eng. Res. Allied Sci.*, 1, 1–5, 2016.
37. Ahmad, T. and Chen, H., Short and medium-term forecasting of cooling and heating load demand in building environment with data-mining based approaches. *Energy Build.*, 166, 460–476, 2018.
38. Jani, D.B., Mishra, M., Sahoo, P.K., Performance prediction of rotary solid desiccant dehumidifier in hybrid air-conditioning system using artificial neural network. *Appl. Therm. Eng.*, 98, 1091–1103, 2016.
39. Bhabhor, K.K. and Jani, D.B., Performance analysis of desiccant dehumidifier with different channel geometry using CFD. *J. Build. Eng.*, 103–021, 2021.
40. Dadi, M. and Jani, D.B., Solar Energy as a Regeneration Heat Source in Hybrid Solid Desiccant – Vapor Compression Cooling System – A Review. *J. Emerg. Technol. Innov. Res.*, 6, 5, 421–425, 2019.
41. Jani, D.B., TRNSYS simulation of desiccant powered evaporative cooling systems in hot and humid climate. *J. Mech. Eng.*, 1, 1, 1–6, 2022.
42. Zhang, Q., Liu, X., Zhang, T., Xie, Y., Performance optimization of a heat pump driven liquid desiccant dehumidification system using exergy analysis. *Energy*, 204, 117891, 2020.
43. Jani, D.B., Mishra, M., Sahoo, P.K., TRNSYS simulation of Desiccant-assisted HVAC System. *Air Cond. Refrig. J.*, 24, 6, 28–33, 2022.
44. Jia, C.X., Dai, Y.J., Wu, J.Y., Wang, R.Z., Use of compound desiccant to develop high performance desiccant cooling system. *Int. J. Refrig.*, 30, 2, 345–353, 2007.

45. Hua, L. and Wang, R., An exergy analysis and parameter optimization of solid desiccant heat pumps recovering the condensation heat for desiccant regeneration and heat transfer enhancement. *Energy*, 238, 121811, 2022.
46. Jani, D.B., A recent development in desiccant assisted HVAC system. *Air Cond. Refrig. J.*, 24, 2, 26–52, 2021.
47. Jani, D.B., Use of artificial neural network in performance prediction of solid desiccant powered vapor compression air conditioning systems. *Instant J. Mech. Eng.*, 3, 3, 5–19, 2021.
48. Jani, D.B., Desiccant cooling as an alternative to traditional air conditioners in green cooling technology. *Instant J. Mech. Eng.*, 1, 1, 1–13, 2019.
49. Halliday, S.P., Beggs, C.B., Sleigh, P.A., The use of solar desiccant cooling in the UK: a feasibility study. *Appl. Therm. Eng.*, 22, 12, 1327–1338, 2002.
50. Kanoğlu, M., Çarpınlıoğlu, M.Ö., Yıldırım, M., Energy and exergy analyses of an experimental open-cycle desiccant cooling system. *Appl. Therm. Eng.*, 24, 5-6, 919–932, 2004.
51. Açıkkalp, E., Caliskan, H., Hong, H., Piao, H., Seung, D., Extended exergy analysis of a photovoltaic-thermal (PVT) module based desiccant air cooling system for buildings. *Appl. Energy*, 323, 119581, 2022.

5

# Eco-Friendly Paint for Sustainable Building Applications to Enhance Thermal Life Comfort

**Vikas Chaubey[1]*, Atul Kumar[2], Aakash Singh[3] and Shekhar Yadav[4]**

[1]*Department of Mechanical Engineering, I.E.T, DSMNRU, Lucknow, Uttar Pradesh, India*
[2]*CoEMTD, I.I.T BHU, Varanasi, Uttar Pradesh, India*
[3]*Department of Applied Science & Humanities, I.E.T, DSMNRU, Lucknow, Uttar Pradesh, India*
[4]*Department of Electronics and Communication Engineering, I.E.T, DSMNRU, Lucknow, Uttar Pradesh, India*

***Abstract***

The present era is constantly focusing on well-furnished buildings and adopting innovation and technology to meet the evolving demands of modern living standards. Researchers and industrialists are focusing on Vedic products like Vedic paints, Vedic colors, Vedic plaster, Vedic bricks, and other Vedic construction materials, which are made from cow dung to enhance the comfort of life. Over the course of history, paints have evolved from Vedic paint (VP) to chemical composition-based paints. This study is exploring the use of Vedic paints by considering technically and economically viable alternatives compared to conventional paints. Vedic paint has the potential to improve thermal properties and reduce total energy consumption in smart buildings. Vedic paint has low thermal conductivity and density, which saves some amount of operational energy as compared to conventional paint in smart buildings. Vedic paint embodies a harmonious blend of time-honored traditions and modern sustainability, providing a practical solution

**Corresponding author*: vikaschaubey0111@gmail.com; ORCID: 0009-0005-7942-8134
Atul Kumar: ORCID: 0000-0002-2986-8458
Aakash Singh: ORCID: 0000-0002-1557-1995
Shekhar Yadav: ORCID: 0009-0001-8557-3729

Ashwani Kumar, Nitesh Dutt and Mukesh Kumar Awasthi (eds.) Heat Transfer Enhancement Techniques: Thermal Performance, Optimization and Applications, (105–118) © 2025 Scrivener Publishing LLC

to contemporary architectural requirements. Vedic paint not only provides aesthetics but also contributes to improving indoor air quality by emitting minimum pollutants, thereby promoting a healthy indoor environment. The primary objective of the present work is to highlight the potential of VP as a sustainable and eco-friendly option for smart buildings. The future of the study is concluded for the construction sector to develop energy efficient paint materials with better sustainability by incorporating natural waste.

***Keywords:*** Vedic paint, life comfort, smart building, eco-friendly, temperature control

## 5.1 Introduction

An effective technical and economical increment has been considered for smart constructing building construction. Housing societies are needed to be leading about to preserve conventional and non-conventional energy resources. It points to vital requirement for sustainable design, development, and implementation in buildings construction to completing housing segment scheme [1]. Worldwide, the consumption of total generated energy by buildings is about 65% and it contributes more in green-house gases [2]. The maximum percentage of energy can be utilized only in conditioning for heating and cooling the house envelop. India has a tropical season; insulation of building walls would result in more energy saving and less green-house gas emissions. It can help to reduce green-house gas and improve sustainable development. Paint is a vital material that is directly exposed to the sun's radiation from the outer wall of a house. It has a prime role in the heat transfer mechanism of the building envelope [3, 4]. Nowadays, synthetic and Prakritik paints are used in India. Furthermore, in rural parts of India, numerous people live in the houses painted and plastered by synthetic and Prakritik products.

Vedic paint is also known as Prakritik paint or natural paint, is made from cow dung, so it is also known as cow dung paint. Cow dung is less expensive and a good thermal insulator that should block emissions. This paint is chemical-free and anti-toxicant paint, and have anti-bacterial and anti-fungal properties. Nowadays, paint is a vital artist in our living spaces, have capabilities to increase life comforts through natural air breathing and allergy-free paint smell. Therefore, Vedic paint can be recommended for patients that have breathing or allergy infection problems. Vedic paint cannot produce a foul smell after painting and no irritation to the eye [5].

Through its vibrant hues, it not only decorates walls but also enhances humors and revitalizes living spaces. As per literature and other resources,

Vedic paint has seven core colors: White, yellow, red, green, pink, grey, and off-orange; integrating the seven colors can make 27 types of sub-colors requirement. The seven prime colors are shown in Figure 5.1.

Vedic paint is an eco-conscious paint solution meticulously crafted not only to adorn smart building walls but also to harmonize with nature's principles, enhancing life comfort in unprecedented ways. Cow dung paint (Vedic paint) (Figure 5.2) is most commonly used in villages from ancient time to present time.

Vedic principles' emphasizing nature and balance and aim is to transform smart buildings, enhancing both environmental sustainability and human well-being. As the world embraces heightened environmental responsibility, Vedic paint emerges not just as a coating but as a holistic

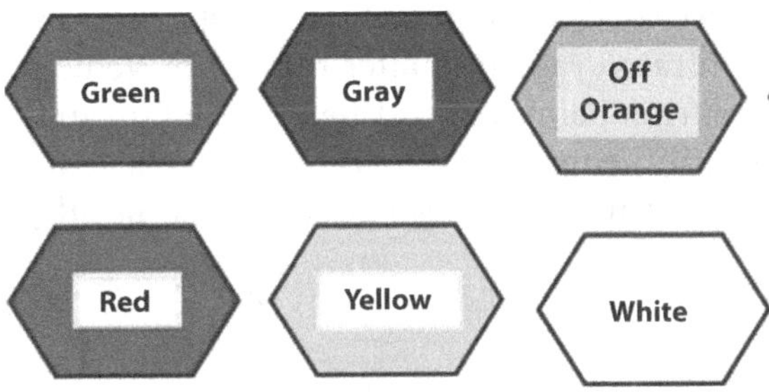


**Figure 5.1** Vedic paint main colors range.

**Figure 5.2** A house painted with cow dung paint (Vedic paint) [8].

approach to creating smarter, greener, and more comfortable living spaces. Come explore how Vedic paint harmonizes with smart building ethos, redefining comfort, sustainability, and the essence of modern living.

The prepared Vedic paint materials has numerous uses, can be applied where non-load bearing, economical lightweight material is required. This paint has good thermal insulator that can protect from sun radiations. So, there are numbers of ways to use Vedic paint over synthetic paint, and some are explored in Table 5.1. Both types of paint have their limitations, but the key distinction lies in their use and their impact on the environment and indoor air quality.

In recent years, cow dung, clay, and other natural materials have been used as alternatives to concrete. Concrete is a major contributor to climate change, as it produces a large amount of carbon footprint and produces waste that cannot be recycled. The cooling characteristics process of wall by solar radiation and infrared radiation after the use of Vedic paint are shown in Figure 5.3. Prakritik paints are certified by the Bureau of Indian Standards. Finally, Prakritik/Vedic paint then is the most efficient and eco-friendly, and a profitable business and also good for the health point of view.

## 5.2 Advantages of Vedic Plaster Over Conventional Plaster

In essence, Vedic paints emerge as a compelling substitute for conventional paints within smart buildings aspiring to prioritize environmental sustainability, occupant health, and long-term cost efficiency while elevating comfort levels within indoor spaces. Some benefits of Vedic paints after comparison of normal paints and Vedic paint are listed as follows:

- Vedic paint is an eco-friendly product and easily offered.
- It is very easy to apply on the surface.
- It saves water because no curing is required.
- It has better thermal characteristics.
- It has good sound resistance and radiation resistance.
- It can be used for ceiling design, gives attractive and smooth finish.
- Vedic paint can be used by adding shell powder to it.

**Table 5.1** Comparison between Vedic paint and synthetic paint.

| Compositions | Parameters | Vedic paint | Synthetic paint |
|---|---|---|---|
| Ingredients | Binders | Practice with plant-based binders like linseed oil, soya oil, or other natural resins | Practice with petroleum-based binders like acrylic, vinyl, or alkyd resins. |
| | Pigments | Can be used with non-toxic pigments from several bases like minerals plants, etc. | Contain chemically manufactured pigments |
| | Fillers | Used with low-impact sources like chalk, clay, or talc. | Contain synthetic fillers |
| | Solvents | Used with citrus or pine-based terpenes. | Contain petrochemical-based solvents. |
| Environmental impact | Environmental impact | Lower environmental impact due to their renewable, biodegradable, and non-toxic ingredient | A higher environmental impact due to their reliance on non-renewable, petroleum-based ingredients. |
| | Indoor air quality | Having better indoor air quality | Have poor indoor air quality and contribute to a range of health issues. |
| Performance | ---- | Allows moisture to escape from the surface, which offer good breathability | They can be less breathable than natural paints but provide a long-lasting finish. |

(*Continued*)

**Table 5.1** Comparison between Vedic paint and synthetic paint. (*Continued*)

| Compositions | Parameters | Vedic paint | Synthetic paint |
|---|---|---|---|
| Durability | ---- | They may require more frequent maintenance and reapplication compared to synthetic paints. | They have issues like peeling, blistering, or mold growth, especially in areas with high humidity or moisture. |
| Cost | ---- | Slightly higher cost. | Lower cost. |

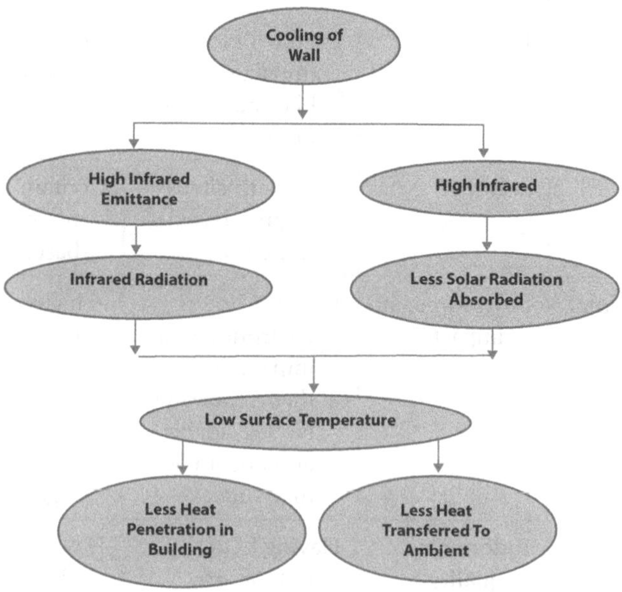


**Figure 5.3** Cooling characteristics of wall [9].

## 5.3 Need for Vedic Paints

The need for Vedic paint, tailored for smart buildings to enhance life comfort, arises from various critical factors aligning with contemporary demands:

- Environmental consciousness: in a global shift towards sustainability, the demand for eco-friendly products has surged.

Vedic paint, derived from natural, non-toxic ingredients, meets this demand by reducing environmental impact during production, application, and disposal [11].

- Health and well-being: traditional paints often release harmful volatile organic compounds (VOCs), contributing to indoor air pollution. Vedic paint addresses this concern by offering low or zero VOC formulas, ensuring improved indoor air quality and promoting healthier living spaces.
- Smart building integration: Vedic paint complements the holistic approach of smart building designs, optimizing energy efficiency, and comfort. Its reflective properties might maintain optimal temperatures, reducing excessive heating, or cooling needs, enhancing overall building efficiency. Cooling of the walls by use of Vedic paint from IR-radiations and sunlight.
- Holistic comfort: going beyond aesthetics, Vedic paint incorporates principles from Vedic wisdom to enhance comfort. Its formulation might include elements resonating with natural energies, potentially creating a positive ambiance and promoting a sense of well-being among occupants.
- Longevity and protection: Vedic paint, with specialized formulations inspired by ancient principles, might offer enhanced durability and protection against weathering, prolonging the lifespan of building materials.
- Ethical and cultural alignment: aligning with societies reevaluating their connection with nature and cultural heritage, Vedic paint resonates with those seeking products honoring ancient wisdom while embracing modern innovation.
- Market demand and differentiation: the growing market for sustainable products offers an opportunity for Vedic paint to stand out. Its unique composition and focus on sustainability cater to an expanding segment of environmentally conscious consumers.

In essence, the need for Vedic paint arises from a convergence of environmental responsibility, health considerations, smart building integration, holistic comfort, cultural alignment, and market demand for innovative, sustainable solutions. Its role transcends traditional paint, shaping eco-conscious, technologically advanced, and comfortable living spaces [6, 7]. Energy input for recycle and reuse of products for maximize building's life block diagram shown in Figure 5.4.

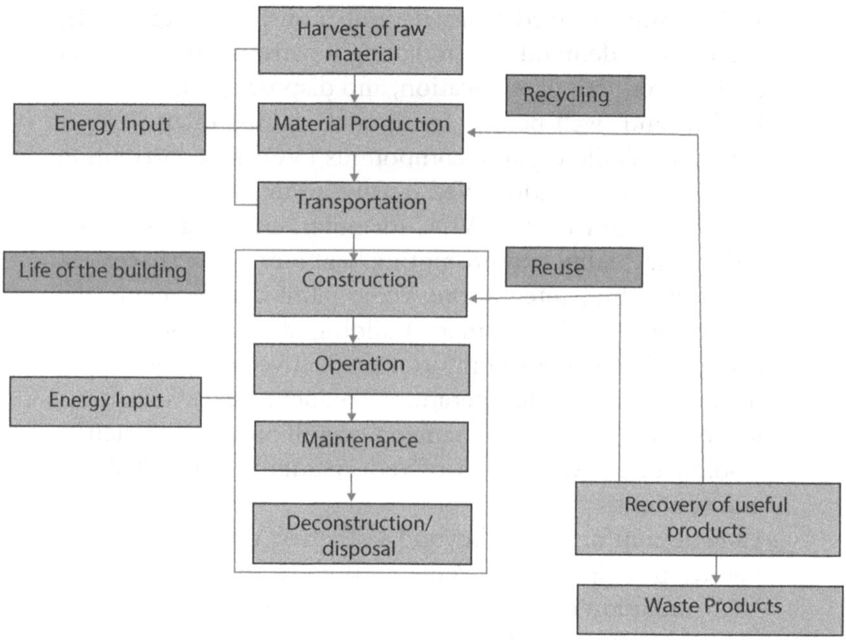


**Figure 5.4** Input for increasing life of wall [10].

## 5.4 Types of Vedic Paints

Vedic paint serves as an environmentally-conscious solution tailored to smart buildings, enriching life comfort while aligning with sustainability, holistic well-being, and eco-consciousness principles. Though it does not conform to traditional "types," Vedic paint embodies various formulations and traits that resonate with its underlying ethos. Here are several aspects that delineate categories within Vedic paint:

- Natural and non-toxic formulations: Vedic paints are primarily composed of natural, organic, or non-toxic elements sourced from plant extracts, minerals, or sustainable origins. These formulations mitigate environmental impact, enhance indoor air quality, and prioritize occupants' health.
- Zero or low VOC paints: volatile organic compounds (VOCs) contribute to air pollution and health concerns. Vedic paints may offer options with minimal or zero VOCs, reducing emissions and fostering healthier indoor environments, in line with eco-friendliness principles.

- Biodegradable paints: emphasizing sustainability, Vedic paints may be biodegradable or designed to decompose naturally over time, lessening their environmental footprint during disposal.
- Energy-efficient and reflective coatings: certain Vedic paint variations may incorporate reflective properties, aiding in maintaining optimal indoor temperatures. These coatings potentially reduce energy consumption for heating or cooling, complementing smart building practices.
- Aura and holistic well-being: Vedic paints may seek to cultivate specific atmospheres by aligning with Vedic principles or traditional wisdom. These endeavors aim to invoke positive energies or an ambiance promoting holistic well-being within spaces.
- Specialized finishes or textures: while upholding eco-friendliness, Vedic paints might offer diverse finishes or textures that harmonize with smart building designs. These could include matte, satin, or textured finishes catering to varied architectural styles.
- Adaptive and protective formulations: aligned with smart building sustainability goals, certain Vedic paint formulations adapt to environmental changes or provide protective layers that extend surface longevity, reducing maintenance requirements.

Vedic paints are custom-tailored for eco-friendly smart buildings, prioritize sustainability, well-being, and environmental harmony [12]. The delineations within this category pivot around formulations and traits that uphold these principles, ensuring they contribute positively to both occupants' comfort and the overall sustainability of built environments.

## 5.5 Chemical Properties of Vedic Paints

Vedic paint, also recognized as eco-friendly paint, boasts distinctive chemical properties that set it apart from conventional paints, rendering it an optimal choice for smart buildings aspiring to elevate life comfort. The chemical characteristics of Vedic paint include:

- Low VOC content: typically, Vedic paint contains minimal to zero VOCs. These compounds tend to evaporate at room temperature, contributing to indoor air pollution. By curbing VOCs, Vedic paint significantly enhances indoor air quality, diminishing health risks and fostering a healthier living environment.

- Natural composition: its formulation incorporates natural constituents like plant extracts, essential oils, natural resins, minerals, and water-based solvents. These components are sustainably sourced, reducing reliance on synthetic or petrochemical-based substances typically found in traditional paints.
- Biodegradability: engineered for biodegradability, Vedic paint decomposes naturally without harming the environment. This property minimizes its impact on landfills and ecosystems, establishing it as an environmentally conscious choice.
- Non-toxic formulation: unlike conventional paints, which may harbor heavy metals or toxic additives, Vedic paint is predominantly free from harmful substances. This attribute ensures safety for human health and the environment alike.
- Mold and mildew resistance: certain variations of Vedic paint integrate natural anti-fungal properties, mitigating mold and mildew growth on painted surfaces. This quality preserves a cleaner and healthier indoor environment.

Collectively, these chemical properties position Vedic paint as a fitting choice for smart buildings, aligning to enhance life comfort while prioritizing sustainability and environmental stewardship [13, 14].

## 5.6 Factors Increasing Comfort

Here, the focus is on factors contributing to enhanced comfort within smart buildings particularly in relation to the use of Vedic paint as an eco-friendly option:

- Low VOC content: Vedic paint, with its minimal or zero VOCs, contributes to better indoor air quality, reducing health risks and enhancing comfort by ensuring occupants breathe cleaner air.
- Natural ingredients: the use of natural ingredients like plant extracts, essential oils, and minerals in Vedic paint promotes

a healthier indoor environment. This natural composition reduces the presence of harmful chemicals, fostering comfort for occupants.

- Biodegradability: Vedic paint being biodegradable minimizes its impact on the environment, aligning with sustainability goals. This aspect contributes to a comfortable living space by reducing the ecological footprint and promoting a sense of environmental well-being.
- Safety and health: the non-toxic formulation of Vedic paint, devoid of heavy metals or harmful additives, ensures a safer and healthier indoor environment. This factor significantly contributes to the overall comfort and well-being of building occupants.
- Durability and longevity: Vedic paint's ability to exhibit good durability and longevity contributes to a comfortable living or working environment by reducing maintenance needs. It ensures the painted surfaces retain their quality for an extended period, minimizing disruptions and maintaining visual comfort.
- Thermal insulation and breathability: some variants of Vedic paint may possess properties contributing to thermal insulation. Maintaining comfortable indoor temperatures and allowing moisture to pass through ensures a more comfortable and pleasant indoor environment.
- Aesthetics and ambiance: Vedic paints often offer a wide range of natural colors and textures, contributing to aesthetically pleasing interiors. A visually appealing environment positively influences comfort and mood within a living or working space.
- Sustainable contribution: the eco-friendly nature of Vedic paint aligns with the sustainable objectives of smart buildings. This factor, coupled with its comfort-enhancing properties, contributes to a sense of well-being and contentment among occupants.

In summary, the use of Vedic paint within smart buildings not only addresses environmental concerns but also significantly contributes to enhancing the overall comfort and well-being of individuals inhabiting these spaces [15].

## 5.7 Conclusion

The study of Vedic paint as an eco-friendly solution for smart buildings underscores its significant role in shaping sustainable and comfortable living spaces. Evolving from ancient wisdom to modern technology integration, Vedic paint epitomizes innovation, aligning seamlessly with environmental consciousness and holistic well-being.

Study reveals the crucial need for Vedic paint in smart buildings, driven by contemporary demands. Exploration of Vedic paint's categories unveils formulations dedicated to sustainability, well-being, and environmental harmony. From natural compositions to adaptive protective traits, each aspect prioritizes occupants' comfort and the enduring sustainability of built environments.

Its low VOC content, natural formulation, biodegradability, non-toxicity, durability, and insulation contribute to a healthier, more sustainable, and cost-efficient built environment.

## 5.8 Future Outlook

The scope of Vedic paint within smart buildings anticipates further growth and enhancement:

- Innovative formulations: ongoing research could lead to more advanced Vedic paint formulations, exploring new natural sources, protective coatings, or enhanced thermal properties.
- Technological integration: further integrating Vedic paint with smart building tech can enhance functionalities, offering coatings adapting to environmental changes or interacting with building systems for efficiency.
- Customization and diversity: exploring varied colors, textures, and finishes can cater to diverse architectural styles, enhancing appeal and practicality.
- Environmental impact studies: deeper lifecycle and environmental impact studies could solidify Vedic paint's position, offering metrics supporting its eco-friendliness.
- Occupant well-being research: continued studies on its effects on health and productivity can bolster its standing as a contributor to healthier indoor environments. And also, focused

educational efforts can increase adoption and support for Vedic paint among consumers, architects, and builders.

Vedic paint's ongoing evolution and resonance with sustainable living demands promise to redefine comfort, sustainability, and modern living within smart buildings.

## References


1. UN, The Global Compact: Corporate Citizenship in the World Economy: A UN Programme, UN, New York, MY, 2003.
2. Fergusson, L., Nidich, S., Bonshek, A., Nidich, R., A concurrent mixed methods study of quality of life and Maharishi Vedic architecture. *J. Maharishi Vedic Res. Inst.*, 17, 75–108, 2021.
3. Rangari, N.T., Kalyankar, T.M., Mahajan, A.A., Lendhe, P.R., Puranik, P.K., Review article: Ayurvastra: Herbal couture technology in textile. *Int. J. Res. Ayurveda Pharm.*, 3, 5, 733–736, 2012.
4. Banupriya, J. and Maheshwari, V., Comparison between herbal and conventional methods in antimicrobial finishes. *Int. J. Fiber Text. Res.*, 4, 2, 41–43, 2014.
5. Sumithra, M. and Vasugi Raaja, N., Herbal coating of denim fabric to enhance the wickability. *Int. J. Biomed. Res.*, 5, 2, 95–97, 2014.
6. Pandey, R.K. and Tripathi, P.S., Vedic values and corporate excellence, in: *Vedic Values and Corporate Excellence*, S.C. Dhamija, and V.K. Singh, (Eds.), pp. 168–71, Gurukul Kaugri University, Haridwar, 2002.
7. Ganapati Sthapati, V., *Building architecture of Sthāpatya Veda*, Dakshinaa Publishing, Chennai, India, 2005.
8. https://www.barrondesigns.com/blog/southwest-interior-design-how-to-get-the-look/
9. Rawat, M. and Singh, R.N., A study on the comparative review of cool roof thermal performance in various regions. *Energy Built Environ.*, 3, 3, 327–347, 2022.
10. Akadiri, P.O., Chinyio, E.A., Olomolaiye, P.O., Design of a sustainable building: A conceptual framework for implementing sustainability in the building sector. *Buildings*, *2*, 2, 126–152, 2012.
11. Swaroop, D., Singh, J., Singh, D., KRISHI VIGYAN KENDRAS–A BOON FOR RESOURCE MANAGEMENT, AND SUSTAINABLE RURAL DEVELOPMENT. *SN*, 333, 2017.
12. Havell, E.B., The symbolism of Indian sculpture and painting. *Burlington Mag. Connoisseurs*, *15*, 78, 331–345, 1909.
13. Solanki, D., Duggal, P., Singh, R., Tomar, R.K., Bhardwaj, U., STUDY AND ANALYSIS OF CHARACTERISTICS OF VEDIC PLASTER.

14. Biegeleisen, L.K., The Chemistry of Consciousness: Consciousness-Based Chemistry and Maharishi Vedic Science, Doctoral dissertation, Maharishi University of Management, 2022.
15. Fergusson, L., Nidich, S., Bonshek, A., Nidich, R., Maharishi Vedic Architecture and quality of life: An international mixed methods study of lived experience. *Int. J. Archit. Urban Devel.*, *10*, 4, 1–18, 2020.

# 6

# Augmentation of Solar, Geothermal, and Earth-Air Heat Exchanger in Sustainable Buildings

**Varun Pratap Singh**[1,2*], **Ashwani Kumar**[3] **and Mukesh Kumar Awasthi**[4]

[1]*Solar Thermal Energy Research Group (STERG), Department of Mechanical Engineering and Mechatronics Engineering, University of Stellenbosch, Stellenbosch, South Africa*
[2]*Department of Mechanical Engineering, School of Advanced Engineering, UPES, Bidholi, Dehradun, Uttarakhand, India*
[3]*Technical Education Department, Uttar Pradesh, Kanpur, India*
[4]*Department of Mathematics, Babasaheb Bhimarao Ambedkar University, Lucknow, India*

***Abstract***

This chapter presents an in-depth exploration of the integration and enhancement of renewable energy technologies in sustainable building practices. The chapter systematically dissects the current state of renewable energy technologies, focusing on solar energy, geothermal energy, and Earth-air heat exchangers (EAHE). It delves into advanced solar technologies, their integration into building design, and assesses their energy efficiency and environmental impacts. The geothermal energy segment examines the extraction of subsurface heat and its applications in space heating, cooling, and power generation, coupled with innovative solutions in this field. The discussion on EAHE emphasizes their principles, functionality, and practical implementation techniques. The chapter culminates in a comprehensive analysis of combined augmentation strategies, highlighting synergies, challenges, and future prospects of integrating these systems in sustainable buildings.

**Corresponding author*: ray_varun@yahoo.com; ORCID: 0000-0002-2148-5604
Ashwani Kumar: ORCID: 0000-0003-4099-935X
Mukesh Kumar Awasthi: ORCID: 0000-0002-6706-5226

Ashwani Kumar, Nitesh Dutt and Mukesh Kumar Awasthi (eds.) Heat Transfer Enhancement Techniques: Thermal Performance, Optimization and Applications, (119–158) © 2025 Scrivener Publishing LLC

The conclusion provides a summary of key findings, implications for sustainable building practices, and recommendations for future research.

***Keywords***: Earth-air heat exchangers, renewable energy technologies, sustainable building, solar energy, geothermal energy

## 6.1 Introduction

In the rapidly evolving realm of sustainable building practices, the integration of renewable energy technologies emerges as a pivotal force in establishing a harmonious balance between human infrastructure and ecological resilience. This chapter undertakes a comprehensive exploration of solar, geothermal, and Earth-air heat exchanger (EAHE) augmentation within the context of sustainable building design. Tracing the historical trajectory and evolution of these technologies provides a foundational understanding, offering insights into the challenges and milestones that have shaped the contemporary landscape. From the nascent days of solar energy to the innovative applications of geothermal systems and EAHE, this historical lens sets the stage for evaluating the present imperatives driving the augmentation of these technologies.

The exploration extends into the overarching objectives of this chapter, transcending technical intricacies to encompass broader goals. These objectives are designed to elucidate the transformative potential of these technologies, such as enhancing energy efficiency, reducing carbon footprints, and fostering adaptability and resilience in the face of evolving environmental dynamics. By outlining these objectives, the chapter seeks to provide a comprehensive roadmap for stakeholders, researchers, and practitioners aiming to navigate the complex landscape of sustainable building practices. The narrative seamlessly transitions into emphasizing the significance of augmenting solar, geothermal, and EAHE in the context of sustainable building practices. Beyond the immediate advantages of reducing reliance on non-renewable energy sources, these augmentation strategies signal a paradigm shift in architectural thinking. The significance lies not only in mitigating environmental impact but also in reshaping the very essence of architectural paradigms towards a more holistic and ecologically conscious approach [1]. Positioned at the nexus of environmental stewardship and technological innovation, the integration of these renewable technologies takes center stage in realizing a future where sustainable buildings seamlessly intertwine with the natural world, embodying resilience, efficiency, and a steadfast commitment to a greener, more sustainable tomorrow [2].

The significance of augmenting solar, geothermal, and EAHE in sustainable building practices cannot be overstated, as these technologies collectively represent a transformative force in shaping the future of the built environment. At its core, this augmentation aligns with the imperative to reduce our reliance on finite, non-renewable energy sources, offering a tangible pathway towards achieving energy independence and mitigating the environmental impact of conventional energy consumption [3]. Beyond this, the integration of these renewable technologies embodies a paradigm shift in architectural thinking, emphasizing a holistic approach that considers not only energy efficiency but also the broader ecological footprint of structures [4]. The significance lies in fostering adaptability and resilience, allowing buildings to harmonize with the natural environment rather than merely coexisting with it. By harnessing solar power, tapping into the Earth's geothermal energy, and utilizing EAHE, sustainable buildings emerge as dynamic entities that actively contribute to a greener, more sustainable tomorrow. This significance extends beyond immediate energy savings, encompassing a commitment to environmental stewardship, a redefinition of architectural paradigms, and a realization of structures that stand as resilient, efficient, and integral components of a sustainable future [5].

## 6.2 Current State of Renewable Energy Technologies

The current state of renewable energy technologies used in buildings reflects a dynamic landscape driven by a growing awareness of environmental sustainability and a desire to reduce dependence on traditional energy sources. Various renewable energy technologies have gained prominence, each offering unique advantages in enhancing the energy efficiency of buildings [6]. Solar photovoltaic (PV) systems, harnessing energy from the sun, have witnessed widespread adoption, with PV panels seamlessly integrated into building designs to generate electricity. Wind turbines, both large and small-scale, are also making strides in urban and rural settings, contributing to on-site power generation. Additionally, geothermal heat pumps tap into the Earth's consistent temperature to provide efficient heating and cooling solutions. Biomass systems utilize organic materials for heating and power generation, while hydropower systems leverage flowing water for electricity production. In the context of buildings, the integration of these technologies varies based on factors like location, climate, and architectural design. For a comprehensive overview, Table 6.1 outlines different types of renewable energy technologies and their applications in buildings [7].

**Table 6.1** Different types of renewable energy technologies and their applications in buildings.

| Renewable energy technology | Building applications |
|---|---|
| Solar photovoltaic (PV) systems | Electricity generation, solar panels integrated into building structures |
| Wind turbines | On-site power generation, urban and rural applications |
| Geothermal heat pumps | Heating and cooling systems utilizing Earth's consistent temperature |
| Biomass systems | Organic material-based heating and power generation |
| Hydropower systems | Electricity generation using flowing water sources |
| Solar thermal collectors | Solar water heating, space heating, and industrial processes |
| Concentrated solar power (CSP) | Electricity generation through concentrated sunlight |
| Passive solar design | Architectural design to maximize natural heating and lighting |
| Ground source heat pumps (GSHP) | Utilizes stable ground temperature for heating and cooling |
| Direct-use geothermal systems | Direct extraction and use of geothermal fluids for heating |
| Earth-air heat exchangers | Passive and active cooling systems utilizing ground temperature |

## 6.3 Solar Augmentation Strategies

Building solar augmentation solutions involve a wide range of techniques and modern technology that attempt to utilize solar energy to improve energy efficiency and reduce environmental harm. An effective approach is incorporating solar PV panels into building structures, harnessing sunlight to produce power. In addition, solar thermal collectors are used for the purpose of heating water and spaces, thereby adding to a full method of Utilizing solar energy. Emerging solar technologies, such as thin-film solar cells and concentrated solar power (CSP), have appeared, providing more efficiency and flexibility. The process of integrating solar components into building design include strategically positioning PV panels on rooftops or facades, as well as constructing structures to maximize the use of natural illumination [8]. This integration not only improves the visual attractiveness but also optimizes energy absorption. The resulting energy efficiency is remarkable, since the use of solar augmentation decreases reliance on traditional power sources, resulting in possible energy conservation and a reduced carbon impact. The environmental effect is additionally reduced by the utilization of clean, renewable energy, hence supporting sustainable construction practices. The potential for savings is significant, as it leads to decreased energy expenses in the long run. Additional benefits encompass freedom from reliance on the power grid, enhanced property valuation, and eligibility for diverse incentives. Nevertheless, restrictions arise due to obstacles like as intermittency, substantial initial expenses, and space constraints. However, current research and technology progress are still working to overcome these restrictions, establishing solar augmentation as a fundamental aspect in the development of energy-efficient and ecologically aware construction methods [9].

During the first phases of design, engineers increasingly prioritized energy consumption, especially after realizing in the early 2000s that conserving energy is crucial in system design. This age signifies a shift in addressing the consequences of the previous century's use of natural resources. The need to tackle environmental repercussions, such as those caused by greenhouse gases and the limited availability of conventional energy sources like fossil fuels, requires a strong dedication to harnessing renewable energies. Efficient energy conservation strategies are of utmost importance throughout the shift from non-renewable to renewable energy sources [10]. The energy consumption throughout building life cycles is mostly attributed to heating, ventilation, and air conditioning (HVAC) systems operating, accounting for roughly 94.4% of the total energy used. In order to address this issue, it is recommended to utilize passive techniques and renewable energy sources to create comfortable circumstances, reducing the need on mechanical equipment. Renewable energy sources, such as solar, wind, wave, tidal, biofuel, geothermal, hydrogen, and ocean energy, offer sustainable options. The diminishing likelihood of traditional energy sources has prompted a worldwide transition towards secure, environmentally friendly, cost-effective, and sustainable energy sources, in line with the urgent need to protect the environment. Solar systems, employing both passive and active tactics, are emerging as a crucial participant in this shift [11].

Utilizing solar energy in buildings requires the use of intricate tactics to optimize its capacity. The main premise is centered on using the thermal energy emitted by the sun through the processes of conduction, convection, and radiation. Building designs utilize these natural processes to produce habitats that can be naturally heated and cooled. Advanced solar technologies, such as thin-film solar cells and CSP, improve efficiency. Incorporating solar components into architectural projects, such as installing PV panels on rooftops or facades, not only enhances energy efficiency but also contributes to the visual attractiveness. This integration decreases reliance on traditional power sources, resulting in energy conservation and a reduced carbon impact. The environmental effect is reduced by utilizing clean, renewable energy, which is in line with sustainable construction methods [12]. The potential for cost reductions is significant, providing enduring advantages through decreased energy expenses. The benefits encompass autonomy from the power grid, enhanced real estate worth, and qualification for diverse incentives. Nevertheless, restrictions arise due to issues such as the intermittent nature of the energy source, the significant initial expenses involved, and the need for ample space. Continual study and technical progress are actively tackling these difficulties, establishing solar augmentation as a crucial component in the development of energy-efficient and ecologically sensitive construction methods [13].

### 6.3.1 Advanced Solar Technologies

Advanced solar technologies play a pivotal role in reshaping the landscape of sustainable buildings, offering innovative solutions to efficiently harness solar energy and seamlessly integrate it into architectural designs. Among the notable technologies driving this transformation are building-integrated photovoltaics (BIPV), where solar PV elements are seamlessly integrated into building materials like glass, roofing, or facades. This not only generates electricity but also serves as a functional part of the building envelope, harmonizing aesthetics with functionality [14].

Thin-film solar cells provide a lightweight and flexible alternative to traditional crystalline silicon solar cells, allowing integration into various surfaces, including curved and irregular ones, thereby expanding the realm of design possibilities. Concentrated solar power (CSP) systems utilize mirrors or lenses to concentrate sunlight onto a small area, generating high-temperature heat for electricity production or storage, making them particularly suitable for large-scale energy generation. Solar thermal collectors capture the sun's heat for applications like water and space heating, employing advanced designs and materials to enhance efficiency and durability. Transparent solar windows embed solar panels in windows, enabling buildings to generate electricity without sacrificing natural light, suitable for both new constructions and retrofits [15].

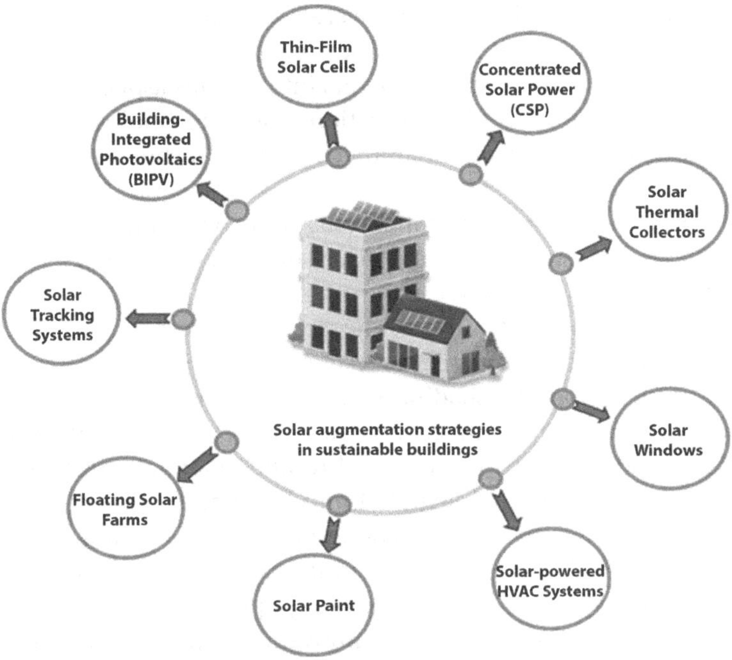


**Figure 6.1** Advanced solar technologies used in sustainable buildings.

Solar shingles replace traditional roofing materials while serving as efficient solar panels, seamlessly blending into a building's structure. Solar paint or PV paint, comprising tiny solar particles, can be applied as a coating on surfaces, turning them into energy-generating structures. Floating solar farms, installed on accessible water bodies, not only generate clean energy but also reduce water evaporation and offer cooling effects. Solar tracking systems dynamically optimize panel orientation to follow the sun's path throughout the day, maximizing energy capture. Advanced solar technologies also extend to HVAC systems, where solar power supplements or entirely powers climate control, reducing reliance on conventional energy sources. Collectively, these technologies exemplify the diverse and innovative ways solar energy is harnessed to create sustainable and energy-efficient buildings [16]. Figure 6.1 discusses the advanced solar technologies used in sustainable buildings.

### 6.3.2 Integration Into Building Design

The integration of advanced solar technologies into building design represents a transformative approach that not only addresses energy needs but also profoundly influences the overall aesthetic and functionality of structures. This integration involves meticulous strategic planning and the seamless incorporation of solar elements into various aspects of the building. Key considerations for the integration of solar technologies into building design include architectural aesthetics, where technologies like BIPV, solar windows, and solar shingles are seamlessly integrated, ensuring that the overall aesthetics of the building remain uncompromised. These technologies serve both functional and visual purposes, enhancing the building's appearance without detracting from its appeal [17].

Another crucial consideration is the strategic placement of solar arrays on rooftops, recognized as prime real estate for solar panels. Factors such as roof pitch, orientation, and shading are carefully considered in the design to optimize energy capture, with technologies like solar shingles and solar tiles blending harmoniously with traditional roofing materials. Solar technologies can also be seamlessly integrated into building facades and cladding, transforming them into energy-generating surfaces. Building-integrated photovoltaic (BIPV) systems embedded in glass or other facade materials contribute not only to the building's energy production but also serve a functional purpose, adding a layer of innovation to the exterior [18].

Transparent solar panels embedded in windows offer a dual function of providing natural light and generating electricity. These solar windows can be thoughtfully incorporated into the building's fenestration design, contributing to both energy efficiency and visual appeal. Solar technologies can further extend into shading elements such as awnings or shades, providing shade to enhance occupant comfort while also capturing solar energy [19]. This approach is particularly effective in creating dual-functionality outdoor spaces. Ground-mounted solar panels integrated into landscaping design, especially in open spaces surrounding buildings, allow for creative placement of solar arrays in harmony with the natural environment. The orientation and layout of the building are carefully considered during the initial design phase to optimize exposure to sunlight, ensuring that solar panels receive maximum sunlight throughout the day [20]. The design also incorporates energy-efficient building envelope elements, such as well-insulated walls and roofs, to complement solar technologies, reducing energy demand and making solar energy a more significant contributor to the building's overall energy needs. Integration of solar technologies often includes the incorporation of smart building systems, utilizing sensors and controls to optimize energy use, and minimize waste through intelligent management. Lastly, building design considers the flexibility and scalability of solar installations, with technologies like thin-film solar cells and modular solar components allowing for adaptability to various building sizes and configurations. The successful integration of solar technologies into building design necessitates a holistic approach that carefully considers architectural, environmental, and energy-related aspects, contributing to the creation of sustainable, energy-efficient, and visually appealing structures that redefine the possibilities of modern architecture.

### 6.3.3 Energy Efficiency and Environmental Impact

The integration of advanced solar technologies into building design significantly enhances both energy efficiency and environmental impact, contributing to sustainable practices [21]. Here is a breakdown of how these aspects are positively influenced:

#### *6.3.3.1 Energy Efficiency*

The integration of advanced solar technologies within sustainable building practices yields a spectrum of benefits that contribute to a more efficient and environmentally conscious built environment. Firstly, through renewable energy generation, technologies like solar PV systems empower buildings to produce clean energy on-site, diminishing reliance on conventional grid-supplied electricity. This shift not only promotes sustainability but also reduces the carbon footprint associated with energy consumption. Moreover, these solar-integrated designs often go hand in hand with optimized building performance. The incorporation of energy-efficient building materials and construction methods enhances the overall energy efficiency of the structure. Features such as solar windows and smart building systems play a crucial role in dynamically managing energy consumption within the building, promoting a holistic approach to sustainable design [22].

Furthermore, the integration of solar technologies contributes to a reduction in energy demand and the alleviation of strain on the electrical grid. By meeting a significant portion of a building's energy demand, especially during peak periods, solar systems play a role in enhancing the overall efficiency of energy distribution. This decentralized energy generation approach not only benefits individual buildings but also contributes to the resilience and sustainability of the broader energy infrastructure. Additionally, with the incorporation of energy storage solutions like batteries, advanced solar systems enable the storage of excess energy generated during peak sunlight hours. This stored energy can be efficiently utilized during periods of low solar activity, ensuring a continuous and reliable power supply. Coupled with smart energy management systems, which monitor and optimize energy consumption in real-time, the integration of advanced solar technologies represents a comprehensive strategy for creating sustainable, energy-efficient, and technologically advanced buildings [23].

#### 6.3.3.2 *Environmental Impact*

The integration of advanced solar technologies into building design is a proactive strategy with multifaceted environmental benefits. Firstly, the utilization of solar energy significantly reduces greenhouse gas emissions, as buildings decrease their reliance on fossil fuels for power generation. This transition plays a vital role in mitigating climate change and lessening the overall environmental impact of energy consumption. Moreover, solar technologies contribute to a lower carbon footprint compared to conventional energy sources. Despite the environmental impact associated with the manufacturing and installation processes of solar panels, the emissions avoided through clean energy generation usually outweigh these costs, making solar a more sustainable choice [24].

Building-integrated solar technologies further emphasize sustainable material choices, favoring construction materials with lower embodied energy and reduced environmental impact. This commitment aligns with broader sustainable building practices, fostering a holistic approach to environmentally conscious design. Additionally, the conservation of natural resources is inherent in the use of solar energy, a renewable resource that can be harnessed continuously without depleting the source. By mitigating urban heat islands through elements like green roofs or reflective surfaces, solar-integrated designs positively impact local microclimates and reduce overall energy demand for cooling. Lastly, the integration of solar technologies contributes to the preservation of ecosystems by decreasing reliance on non-renewable energy sources. This shift away from extractive industries, such as coal or oil extraction, helps mitigate severe environmental consequences and promotes a more harmonious coexistence with nature, aligning with global efforts toward a sustainable future [25].

Top of Form

## 6.4 Geothermal Energy in Building Systems

Geothermal energy integration in building systems is a sustainable approach that leverages the Earth's natural heat to provide heating, cooling, and even electricity. Geothermal systems offer a reliable and environmentally friendly alternative to traditional HVA systems. Here's an overview of how geothermal energy is utilized in building systems [26, 27]:

Ground source heat pumps (GSHP): GSHP systems excel in providing both heating and cooling by tapping into the Earth's stable temperature. During winter, these systems extract heat from the ground to warm the building, while in summer, the process reverses, transferring excess heat from the building to the ground. Notably, GSHP systems exhibit high energy efficiency, demanding less electricity to move heat compared to conventional HVAC systems. This heightened efficiency translates into lower energy costs for users and contributes to a decreased environmental footprint, making GSHP a sustainable and economically favorable choice for year-round temperature control [28, 29].

Direct-use geothermal systems: direct-use geothermal systems harness the Earth's subsurface heat for direct heating applications. This involves extracting hot water or steam, which is then circulated through a distribution system to provide warmth to buildings using radiators, underfloor heating, or other devices. Additionally, these systems find application in agriculture, specifically in greenhouses, where they contribute to maintaining optimal growing conditions for plants by supplying the necessary heat. The versatility of direct-use geothermal systems makes them a sustainable and effective solution for both residential and agricultural heating needs.

Environmental impact: geothermal systems contribute significantly to environmental sustainability by reducing greenhouse gas emissions. By diminishing reliance on fossil fuels for heating and cooling, these systems play a crucial role in creating a cleaner and more sustainable energy mix. Despite some emissions associated with manufacturing and installation, the long-term operational emissions reduction outweighs the initial impact, resulting in a low overall carbon footprint.

Closed-loop and open-loop systems: geothermal systems come in two main configurations—closed-loop and open-loop. Closed-loop systems circulate a heat transfer fluid through a closed loop of pipes buried in the ground, absorbing heat in the winter and releasing it in the summer. On the other hand, open-loop systems directly pump groundwater from an aquifer or well, circulating it through the heat pump before discharge. The choice between these systems depends on factors such as water availability and geological conditions.

Integration into building design: geothermal systems offer efficient space utilization, requiring less space than traditional HVAC systems. This makes them versatile and suitable for various building sizes and designs. Additionally, these systems seamlessly integrate with radiant heating systems, where warm water circulates beneath the floor through pipes or tubes, providing even and efficient heating.

Financial incentives: the adoption of geothermal systems is further incentivized by various financial benefits. Governments often offer financial incentives, tax credits, or rebates to encourage the installation of geothermal systems in buildings. These incentives make geothermal systems more economically viable for homeowners and businesses, promoting their widespread adoption and contributing to sustainable energy practices.

### 6.4.1 Harnessing Subsurface Heat

Harnessing subsurface heat is a compelling and sustainable energy solution that taps into the thermal energy stored beneath the Earth's surface for diverse applications. Geothermal energy extraction principles form the foundation of this process, offering a range of methods to harness subsurface heat. Geothermal power plants employ both enhanced geothermal systems (EGS) and hydrothermal power plants to generate electricity by creating engineered reservoirs or utilizing natural geothermal reservoirs, respectively. Direct heating systems, such as GSHP, transfer heat to and from the ground for space heating and cooling, while district heating systems distribute subsurface heat through a network of pipes for residential and commercial heating. Industrial applications leverage subsurface heat for various processes, emphasizing its value for industries requiring high-temperature resources [30].

Subsurface heat exchange systems, employing closed-loop and open-loop methods, facilitate the transfer of heat between the subsurface and buildings. From an environmental perspective, harnessing subsurface heat results in a lower carbon footprint compared to fossil fuel alternatives, contributing to global efforts to reduce greenhouse gas emissions. Additionally, geothermal energy stands as a sustainable and renewable resource, offering a continuous and reliable source of heat and power with proper management. Ongoing technological advancements, research, and development focus on improving efficiency, expanding viable geothermal resources, and enhancing drilling techniques and reservoir engineering. Site-specific considerations, including geological assessments, are crucial for successful subsurface heat harnessing, and the economic viability of such endeavors depends on factors such as resource quality, drilling costs, and supportive policies [31]. Overall, the continued advancement of subsurface heat harnessing systems holds significant promise for a sustainable and diverse global energy mix [32].

### 6.4.2 Applications in Space Heating, Cooling, and Power Generation

Geothermal energy is harnessed for diverse applications in building systems, providing sustainable solutions for space heating, cooling, and power generation. In space heating, GSHP extract thermal energy from the Earth's subsurface for distribution through a building's heating system, and district heating systems centrally distribute hot water to multiple buildings. Radiant heating systems, integrating geothermal energy, circulate warm water beneath the floor for efficient heating. For space cooling, GSHP systems operate in reverse during summer, and some regions directly use geothermal energy for cooling. Cogeneration systems combine geothermal energy for simultaneous electricity and heat production. Geothermal power plants, EGS, and binary cycle power plants generate electricity from high to low-temperature geothermal resources. Geothermal energy finds applications in industrial processes requiring high-temperature heat. Environmentally, it reduces carbon emissions and offers economic viability through lower operating costs and government incentives [33, 34]. Applications of geothermal energy in building systems are shown in Figure 6.2.

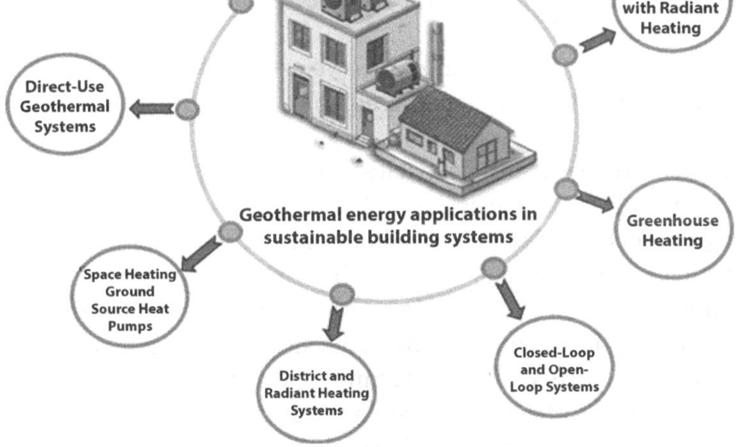


**Figure 6.2** Applications of geothermal energy in building systems.

Space heating: in the context of space heating, GSHP serve as a widely adopted solution for residential, commercial, and industrial buildings. These systems extract thermal energy from the Earth's subsurface during winter, compress the heat, and distribute it at an elevated temperature through the building's heating system. District heating systems, prevalent in regions rich in geothermal resources, efficiently deliver hot water from geothermal reservoirs to multiple buildings, offering centralized warmth for entire neighborhoods. Additionally, geothermal energy seamlessly integrates into radiant heating systems, where warm water circulates beneath the floor through pipes or tubes, ensuring even and energy-efficient heating, known for its comfort and effectiveness.

Space cooling: for space cooling, GSHP exhibit a versatile approach by operating in reverse during the summer. These systems extract heat from the building and transfer it into the cooler subsurface, providing an efficient and sustainable alternative to traditional air conditioning systems. Additionally, in regions with high-temperature geothermal reservoirs, geothermal cooling systems directly utilize geothermal energy for cooling purposes. Chilled water sourced from the subsurface is circulated through cooling systems, contributing to the maintenance of comfortable indoor temperatures in an environmentally friendly manner.

Combined heating and cooling: in the realm of combined heating and cooling, geothermal energy seamlessly integrates into cogeneration or combined heat and power (CHP) systems. These sophisticated systems achieve a dual purpose by simultaneously generating electricity and extracting useful heat for space heating or cooling. Cogeneration enhances the overall efficiency of geothermal energy utilization, making it a sustainable and resourceful solution for meeting both electricity and thermal needs in various applications.

Power generation: in the realm of power generation, geothermal energy employs diverse methods to harness Earth's heat. Geothermal power plants utilize high-temperature reservoirs, drawing heat from the subsurface through wells to generate electricity via turbines connected to generators. Enhanced geothermal systems (EGS) extend the reach of geothermal power by engineering reservoirs in hot rock formations, injecting water to produce steam for power generation. In regions with lower-temperature geothermal resources, binary cycle power plants use a secondary fluid with a lower boiling point than water to create steam, driving turbines for electricity generation. This array of technologies highlights the adaptability and versatility of geothermal energy in producing sustainable power.

Industrial applications: geothermal energy finds valuable application in the industrial sector, particularly in providing process heat for various operations. Industries leverage geothermal heat for high-temperature processes like drying, desalination, and various manufacturing applications. The reliability and consistency of geothermal heat make it a suitable and sustainable solution for fulfilling the thermal requirements of industrial processes. This not only contributes to the efficiency of industrial operations but also aligns with the broader goal of reducing reliance on conventional energy sources with higher environmental impacts.

Environmental and economic considerations: geothermal energy in building systems presents compelling environmental and economic considerations. Firstly, it significantly contributes to carbon emission reduction when compared to conventional fossil fuel-based systems, aligning with global initiatives to combat climate change and promote sustainability. Secondly, geothermal systems demonstrate economic viability over the long term by offering cost savings through lower operating expenses and reduced dependency on external energy sources. Government incentives and supportive policies further bolster the economic attractiveness of geothermal projects, making them not only environmentally friendly but also economically sound choices for building energy solutions.

The geothermal energy plays a crucial role in building systems by providing sustainable solutions for space heating, cooling, and power generation. Its versatility, environmental benefits, and economic advantages make it a valuable component of the transition towards cleaner and more efficient energy use in the built environment.

### 6.4.3 Innovative Geothermal Solutions

Innovative geothermal solutions are continuously evolving to enhance the efficiency, accessibility, and versatility of geothermal energy applications. These solutions aim to address challenges, expand the use of geothermal energy, and contribute to sustainable practices. Here are some innovative geothermal solutions [35, 36]:

Advanced drilling technologies: advanced drilling techniques, including directional drilling, allow for the creation of complex geothermal well systems. This enhances the efficiency of heat extraction from subsurface reservoirs.

Closed-loop ground source heat pumps: advancements in GSHP technology include closed-loop systems that circulate a heat transfer fluid through a closed loop of pipes buried in the ground. This reduces the environmental impact and increases the efficiency of geothermal heating and cooling systems.

Enhanced geothermal systems (EGS): EGS involves creating engineered reservoirs in hot rock formations by injecting water. Innovations in reservoir engineering and stimulation techniques aim to improve the productivity of EGS projects, expanding their feasibility in various geological settings.

Nano-fluids for heat transfer: the use of nano-fluids, which contain nanoparticles, in heat exchangers can enhance the heat transfer efficiency of geothermal systems. This innovation improves the overall performance of geothermal heat exchange processes.

Smart geothermal systems: smart geothermal systems leverage internet of things (IoT) technologies for real-time monitoring and control. This allows for optimization of geothermal heat pump performance, energy usage, and predictive maintenance, improving overall system efficiency.

Geothermal combined with other renewable technologies: integrating geothermal energy with other renewable technologies, such as solar or wind, creates hybrid systems. These systems provide a more consistent and reliable energy supply by compensating for variations in renewable resource availability.

Low-temperature geothermal applications: innovations in technologies that can efficiently extract heat from low-temperature geothermal reservoirs expand the geographical reach of geothermal applications. This is particularly relevant for regions with lower subsurface temperatures.

Deep direct-use systems: deep direct-use geothermal systems explore the utilization of higher temperatures found at greater depths. These systems aim to supply heat directly to industrial processes or district heating systems, offering a sustainable alternative to conventional heating methods.

Microscale geothermal systems: compact geothermal systems designed for individual buildings or small communities are emerging. These microscale systems offer decentralized and efficient solutions for localized heating and cooling needs.

Next-generation power plants: supercritical binary cycle power plants operate at higher temperatures and pressures, improving the efficiency of electricity generation from geothermal resources. These next-generation power plants enhance the economic viability of geothermal electricity production.

Geothermal heat storage: innovations in STES allow for the storage of excess geothermal heat during periods of low demand. This stored heat can be retrieved during peak demand seasons, ensuring a consistent and reliable energy supply.

Incorporating these innovative solutions into geothermal projects contributes to the continued growth and adoption of geothermal energy as a sustainable and reliable energy source. As technology advances, the potential for geothermal applications across various sectors continues to expand, contributing to a more sustainable energy future.

## 6.5 Earth-Air Heat Exchangers: Passive and Active Cooling

Earth-air heat exchangers (EAHE) are innovative systems that leverage the stable temperature of the Earth to provide passive and active cooling for buildings. These systems take advantage of the fact that a few feet below the Earth's surface, the temperature remains relatively constant throughout the year. How EAHE works is discussed in Figure 6.3 and an overview of both passive and active cooling approaches using EAHE are discussed herewith [37]:

Passive cooling with Earth-air heat exchangers: Passive cooling with EAHE is a sustainable method that harnesses natural convection and the temperature differential between outdoor air and the subsurface to cool indoor spaces. In this system, underground ducts or pipes are strategically buried below the building or in close proximity, serving as heat exchangers

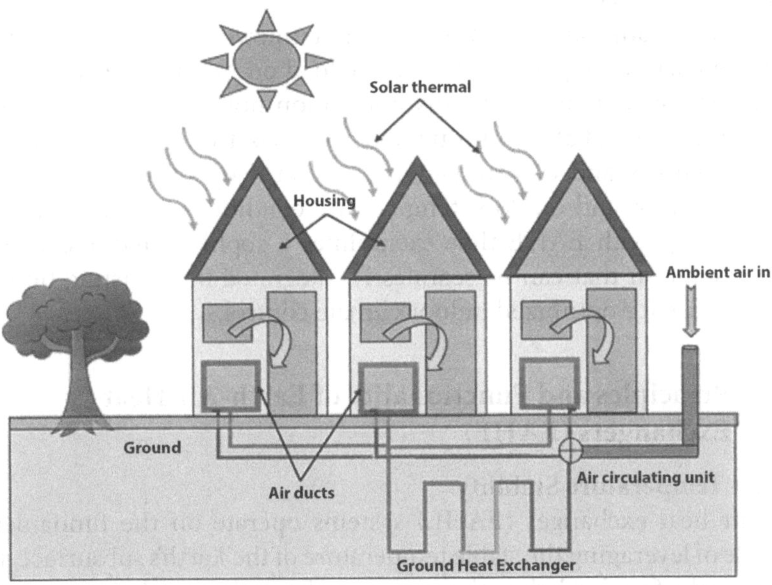


**Figure 6.3** Earth-air heat exchangers (EAHE) working.

that facilitate the exchange of heat between the air inside the ducts and the surrounding subsurface. During warmer periods, outdoor air is drawn through these underground ducts, and as it traverses the cool subsurface, it sheds heat, resulting in a cooling effect. The now-cooled air is then directed into the building, effectively reducing indoor temperatures. Another passive cooling strategy involves night flushing, where cooler outdoor air is drawn through the underground ducts during the night. This process pre-cools the building mass, leveraging the thermal inertia of the structure itself to absorb and retain the coolness generated, thereby contributing to maintaining a comfortable indoor environment during the day. The incorporation of passive cooling with EAHE not only aligns with sustainable building practices but also utilizes the inherent properties of the subsurface and building mass to create an energy-efficient and naturally cooled indoor environment [38].

Active cooling with Earth-air heat exchangers: Active cooling with EAHE represents an innovative approach that employs fans or blowers to actively circulate air through an underground heat exchanger, offering enhanced control over the cooling process. In this method, fans or blowers draw outdoor air through the underground ducts, actively circulating it into the building after being cooled by the subsurface. The use of active systems enables precise temperature control of the air entering the building, allowing for adjustments based on specific cooling requirements and variations in outdoor conditions. Furthermore, the integration of active EAHE systems with traditional HVAC systems is a noteworthy feature, providing supplementary cooling and reducing the load on mechanical cooling systems during specific periods. This integration not only enhances overall energy efficiency but also introduces dynamic control mechanisms, incorporating sensors and control systems to adjust operations dynamically based on indoor and outdoor temperature conditions. The utilization of active cooling with EAHE thus exemplifies a sophisticated and energy-efficient solution that can be seamlessly integrated into existing building infrastructure for optimized indoor climate control.

### 6.5.1 Principles and Functionality of Earth-Air Heat Exchangers (EAHE)

**Ground Temperature Stability**

Earth-air heat exchanger (EAHE) systems operate on the fundamental principle of leveraging the stable temperature of the Earth's subsurface, typically found a few feet below the surface and remaining relatively constant throughout the year. The functionality of EAHE involves the use of buried pipes or ducts as heat exchangers, facilitating the exchange of heat between the air flowing through the system and the surrounding subsurface. This principle serves as the foundation for both passive and active cooling strategies within EAHE systems. In passive cooling, outdoor air naturally circulates through the underground ducts, benefiting from the sub surface's cooling effect, while in active cooling, fans or blowers actively draw outdoor air through the underground heat exchanger, providing a controlled and enhanced cooling process. Overall, the principle and functionality of EAHE systems harness the Earth's consistent subsurface temperature to contribute to energy-efficient cooling solutions for buildings.

**Passive Cooling**

In passive cooling, EAHE systems operate on the principle of utilizing natural convection and the temperature contrast between outdoor air and the subsurface to cool indoor spaces. The functionality involves the natural circulation of outdoor air through underground ducts during warm periods. As this air traverses through the cool subsurface, heat exchange occurs, resulting in the cooling of the air. Subsequently, the cooled air is directed into the building, thereby facilitating a natural cooling process without the necessity of mechanical systems. This approach harnesses the inherent thermal properties of the subsurface to achieve energy-efficient and environmentally friendly cooling within structures.

**Active Cooling**

In active cooling, the EAHE systems operate based on the principle of employing fans or blowers to actively circulate outdoor air through the underground heat exchanger, offering enhanced control over the cooling process. The functionality involves the use of fans or blowers to draw outdoor air through underground ducts. As the outdoor air passes through the subsurface, it undergoes cooling, facilitated by the stable temperature of the subsurface. The actively cooled air is then directed into the building. This method allows for precise temperature control, providing flexibility in adjusting the cooling process according to specific requirements and variations in outdoor conditions. Furthermore, active EAHE systems can be seamlessly integrated with traditional HVAC systems, offering supplementary cooling and contributing to overall energy efficiency by reducing the load on mechanical cooling systems during specific periods. The underground ducts or pipes form a crucial component of this system, serving as conduits for the exchange of heat between the circulating air and the subsurface.

**Underground Ducts or Pipes**

The principle behind EAHE systems employing underground ducts or pipes lies in utilizing these conduits as the medium for air exchange with the subsurface. These ducts serve as essential components for the functionality of the system, enabling the exchange of heat between the circulating air and the subsurface. The design and strategic placement of these underground elements are pivotal factors for ensuring effective heat exchange and optimizing the overall performance of the EAHE system. The underground ducts or pipes act as conduits through which the outdoor air, drawn into the system, interacts with the stable temperature of the subsurface, facilitating the desired cooling effect for buildings.

**Thermal Inertia**

Thermal inertia in the context of EAHE systems involves the principle of building materials possessing the ability to store and release heat over time. The functionality of thermal inertia is realized through the mass of the building, which acts as a thermal storage medium. In the process of passive cooling with EAHE, the building mass efficiently absorbs and retains the coolness generated during the night or cooler periods. This stored thermal energy becomes instrumental in maintaining a comfortable indoor environment during subsequent warmer periods. The concept of thermal inertia showcases how the inherent properties of building materials contribute to the overall energy dynamics of the EAHE system, enhancing its effectiveness in providing sustainable and energy-efficient cooling solutions.

**Night Flushing**

Night flushing, as applied in the context of EAHE systems, operates on the principle of utilizing cooler outdoor air during the night to pre-cool the building and its mass. The functionality of night flushing in passive EAHE systems is designed to harness lower night-time temperatures effectively. During this process, the system draws in cooler outdoor air through the underground ducts. This pre-cooled air is then circulated within the building, contributing to the reduction of indoor temperatures. By strategically leveraging night flushing, passive EAHE systems optimize the thermal performance of the building mass, enhancing overall energy efficiency and occupant comfort during warmer daytime conditions.

**Integration with HVAC Systems**

Integration with HVAC systems, in the context of EAHE systems, operates on the principle of seamlessly combining active EAHE systems with traditional HVAC systems. The functionality of this integration is designed to provide supplementary cooling to the building. The EAHE system, driven by fans or blowers, actively circulates outdoor air through the underground heat exchanger. This actively cooled air, having undergone the temperature exchange with the subsurface, is then directed into the building. By integrating with HVAC systems, the EAHE system reduces the load on mechanical cooling systems during specific periods. This integration not only enhances overall energy efficiency but also contributes to cost savings in the operational aspects of the building's climate control systems.

**Dynamic Control**

Dynamic control in EAHE systems revolves around the principle of utilizing sensors and control systems to dynamically adjust the operation of the system based on real-time conditions. The functionality of dynamic control in active EAHE systems is crucial for optimizing performance and maximizing energy savings. These systems incorporate sensors that continuously monitor both indoor and outdoor temperatures. The control system then responds to these temperature variations, making real-time adjustments to the operation of the EAHE system. This dynamic control mechanism ensures that the system operates efficiently and effectively in response to the prevailing environmental conditions, contributing to the overall energy efficiency of the building's cooling system.

### 6.5.2 Benefits of Earth-Air Heat Exchanger (EAHE) Systems

Earth-air heat exchanger (EAHE) systems offer several benefits, making them a valuable addition to building designs. One key advantage is their contribution to energy efficiency, as these systems leverage the Earth's stable temperature to provide cooling, reducing the need for extensive reliance on mechanical cooling systems and ultimately lowering overall energy consumption. This aligns with sustainable building practices, promoting environmental sustainability by decreasing dependence on traditional air conditioning methods, resulting in a reduced environmental footprint [39]. The operational efficiency of EAHE systems also translates into potential cost savings, particularly in regions with moderate climate conditions, as lower energy consumption leads to reduced utility costs over the system's lifecycle. Moreover, the principles of EAHE systems contribute to maintaining a stable and comfortable indoor environment by effectively regulating indoor temperatures throughout the year. In summary, EAHE embody sustainable practices, offering not only energy efficiency and environmental responsibility but also economic advantages and improved indoor comfort, making them a compelling choice for environmentally conscious and cost-effective building solutions [40].

Top of Form

### 6.5.3 Practical Implementation Techniques for Earth-Air Heat Exchangers (EAHE)

Practical implementation techniques for EAHE involve a systematic approach to ensure optimal performance and energy efficiency. The first step is a comprehensive site assessment, considering factors such as local climate, soil composition, and available space. Conducting soil tests to determine thermal conductivity, assessing the water table depth, and identifying potential obstacles are crucial. The chosen installation location should maximize the effectiveness of the EAHE system [41–43].

The design of underground ducts is a critical aspect of implementation. Optimization of layout and depth based on the building's heating and cooling requirements, considering factors like soil type and available space for excavation, ensures effective heat exchange with the subsurface. The selection of a suitable heat exchange medium, proper insulation of ducts, and incorporation of passive cooling strategies, such as natural convection, contribute to system efficiency. Active cooling integration with fans or blowers, dynamic control systems with sensors for real-time adjustments, and utilization of thermal inertia by leveraging the building's mass as a thermal storage medium enhance control over the cooling process.

Night flushing design, where cooler outdoor air is used during the night to pre-cool the building, can be facilitated through system design and automated controls. Integration with HVAC systems allows for supplementary cooling, reducing the load on mechanical cooling during specific periods. Regular maintenance, including routine inspections, cleaning of ducts, and checking the functioning of fans and blowers, ensures continued efficiency. Monitoring and optimization through installed systems provide ongoing insights for peak performance. Compliance with local regulations, building codes, and environmental standards is essential for successful implementation. Community education programs foster awareness and acceptance of EAHE technology, addressing concerns and encouraging sustainable building practices. The combination of these implementation techniques ensures the effective and sustainable deployment of EAHE systems in building designs [44, 45].

## 6.6 Combined Augmentation Strategies for Sustainable Buildings

In the pursuit of creating truly sustainable buildings, a holistic approach that integrates multiple renewable energy technologies is essential. This chapter explores combined augmentation strategies, focusing on synergies among solar, geothermal, and Earth-air systems to optimize energy efficiency and environmental impact. The integration of these technologies offers a comprehensive solution for both heating and cooling needs, ensuring a resilient and eco-friendly built environment [46–50].

Integrated system design: the foundation of combined augmentation lies in the thoughtful design of an integrated energy system that harmoniously incorporates solar PV, solar thermal, geothermal heat pumps, and EAHE. This approach ensures that each technology complements the others, maximizing overall system efficiency.

Complementary energy generation: by combining solar PV for electricity generation with solar thermal systems for heating, the building benefits from complementary energy outputs. Solar panels harness sunlight for electricity, while solar thermal collectors capture and store heat energy for space heating and water heating, creating a balanced and continuous energy supply.

Seasonal thermal energy storage (STES): STES becomes a crucial component in the combined strategy, allowing for the storage of excess energy during peak solar periods. The geothermal system can play a role in storing this surplus energy in the ground, providing a reliable reservoir for thermal needs during less sunny periods or higher demand.

Geothermal heat pump integration: integrating geothermal heat pumps with solar thermal systems enhances efficiency. Solar thermal collectors preheat the fluid used in geothermal heat pumps, reducing the load on the system, and optimizing the utilization of both solar and geothermal resources.

Earth-air heat exchanger for cooling support: EAHE contribute to the strategy by offering passive or active cooling support. During warmer periods, these systems exchange heat with the cooler subsurface, reducing the reliance on mechanical cooling and promoting energy savings.

Hybrid renewable systems: the combined augmentation strategy results in a hybrid renewable energy system. Solar energy caters to peak demand, geothermal systems provide consistent thermal control, and EAHE systems contribute to passive cooling. This combination ensures a continuous and reliable energy supply throughout the year.

Smart energy management: advanced energy management systems are implemented to orchestrate the combined system's operation intelligently. Utilizing real-time data and considering various factors such as weather conditions and building occupancy, smart management optimizes the performance of each technology within the integrated system.

Environmental impact reduction: the overarching goal of these combined augmentation strategies is to significantly reduce the environmental impact of building energy systems. By relying on renewable sources, minimizing energy waste, and embracing advanced technologies, the integrated system aligns with sustainable building practices and environmental stewardship.

In conclusion, the chapter delves into the intricacies of combining solar, geothermal, and Earth-air systems to create a unified and efficient energy strategy for sustainable buildings. This integrated approach not only addresses the challenges of heating and cooling but also sets a benchmark for environmentally conscious and resilient building design.

### 6.6.1 Synergies Among Solar, Geothermal, and Earth-Air Systems

The combined augmentation of solar, geothermal, and Earth-Air systems presents a holistic and integrated strategy for enhancing the sustainability and efficiency of building energy systems. The integration of these renewable technologies involves designing a comprehensive energy system that optimally combines the strengths of each. This integrated design ensures efficiently meeting the energy demands for both heating and cooling. Solar energy contributes to space heating and electricity generation, while geothermal and Earth-air systems provide stable and consistent thermal control. The complementary nature of these systems ensures a continuous and balanced energy supply throughout the year. Solar panels generate electricity, while geothermal and Earth-air systems provide thermal energy for heating and cooling, ensuring a reliable supply even during periods of low solar availability [51].

Seasonal thermal energy storage (STES) plays a key role in the synergies among these systems by storing excess energy generated during peak solar periods or periods of low demand. This stored energy is then retrieved during periods of high demand or low solar availability, contributing to a reliable and continuous energy supply. The integration of geothermal heat pumps with solar thermal systems enhances overall efficiency by preheating the fluid used in geothermal heat pumps, reducing the energy required for heating. Earth-air heat exchangers (EAHE) further complement solar and geothermal systems by providing passive or active cooling. During warmer periods, EAHE systems cool indoor spaces by exchanging heat with the cooler subsurface, reducing the load on mechanical cooling systems and enhancing energy savings [52].

The resulting hybrid renewable energy system offers a reliable and resilient energy solution. It harnesses solar energy when sunlight is abundant, maintains consistent thermal control through geothermal systems, and contributes to passive cooling with EAHE systems. Advanced energy management systems optimize the operation of these integrated systems based on real-time data, considering factors such as weather conditions, building occupancy, and energy demand. This ensures the efficient utilization of available resources and maximizes the benefits of the integrated system. Moreover, the combined augmentation strategy aims to reduce the environmental impact of building energy systems by relying on renewable sources, maximizing energy efficiency, and contributing to lower carbon emissions. Overall, the synergies among solar, geothermal, and Earth-air systems demonstrate a robust and versatile approach to sustainable building practices, aligning with the broader goal of creating energy-efficient, environmentally friendly, and resilient built environments [53].

### 6.6.2 Challenges and Considerations in the Integration of Solar, Geothermal, and Earth-Air Systems for Sustainable Buildings

While the integration of solar, geothermal, and Earth-air systems offers significant advantages in creating energy-efficient and sustainable buildings, it comes with its own set of challenges and considerations. Addressing these factors is crucial for the successful implementation and optimization of a combined augmentation strategy. Few key challenges and considerations in the integration process are discussed in this section [53–57]:

**System Compatibility**

Achieving seamless compatibility among diverse renewable energy technologies, including solar panels, geothermal heat pumps, and EAHE, poses a significant challenge. To address this challenge, prioritizing technologies with built-in interoperability features is crucial. It is essential to ensure that the components of each system can work cohesively to maximize efficiency. Thorough feasibility studies during the design phase play a pivotal role in identifying and addressing potential compatibility issues, ensuring a harmonized integration of these technologies into a cohesive and efficient renewable energy system.

**Space and Site Constraints**

Limited space and site-specific restrictions present challenges for installing multiple systems, particularly geothermal ground loops or extensive solar arrays. To address this challenge, efficient system design and creative architectural solutions are essential. Utilize rooftop spaces for solar panels, explore vertical installations, and carefully plan geothermal borehole locations to maximize efficiency within the constraints of available space. Strategic planning and innovative design considerations play a crucial role in overcoming the limitations posed by spatial constraints, ensuring the effective integration of renewable energy technologies.

**Cost Implications**

The upfront costs associated with installing multiple renewable energy systems can be a significant challenge, potentially acting as a financial barrier for some projects. To address this challenge, it is crucial to conduct a comprehensive cost-benefit analysis, evaluating the long-term savings and return on investment provided by the integrated systems. Additionally, exploring available incentives, grants, or financing options can help alleviate the financial burden and promote the adoption of sustainable technologies. Careful financial planning and leveraging available support mechanisms are key considerations for overcoming cost-related challenges in implementing a combined augmentation strategy for renewable energy systems.

**Maintenance Requirements**

Addressing the maintenance needs of multiple renewable energy systems poses a challenge due to their unique requirements. To navigate this complexity, a detailed maintenance plan should be developed, considering the specific demands of each system. Regular inspections, monitoring, and timely repairs are essential to ensure the longevity and efficiency of the integrated system.

**Energy Storage and Management**

Coordinating the storage and management of energy generated by diverse renewable energy systems is a challenge. To overcome this, implement energy storage solutions such as batteries or thermal storage systems to store excess energy during peak production times. Additionally, deploy advanced energy management systems that optimize energy distribution based on real-time demand and availability.

**Climate and Geographic Variability**

The performance of renewable energy systems is influenced by climate conditions, presenting a challenge. To address this, conduct a thorough climate analysis and site assessment to tailor the integrated system to local conditions. Consider the geographic-specific benefits and limitations of solar, geothermal, and Earth-air technologies.

**Regulatory and Permitting Challenges**

Navigating compliance with local building codes, zoning regulations, and environmental permits can be challenging for integrated renewable energy systems. Collaborate with local authorities early in the planning process to ensure alignment with regulatory requirements. Obtain necessary permits before construction and engage with community stakeholders to build support for sustainable building practices.

**Technological Advancements and Updates**

The rapid evolution of renewable energy technologies poses a challenge for system compatibility over time. Stay informed about technological developments and plan for future upgrades or retrofits. Design the integrated system with flexibility to accommodate advancements, ensuring adaptability and resilience over time.

**Occupant Comfort and Behavior**

Occupant comfort preferences and behavior can impact the effectiveness of integrated systems, especially those relying on passive strategies. To address this challenge, incorporate occupant education programs and feedback mechanisms to promote sustainable behaviors. Design the building to provide comfort through user-friendly interfaces, ensuring occupants are supportive and engaged in energy-saving practices.

**Long-Term Performance Monitoring**

Ensuring the long-term performance of integrated systems requires continuous monitoring and data analysis. Implement robust monitoring systems that track the performance of each technology. Regularly analyze data to identify trends, troubleshoot issues, and make informed decisions for system optimization.

Addressing these challenges and considerations in the integration of solar, geothermal, and Earth-air systems is essential for the success of sustainable building projects. A thorough understanding of these factors allows for strategic planning, effective problem-solving, and the creation of resilient and high-performing integrated systems.

### 6.6.3 Future Trends and Prospects in the Integration of Solar, Geothermal, and Earth-Air Systems for Sustainable Buildings

The integration of solar, geothermal, and Earth-air systems in sustainable buildings continues to evolve, driven by advancements in technology, increasing environmental awareness, and the need for resilient and energy-efficient structures. Looking toward the future, several trends and prospects emerge that shape the trajectory of integrated renewable energy systems [58–60]:

**Smart Building Integration**
The integration of artificial intelligence (AI) and smart building technologies is poised to become more prevalent, ushering in an era of real-time monitoring, predictive analytics, and adaptive control for solar, geothermal, and Earth-air systems. The prospect of smart buildings optimizing energy consumption by dynamically adjusting system parameters based on occupant behavior, weather conditions, and energy demand holds the promise of enhancing overall efficiency and contributing significantly to sustainable practices.

**Energy Storage Innovations**
Advances in energy storage technologies, particularly improved battery systems and innovative thermal storage solutions, are set to play a crucial role in managing intermittent energy generation from solar sources. The prospect of enhanced energy storage capabilities will empower buildings to store surplus energy during peak production times, ensuring a consistent and reliable energy supply even during periods of low renewable energy generation.

**Hybrid Renewable Energy Systems**
A discernible trend toward developing hybrid renewable energy systems, combining solar, geothermal, and other renewable sources, is gaining momentum to create more resilient and diversified energy portfolios. The prospect of hybrid systems optimizing energy production, minimizing dependency on a single energy source, and providing flexibility to adapt to changing environmental conditions holds the potential to improve overall energy security.

**Technological Convergence**
The continued convergence of technologies, exemplified by BIPV and advanced HVAC systems, is poised to simplify the integration process and enhance the aesthetics of sustainable buildings. The prospect of integrated

technologies becoming more seamless, blurring the lines between architectural elements and energy systems, will likely contribute to the mainstream adoption of sustainable building practices.

**Community Microgrids**

A notable trend in urban planning and sustainable development is the development of community microgrid, where multiple buildings share and exchange energy. The prospect of community microgrid allowing for efficient energy distribution, sharing excess energy among neighboring buildings, and increasing overall resilience during grid disruptions is becoming a prominent aspect of sustainable urban development.

**Lifecycle Assessment Emphasis**

There is a discernible trend towards an increasing emphasis on the lifecycle assessment of integrated systems, shaping decision-making with a focus on long-term sustainability, environmental impact, and end-of-life considerations. The prospect of lifecycle assessments guiding the selection of materials, technologies, and maintenance practices ensures that integrated systems align with circular economy principles, contributing positively to environmental goals.

**Global Policy Support**

Growing global awareness of climate change and the need for sustainable practices is expected to lead to more supportive policies and incentives for the integration of renewable energy systems in buildings. The prospect of increased policy support is likely to drive widespread adoption, making sustainable building practices more accessible and financially viable for a broader range of projects.

**Innovations in Geothermal Technologies**

Ongoing research and development in geothermal technologies, encompassing enhanced geothermal systems and geothermal heat pump innovations, are expected to broaden the applicability and efficiency of geothermal integration. The prospect of innovations in geothermal technologies making systems more adaptable to diverse geological conditions and increasing their overall effectiveness in heating and cooling applications is a noteworthy development.

**Energy Performance Certifications**

There is a discernible trend emphasizing energy performance certifications, such as LEED (Leadership in Energy and Environmental Design) and

BREEAM (Building Research Establishment Environmental Assessment Method), driving the demand for integrated renewable energy systems in commercial and residential buildings. The prospect of achieving high-performance certifications becoming a standard practice is influencing construction practices and encouraging the widespread implementation of integrated systems.

**Educational and Public Awareness Initiatives**

An observable trend involves increasing educational initiatives and public awareness campaigns promoting the benefits of integrated renewable energy systems, fostering a greater understanding of sustainable building practices. The prospect of informed building owners, developers, and communities being more likely to invest in integrated systems is contributing to a cultural shift towards sustainable and environmentally conscious construction.

In conclusion, the integration of solar, geothermal, and Earth-air systems for sustainable buildings is poised for significant advancements. Future trends and prospects indicate a shift towards smarter, more resilient, and community-oriented approaches, reinforcing the role of integrated renewable energy systems in creating a sustainable built environment.

## 6.7 Conclusion

The integration of solar, geothermal, and Earth-air systems in sustainable buildings represents a pivotal approach in the quest for energy-efficient and environmentally conscious construction. This chapter has delved into various aspects of this integration, exploring methodologies, challenges, considerations, and future trends. Here is a concise summary of key findings:

**Holistic Integration Approach**

The chapter emphasized the significance of adopting a holistic approach in integrating solar, geothermal, and Earth-air systems. This comprehensive strategy addresses both heating and cooling needs, promoting resilience and sustainability in building design.

**Challenges and Considerations**

Challenges in system compatibility, space constraints, cost implications, and maintenance requirements were identified. Successful integration requires careful consideration of these factors to ensure optimal performance and longevity.

**Smart Technologies and Innovations**

Future trends highlighted the increasing integration of smart technologies, advancements in energy storage, and the convergence of technologies. These trends aim to simplify integration processes, enhance efficiency, and contribute to the aesthetic appeal of sustainable buildings.

**Community and Global Perspectives**

The emergence of community microgrid, global policy support, and a growing emphasis on lifecycle assessments underscored the importance of considering broader perspectives in sustainable building practices. Collaboration and global initiatives will play a key role in driving widespread adoption.

**Geothermal Innovations and Certifications**

Ongoing innovations in geothermal technologies, coupled with the growing importance of energy performance certifications, highlighted the role of continuous research and industry standards in shaping the future of integrated systems.

**Public Awareness and Education**

The chapter underscored the importance of public awareness and educational initiatives in promoting sustainable building practices. Informed stakeholders are more likely to invest in integrated renewable energy systems, contributing to a cultural shift towards environmentally conscious construction.

**Prospects for Sustainable Building Practices**

The prospects for sustainable building practices are optimistic, with a trajectory towards smarter, more resilient, and community-oriented approaches. The integration of solar, geothermal, and Earth-air systems aligns with the broader goals of creating energy-efficient and environmentally friendly built environments.

The integration of solar, geothermal, and Earth-air systems offers a multifaceted solution to the challenges of sustainable building practices. As technology evolves, challenges are addressed, and awareness grows, the future holds promise for more widespread adoption and innovation in integrated renewable energy systems, contributing to a sustainable and resilient built environment.

### 6.7.1 Implications for Sustainable Building Practices

The integration of solar, geothermal, and Earth-air systems holds profound implications for sustainable building practices. The key findings of this chapter underscore several implications that can shape the future of construction and contribute to a more sustainable built environment [61]:

**Enhanced Energy Efficiency**
The holistic integration of renewable energy technologies enables buildings to achieve a higher level of energy efficiency. This has significant implications for reducing carbon footprints and promoting environmentally responsible practices in the construction industry.

**Resilient and Adaptive Design**
Integrated systems provide buildings with resilience and adaptability to varying environmental conditions. This adaptability is crucial in the face of climate change impacts, ensuring that structures can efficiently respond to changing energy needs and environmental factors.

**Cost-Effective Sustainability**
While upfront costs may be a consideration, the long-term implications of reduced energy bills, potential incentives, and positive environmental impact position integrated systems as cost-effective solutions for sustainable building practices.

**Positive Impact on Occupant Well-Being**
Sustainable buildings, with integrated renewable energy systems, contribute to enhanced indoor air quality, thermal comfort, and overall occupant well-being. These positive impacts can create healthier living and working environments.

**Community-Level Sustainability**
The trend towards community microgrid and collaborative approaches fosters a sense of community-level sustainability. Shared energy resources and a collective commitment to environmental responsibility contribute to more resilient and sustainable neighborhoods.

**Influence on Regulatory Standards**
The growing adoption of integrated renewable energy systems is likely to influence regulatory standards and certifications. Building codes may evolve to incentivize or mandate the integration of solar, geothermal, and Earth-air systems, further promoting sustainable building practices.

### 6.7.2 Recommendations for Future Research

As the field of integrated renewable energy systems in sustainable buildings continues to advance, future research efforts should explore key areas to drive innovation and address existing challenges. The following recommendations outline avenues for future research [62]:

**Technological Advancements in Geothermal Systems**
Research into further advancements in geothermal technologies, including enhanced geothermal systems and novel heat pump designs, can improve the efficiency and applicability of geothermal integration in various geological conditions [63–64].

**Long-Term Performance Monitoring Studies**
Longitudinal studies on the performance of integrated systems in real-world settings will provide valuable insights into their long-term effectiveness. This includes monitoring energy production, system maintenance needs, and occupant satisfaction over extended periods.

**Human-Centric Design Considerations**
Research should delve into the human-centric design aspects of sustainable buildings. Understanding occupant behavior, preferences, and the impact of integrated systems on daily living can inform design strategies that prioritize both sustainability and user comfort.

**Circular Economy Approaches**
Future research should explore circular economy approaches in the construction and operation of sustainable buildings. This includes studying material life cycles, recyclability, and the potential for repurposing components of integrated systems at the end of their service life.

**Urban Planning and Community-Level Integration**
Studies focused on urban planning and the implementation of integrated systems at a community level can provide insights into the scalability and benefits of community microgrids. Understanding the social, economic, and environmental implications of such approaches is critical.

**Policy and Regulatory Impact Analysis**
Research efforts should analyze the impact of policies and regulatory frameworks on the adoption of integrated renewable energy systems. This includes assessing the effectiveness of existing incentives, identifying

barriers to adoption, and proposing policy recommendations to further encourage sustainable building practices.

**Innovations in Energy Storage**

Continued research into energy storage innovations is crucial for addressing the intermittency of renewable energy sources. Exploring advancements in battery technologies, thermal storage, and grid-scale storage solutions can contribute to more reliable and resilient integrated systems [65–67].

**Cross-Disciplinary Collaboration**

Encouraging cross-disciplinary collaboration between architects, engineers, environmental scientists, and social scientists is essential. Collaborative research can yield comprehensive insights into the holistic integration of renewable energy systems and their broader implications.

By focusing on these recommendations, the research community can contribute to the ongoing evolution of sustainable building practices, ensuring that integrated renewable energy systems continue to play a central role in creating a more sustainable and resilient built environment.

## References

1. Pearce, A.R. and Ahn, Y.H., *Sustainable Buildings and Infrastructure*, Routledge, 2012.
2. Ramalho, J.V.A., Fernando, H.J., Brum, R.S., Domingues, A.M.B., Navarro Pastor, N.R., Burlón Olivera, M.R., Accessing the thermal performance of Earth–air heat exchangers surrounded by galvanized structures. *Sustain. Energy Technol. Assess.*, 54, 102838, 2022, doi: https://doi.org/10.1016/j.seta.2022.102838.
3. Pombo, O., Rivela, B., Neila, J., The challenge of sustainable building renovation: assessment of current criteria and future outlook. *J. Clean. Prod.*, 123, 88–100, 2016.
4. Kaushal, M., Performance analysis of clean energy using geothermal earth to air heat exchanger (GEAHE) in Lower Himalayan Region – Case study scenario. *Energy Build.*, 248, 111166, 2021, doi: https://doi.org/10.1016/j.enbuild.2021.111166.
5. Afrand, M., Shahsavar, A., Sardari, P.T., Sopian, K., Salehipour, H., Energy and exergy analysis of two novel hybrid solar photovoltaic geothermal energy systems incorporating a building integrated photovoltaic thermal system and an earth air heat exchanger system. *Sol. Energy*, 188, 83–95, 2019, doi: https://doi.org/10.1016/j.solener.2019.05.080.

6. Soares, N., Rosa, N., Monteiro, H., Costa, J.J., Advances in standalone and hybrid earth-air heat exchanger (EAHE) systems for buildings: A review. *Energy Build.*, 253, 111532, 2021, doi: https://doi.org/10.1016/j.enbuild.2021.111532.
7. Shahsavar, A., Talebizadehsardari, P., Arıcı, M., Comparative energy, exergy, environmental, exergoeconomic, and enviroeconomic analysis of building integrated photovoltaic/thermal, earth-air heat exchanger, and hybrid systems. *J. Clean. Prod.*, 362, 132510, 2022, doi: https://doi.org/10.1016/j.jclepro.2022.132510.
8. Sharma, A.K., Nayal, M., Jain, S., Singh, V.P., Chapter 07 Optimization Techniques of Solar Thermal and Hybrid Energy Systems, in: *Highly Efficient Thermal Renewable Energy Systems Design, Optimization and Applications*, 1st ed., R.W.V. Verma, S. Thangavel, N. Dutt, A. Kumar (Eds.), pp. 1–14, CRC Press, 2024.
9. Kushwaha, P.K., Sharma, N.K., Kumar, A., Meena, C.S., Recent Advancements in Augmentation of Solar Water Heaters Using Nanocomposites with PCM: Past, Present, and Future. *Buildings*, 13, 1, 2023, doi: 10.3390/buildings13010079.
10. Singh, V.P. and Dwivedi, G., Technical Analysis of a Large-Scale Solar Updraft Tower Power Plant. *Energies*, 16, 1, 103325, 2023, doi: 10.3390/en16010494.
11. Guno, C.S., Agaton, C.B., Villanueva, R.O., Villanueva, R.O., Optimal investment strategy for solar pv integration in residential buildings: A case study in the Philippines. *Int. J. Renew. Energy Dev.*, 10, 1, 79–89, 2021, doi: 10.14710/ijred.2021.32657.
12. Kundu, A., Kumar, A., Singh, V.P., Meena, C.S., Dutt, N., Chapter 1: Introduction to Thermal Energy Resources and Their Smart Applications, in: *Thermal Energy Systems:Design, Computational Techniques, and Applications*, pp. 1–15, 2023.
13. Kumar, A., Singh, V.P., Meena, C.S., Dutt, N., *Thermal Energy Systems: Design, Computational Techniques, and Applications*, 1st ed., CRC Press, 2023.
14. Finnigan, S.M., Clear, A.K., Farr-Wharton, G., Ladha, K., Comber, R., Augmenting audits: Exploring the role of sensor toolkits in sustainable buildings management. *Proc. ACM Interact. Mobile Wearable Ubiquitous Technol.*, 1, 2, 1–19, 2017, doi: 10.1145/3090075.
15. Singh, V.P. and Jain, S., Economic analysis of a large scale solar updraft tower power plant. *Sustain. Energy Technol. Assess.*, 58, 103325, 2023, doi: https://doi.org/10.1016/j.seta.2023.103325.
16. Rathore, G.S., S.R., A.N., Karn, A., Singh, V.P., Recent Progress in Solar Dryers using Phase Changing Material : A Review. *Int. J. Energy Resour. Appl.*, 2, 1, 57–77, 2023, doi: 10.56896/IJERA.2023.2.1.005.
17. Frontini, F., Manfren, M., Tagliabue, L.C., A Case Study of Solar Technologies Adoption: Criteria for BIPV Integration in Sensitive Built Environment. *Energy Procedia*, 30, 1006–1015, 2012, doi: https://doi.org/10.1016/j.egypro.2012.11.113.
18. Bushra, N., A comprehensive analysis of parametric design approaches for solar integration with buildings: A literature review. *Renew. Sustain. Energy Rev.*, 168, 112849, 2022, doi: https://doi.org/10.1016/j.rser.2022.112849.

19. Pant, G., Meena, C.S., Saxena, A., Kumar, A., Singh, V.P., Dutt, N., Study the Temperature Variation in Alternate Coils of Insulated Condenser Cum Storage Tank: Experimental Study, in: *Lecture Notes in Mechanical Engineering*, pp. 627–638, Springer, Singapore, 2023, doi: https://doi.org/10.1007/978-981-99-2382-3_52.
20. Dutt, N., Jageshwar, A., Ashwani, H., Mukesh, K., Awasthi, K., Pratap, V., Thermo hydraulic performance of solar air heater having discrete D shaped ribs as artificial roughness. *Environ. Sci. Pollut. Res.*, 2023, doi: 10.1007/s11356-023-28247-9.
21. Chel, A. and Kaushik, G., Renewable energy technologies for sustainable development of energy efficient building. *Alex. Eng. J.*, 57, 2, 655–669, 2018, doi: https://doi.org/10.1016/j.aej.2017.02.027.
22. Saxena, A., Prajapati, A., Pant, G., Meena, C., Kumar, A., Singh, V., Water Consumption Optimization of Hybrid Heat Pump Water Heating System, in: *Lecture Notes in Mechanical Engineering- Recent Advances in Mechanical Engineering Select Proceedings of FLAME 2022*, A. Shukla, B. Sharma, A. Arabkoohsar, P. Kumar (Eds.), pp. 721–732, Springer Nature Singapore, 2023.
23. Yang, L., He, B., Ye, M., The application of solar technologies in building energy efficiency: BISE design in solar-powered residential buildings. *Technol. Soc.*, 38, 111–118, 2014, doi: https://doi.org/10.1016/j.techsoc.2014.03.002.
24. Bawankar, S. *et al.*, Environmental impact assessment of lithium ion battery employing cradle to grave. *Sustain. Energy Technol. Assess.*, 60, May, 2023, doi: 10.1016/j.seta.2023.103530.
25. Vujanović, M., Wang, Q., Mohsen, M., Duić, N., Yan, J., Recent progress in sustainable energy-efficient technologies and environmental impacts on energy systems. *Appl. Energy*, 283, 116280, 2021, doi: https://doi.org/10.1016/j.apenergy.2020.116280.
26. Shortall, R., Davidsdottir, B., Axelsson, G., Geothermal energy for sustainable development: A review of sustainability impacts and assessment frameworks. *Renew. Sustain. Energy Rev.*, 44, 391–406, 2015, doi: https://doi.org/10.1016/j.rser.2014.12.020.
27. Cunha, R.P. and Bourne-Webb, P.J., A critical review on the current knowledge of geothermal energy piles to sustainably climatize buildings. *Renew. Sustain. Energy Rev.*, 158, 112072, 2022, doi: https://doi.org/10.1016/j.rser.2022.112072.
28. Sivasakthivel, T., Verma, V., Tarodiya, R., Meena, C.S., Singh, V.P., Kumar, R., Chapter 11: Analysis of Optimum Operating Parameters for Ground Source Heat Pump System for Different Cases of Building Heating and Cooling Mode Operations, in: *Thermal Energy Systems: Design, Computational Techniques, and Applications*, 1st ed., A. Kumar, N. Dutt, V. Singh, C. Meena (Eds.), pp. 183–207, CRC Press, 2023.
29. Kumar, A., Singh, V.P., Meena, C.S., Dutt, N., Aim and Scope of Thermal Energy Systems: Design, Computational Techniques, and Applications, in:

*Thermal Energy Systems: Design, Computational Techniques, and Applications*, A. Kumar, V.P. Singh, C.S. Meena, N. Dutt (Eds.), pp. 9–10, CRC Press, 2022.
30. Kumar, A., Singh, V.P., Meena, C.S., Dutt, N., Preface of Thermal Energy Systems, in: *Thermal Energy Systems: Design, Computational Techniques, and Applications*, pp. x–xiii, Technical Education Department, Kanpur, Uttar Pradesh, India: CRC Press, 2023.
31. Nayal, M., Sharma, A.K., Jain, S., Singh, V.P., Chapter 11 Green Hydrogen Production Methods, Designs and Applications, in: *Highly Efficient Thermal Renewable Energy Systems Design, Optimization and Applications*, 1st ed., R.W.V. Verma, S. Thangavel, N. Dutt, A. Kumar (Eds.), pp. 178–193, CRC Press, 2024.
32. Curtis, R., Lund, J., Sanner, B., Rybach, L., Hellström, G., Ground source heat pumps–geothermal energy for anyone, anywhere: current worldwide activity, in: *Proceedings World Geothermal Congress*, vol. 24, p. 175, 2005.
33. Aggarwal, V. *et al.*, Potential and Future Prospects of Geothermal Energy in Space Conditioning of Buildings: India and Worldwide Review. *Sustainability*, 12, 20, 2020, doi: 10.3390/su12208428.
34. Soltani, M. *et al.*, A comprehensive study of geothermal heating and cooling systems. *Sustain. Cities Soc.*, 44, 793–818, 2019, doi: https://doi.org/10.1016/j.scs.2018.09.036.
35. DeLovato, N., Sundarnath, K., Cvijovic, L., Kota, K., Kuravi, S., A review of heat recovery applications for solar and geothermal power plants. *Renew. Sustain. Energy Rev.*, 114, 109329, 2019, doi: https://doi.org/10.1016/j.rser.2019.109329.
36. Anderson, A. and Rezaie, B., Geothermal technology: Trends and potential role in a sustainable future. *Appl. Energy*, 248, 18–34, 2019, doi: https://doi.org/10.1016/j.apenergy.2019.04.102.
37. Rosa, N., Soares, N., Costa, J.J., Santos, P., Gervásio, H., Assessment of an earth-air heat exchanger (EAHE) system for residential buildings in warm-summer Mediterranean climate. *Sustain. Energy Technol. Assess.*, 38, 100649, 2020, doi: https://doi.org/10.1016/j.seta.2020.100649.
38. Kundu, A., Kumar, A., Dutt, N., Singh, V.P., Meena, C.S., Chapter 7: Modelling and Simulation of Thermal Energy System for Design Optimization, in: *Thermal Energy Systems: Design, Computational Techniques, and Applications*, A. Kumar, N. Dutt, V. Singh, C. Meena (Eds.), pp. 103–137, CRC Press, 2023.
39. Kumar, A., Gori, Y., Meena, C.S., Singh, V.P., Sharma, V.K., Design and analysis of heavy vehicle medium duty transmission gearbox system, in: *Addit. Manuf. Ind. 4.0 Methods, Tech. Model. Nano Asp.*, pp. 67–86, 2022.
40. Mihalakakou, G. *et al.*, Applications of earth-to-air heat exchangers: A holistic review. *Renew. Sustain. Energy Rev.*, 155, 111921, 2022, doi: https://doi.org/10.1016/j.rser.2021.111921.
41. Bordoloi, N., Sharma, A., Nautiyal, H., Goel, V., An intense review on the latest advancements of Earth Air Heat Exchangers. *Renew. Sustain. Energy Rev.*, 89, 261–280, 2018, doi: https://doi.org/10.1016/j.rser.2018.03.056.

42. Chauhan, K. and Singh, V.P., Prospect of biomass to bioenergy in India : An overview. *Mater. Today Proc.*, 2023, doi: 10.1016/j.matpr.2023.01.419.
43. Singh, A.R., Singh, S.K., Singh, V.P., Process parameters optimization of carbon nano tube based catalytic transesterification of algal oil. *Mater. Today Proc.*, 2023, doi: 10.1016/j.matpr.2023.01.418.
44. Jain, S., Kumar, N., Singh, V.P., Mishra, S., Sharma, N.K., Transesterification of Algae Oil and Little Amount of Waste Cooking Oil Blend at Low Temperature in the Presence of NaOH. *Energies*, 16, 1, 1–12, 2023, doi: https://doi.org/10.3390/ en16031293.
45. Singh, V.P., Meena, C.S., Kumar, A., Dutt, N., Double Pass Solar Air Heater : A Review. *Int. J. Energy Resour. Appl.*, 1, 2, 22–43, 2022, doi: 10.56896/IJERA.2022.1.2.009.
46. Debrah, C., Chan, A.P.C., Darko, A., Artificial intelligence in green building. *Autom. Constr.*, 137, 104192, 2022, doi: https://doi.org/10.1016/j.autcon.2022.104192.
47. Nayal, M., Sharma, A.K., Jain, S., Singh, V.P., Chapter 06 Design and Modelling of Solar, Geothermal and Hybrid Energy Systems.pdf, in: *Highly Efficient Thermal Renewable Energy Systems Design, Optimization and Applications*, 1st ed., R.W.V. Verma, S. Thangavel, N. Dutt, A. Kumar (Eds.), pp. 1–15, CRC Press, 2024.
48. Kumar, R. *et al.*, Experimental and RSM-Based Process-Parameters Optimisation for Turning Operation of EN36B Steel. *Mater. (Basel)*, 16, 1, 1–18, 2023, doi: 10.3390/ma16010339.
49. Singh, V.P., Karn, A., Dwivedi, G., Alam, T., Kumar, A., Experimental Assessment of Variation in Open Area Ratio on Thermohydraulic Performance of Parallel Flow Solar Air Heater. *Arab. J. Sci. Eng.*, 41, 12, 1–17, 2022, doi: 10.1007/s13369-022-07525-7.
50. Singh, V.P. *et al.*, Heat transfer and friction factor correlations development for double pass solar air heater artificially roughened with perforated multi-V ribs. *Case Stud. Therm. Eng.*, 39, September, 102461, 2022, doi: https://doi.org/10.1016/j.csite.2022.102461.
51. Singh, V.P., Jain, S., Gupta, J.M.L., Analysis of the effect of perforation in multi-v rib artificial roughened single pass solar air heater: - Part A. *Exp. Heat Transfer*, 1–20, Oct. 2021, doi: 10.1080/08916152.2021.1988761.
52. Singh, V.P. *et al.*, Nanomanufacturing and Design of High-Performance Piezoelectric Nanogenerator for Energy Harvesting, in: *Nanomanufacturing and Nanomaterials Design: Principles and Applications*, pp. 241–272, 2022.
53. Hwang, B.-G. and Tan, J.S., Green building project management: obstacles and solutions for sustainable development. *Sustainable Dev.*, 20, 5, 335–349, Sep. 2012, doi: https://doi.org/10.1002/sd.492.
54. Ashwani Kumar, Y.G., Gangwar, A.K.S., Kumar, A., Meena, C.S., Singh, V.P., Dutt, N., Prasad, A., Biomedical study of femur bone fracture and healing, in: *Advanced Materials for Biomedical Applications*, pp. 283–298, 2022.

55. Datta, A., Kumar, A., Kumar, A., Kumar, A., Singh, V.P., Advanced materials in biological implants and surgical tools, in: *Advanced Materials for Biomedical Applications*, 1st ed., N.D.A. Kumar, Y. Gori, A. Kumar, C.S. Meena (Eds.), pp. 283–298, 2022.
56. Shubham Srivastava, R.K., Verma, D., Thusoo, S., Kumar, A., Singh, V.P., Nanomanufacturing for Energy Conversion and Storage Devices, in: *Nanomanufacturing and Nanomaterials Design: Principles and Applications*, pp. 165–174, 2022.
57. Dutt, N., Binjola, A., Hedau, A.J., Kumar, A., Singh, V.P., Meena, C.S., Comparison of CFD Results of Smooth Air Duct with Experimental and Available Equations in Literature. *Int. J. Energy Resour. Appl.*, 1, 1, 40–47, 2022, doi: https://doi.org/10.56896/IJERA.2022.1.1.006.
58. Fan, C., Chen, M., Tang, R., Wang, J., A novel deep generative modeling-based data augmentation strategy for improving short-term building energy predictions. *Build. Simul.*, 15, 2, 197–211, 2022, doi: 10.1007/s12273-021-0807-6.
59. Saini, M., Sharma, A., Singh, V.P., Dwivedi, G., Jain, S., Solar Thermal Receivers – A Review, in: *Adv. Mater. Manuf. Energy Eng. Vol. II*, vol. II, pp. 310–325, 2022.
60. Singh, V.P., Jain, S., Kumar, A., Establishment of correlations for the Thermo-Hydraulic parameters due to perforation in a multi-V rib roughened single pass solar air heater. *Exp. Heat Transfer*, 35, 5, 1–20, 2022, doi: 10.1080/08916152.2022.2064940.
61. Fang, H., Tan, H., Kosonen, R., Yuan, X., Jiang, K., Ding, R., Study of the Data Augmentation Approach for Building Energy Prediction beyond Historical Scenarios. *Buildings*, 13, 2, 2023, doi: 10.3390/buildings13020326.
62. Chi, H.-L., Kang, S.-C., Wang, X., Research trends and opportunities of augmented reality applications in architecture, engineering, and construction. *Autom. Constr.*, 33, 116–122, 2013, doi: https://doi.org/10.1016/j.autcon.2012.12.017.
63. Verma, V., Thangavel, S., Dutt, N., Kumar, A., Weerasinghe, R., Development of Thermal Energy Storage: Solar, Geothermal and Hydrogen Energy, in: *Highly Efficient Thermal Renewable Energy Systems: Design, Design, Optimization and Applications*, V. Verma, S. Thangavel, N. Dutt, A. Kumar, R. Weerasinghe (Eds.), CRC Press, Boca Raton, FL, USA, 2024, Chapter 01; ISBN 9781032595641.
64. Ahmadizadeh, M., Heidari, M., Thangavel, S., Naamani, E.A., Khashehchi, M., Verma, V., Kumar, A., Technological Advancements in Sustainable and Renewable Solar Energy Systems, in: *Highly Efficient Thermal Renewable Energy Systems: Design, Design, Optimization and Applications*, V. Verma, S. Thangavel, N. Dutt, A. Kumar, R. Weerasinghe (Eds.), CRC Press, Boca Raton, FL, USA, 2024, Chapter 02; ISBN 9781032595641.
65. Chitt, M., Thangavel, S., Verma, V., Kumar, A., Green Hydrogen Productions: Methods, Designs and Smart Applications, in: *Highly Efficient Thermal Renewable Energy Systems: Design, Design, Optimization and Applications*, V.

Verma, S. Thangavel, N. Dutt, A. Kumar, R. Weerasinghe (Eds.), CRC Press, Boca Raton, FL, USA, 2024, Chapter 16; ISBN 9781032595641.

66. Khashehchi, M., Thangavel, S., Rahmanivahid, P., Heidari, M., Moazzeni, T., Verma, V., Kumar, A., Solar Desalination Techniques: Challenges and Opportunities, in: *Highly Efficient Thermal Renewable Energy Systems: Design, Design, Optimization and Applications*, V. Verma, S. Thangavel, N. Dutt, A. Kumar, R. Weerasinghe (Eds.), CRC Press, Boca Raton, FL, USA, 2024, Chapter 19; ISBN 9781032595641.
67. *Highly Efficient Thermal Renewable Energy Systems: Design, Design, Optimization and Applications*, V. Verma, S. Thangavel, N. Dutt, A. Kumar, R. Weerasinghe (Eds.),CRC Press, Boca Raton, FL, USA, 2024, ISBN 9781032595641.

7

# CFD Numerical Investigation of Thermal Performance of Diamond Shape Micro Rectangular Heat Exchanger

**Jaideep[1], Pritosh Tomar[2] and Ashwani Kumar[3]***

*[1]Department of Mechanical Engineering, Quantum University, Roorkee, Uttarakhand, India*
*[2]Department of Mechanical Engineering, Womens Institute of Technology, Veer Madho Singh Bhandari Uttarakhand Technical University, Dehradun, Uttarakhand, India*
*[3]Department of Mechanical Engineering, Technical Education Department, Uttar Pradesh, Kanpur, India*


## *Abstract*

The present study examines the thermohydraulic behavior of micro pin fin heat sinks with diamond-shaped pin fins with a constant surface area, with air at 4% volumetric concentration at similar boundary conditions and with the values of Reynolds number at 1500, 2000, 2500, 3000, 3500, and 4000 for further performance optimization. A numerical study has been carried out to examine the heat transfer and fluid flow characteristics of open microchannel heat sinks consisting of diamond-shaped pin fins. The variation of fin height involves an open configuration of the microchannel having open spaces between the fin top surface and the top wall of the heat sink. Comparative analysis has been done among seven different heat sinks of varying fin height, $H_f$ = 1.5, 2, 2,5, and 3 mm with an increment of heat transfer rate at $H_f$ = 3 mm, and friction decrement of 1.5 mm. Uniform heat flux at 1000 kW/m$^2$ is supplied to heat the microchannel. The heat transfer rate improved 5.895 times as compared to the smooth channel at $H_f$ = 2.5, and the maximum thermal performance improved 340% as compared to smooth channel.


**Corresponding author*: drashwanikumardte@gmail.com; ORCID: 0000-0003-4099-935X
Jaideep: ORCID: 0009-0000-7264-783X
Pritosh Tomar: ORCID: 0000-0003-3856-5982

Ashwani Kumar, Nitesh Dutt and Mukesh Kumar Awasthi (eds.) Heat Transfer Enhancement Techniques: Thermal Performance, Optimization and Applications, (159–192) © 2025 Scrivener Publishing LLC

***Keywords:*** Diamond shape, HTE, heat exchanger

## 7.1 Introduction

All industrial and solar panel firms use the most efficient microchannel heat exchanger sink (MCHS). These tiny channels are commonly used as fin and small tube heat exchangers that transport heat from the recovery system to the environment, saving energy and time while also enhancing the life of the convection, conduction, and radiation component system. In the model, the MCHS (microchannel heat exchanger sink) acts as fin roughness. Fins are expanded surfaces that are meant to either enhance the transfer rate for a certain surface temperature or decrease the surface temperature for a set heat transfer rate. These are often utilized to accelerate the rate of heat transmission to the environment. Most of the time, controlling the temperature difference is not feasible, and increasing the heat transfer coefficient may necessitate the installation of a pump or a fan, or replacing the existing one with a new one with capacity; an alternative is to increase the effective surface area through extended surfaces or fins. Any heat exchanger model requires this microchannel heat sink. A heat exchanger is essential for heat storage and recovery. Heat transfer augmentation techniques are widely used in a variety of settings, including industries, heating and cooling in evaporators, thermal power plants, air conditioning (AC), refrigerators, and radiators for space vehicles and cars. enhance the surface area, roughness, and boundary conditions to enhance the heat transfer rate.

In the chapter, state-of-the-art literature evaluations were employed to examine current advancements. Although much study has been conducted in order to improve the heat transfer rate utilizing passive enhancement approaches, it still has a broad field of research due to its numerous uses and direct impact on the power industry. The list below summarizes the important studies on heat transfer increases on pipes utilizing passive approaches. These studies establish the groundwork for validating our smooth channel results and determining what additional improvements might be done to improve the heat transfer rate in a heat exchanger.

Prabhakar Bhandari, Yogesh K. Prajapati [1] investigated thermal performance of open Microchannel heat sink with variable pin fin height it can be observed that increasing the fin height in heat sink increases the heat transfer rate however, it persists only up to 1.5 mm of fin height. Beyond this height of the fin, i.e., for 1.75 mm and 2.0 mm, heat dissipation capacity of the heat sink drops. Moreover, heat sinks of shorter fins (0.5–1.00 mm) depict considerably low thermal performance nevertheless; pressure drop is also lower in such configuration and increases with fin height.

Masomeh Ebrahim Qomi *et al.* [2] investigated heat transfer enhancement in a microchannel using a pulsating magnetohydrodynamics (MHD) hybrid nanofluid flow. The results show heat transfer augmentation by raising the strength of magnetic field or Reynolds number. This is such that the heat transfer is intensified by a factor of 3.5 and 1.2, respectively, if the Reynolds number rises from 100 to 500 and Hartmann number from 0 to 50. Increasing Strouhal number from 0.5 to 2 makes about 20% higher heat transfer with lower oscillation amplitude. Pressure drop through the microchannel is also calculated to evaluate the hydrodynamic performance.

Prabhakar Bhandari, Yogesh K. Prajapati [3] changed the balance tallness of rectangular equal microchannel and found that open design of warmth sink with 75–80% balance stature has ideal warmth move rate contrasted with totally shut warmth sink. Creator has additionally recorded different advantages of open microchannel heat sink. Besides, work on open microchannel with pin balance setup has seldom reported.

Sakine Mehrabi, Saeid Kheradmand, Omid R. Farivar [4] investigated the "numerical simulation of thermal and hydraulic performance of a micro plate-pin fin heat sink," which was investigated and published. To evaluate overall performance, the heat sink effectiveness factor which represents a combination of the hydraulic and thermal performance was defined and corresponding figures were provided to derive an optimum geometry.

Rajnesh Kumar *et al.* [5] investigated the uses of forward chamfered ribbed roughness geometry for his work. In this work, the different values of the ratio of e/D were taken which helped to enhance the Nusselt number and the friction factor was found to be maximum in case of e/D at the value of 0.043.

Prabhakar Bhandari, Yogesh K. Prajapati [6] investigated the two-phase flow boiling and its suppression techniques. Emphasis has been given to the recent literatures addressing the issues of flow boiling instabilities in microchannels. The broad aspects that have been covered are overview of instabilities, their causes, consequences and suppression techniques. Major attention has been paid to highlight the promising solutions that have been investigated to mitigate the instabilities. Study revealed that inception of instabilities is inherent during flow boiling.

Sanjay K. Sharma *et al.* [7] investigated heat transfer enhancement using thin heat sink. The Nusselt number was found to be very high and was 49.28 in case 1 and friction factor was also found to be high up to the range of 2.88–7.18 in case 4.

L. Varshney *et al.* [8] investigated heat transfer enhancement using rectangular sectioned tapered rib. There were 12 different types of tapered ribs at a taper of 1.60, 2.30, and 3.20 at the pitch of 10, 15, 20, and 25 mm which were analyzed at which maximum thermal enhancement was obtained at 1.60.

Promthaisong Pitak *et al.* [9] investigated heat transfer enhancement using broken V-ribs. Broken v-rib was compared generally without the V-rib at the following parameter, d/H = 0, 0.01, 0.02,0.03, 0.04, and 0.05, and this rib produced the thermal enhancement factor about 346–539%.

Ganesan Narendran, Nagarajan Gnanasekaran and Dharmaraj A. Perumal [10] reviewed the various geometries of a microchannel that has been reported for the last two decades in literature majorly for the application of high power devices. Recent research on microchannel is more focused on numerical and experimental work with various configurations of the heat sink. In this article, a comparative work on different flow geometries used in the microchannel and their influence on heat transfer and pressure drop is investigated with the brief representation of different working fluids used in microchannel heat sink for the purpose of electronic cooling and their associated performance characteristics with various examined parameters.

Montelpare *et al.* [11] investigated heat transfer enhancement using experimental infrared (IR) thermographic method for the evaluation of the heat transfer coefficient of liquid-cooled short pin fins sink arranged in line. The results show a dependence for the Nusselt trends on the pin position; this is due to the shadow effect of the pin upstream with respect to the one downstream. This behavior is strongly influenced on the pin geometry and particularly there is a better behavior for the triangular and rhomboidal pins.

S. Alfarawi *et al.* [12] studied the enhancement of heat transfer factors by using roughened hybrid ribs rectangular duct and main factors of heat and frictions measurement were observed when three different ribs roughness (semi-circle, rectangular, and two hybrid ribs) were applied in rectangular channel for showing the complete turbulence flow. In this research rectangular channel designed by length of 320 mm, height 120 mm, relative pitch to height ratio from 6.6–53.3, and Reynolds 12,500–86,500.

Andallib Tariq *et al.* [13] carried out an investigation of heat transfer improvement in ferment of mounted slit rib on model surface. In this experimental study, the results were examined with use of slit rib mounted on surface and study was taken in from of downstream flowing flow and heat transfer rate. This article observed maximum 10%, 20%, 30%, 40%, and 50% improvement compared to complete area of rib.

Abhilash Kumar, R. Saravana, Sathiya Prabhahar [14] investigated heat transfer augmentation of internal ribs square channel. In the study, square channel is used with different rib roughness like plain rib, perforated round corner of rectangular rib, and perforated elliptical ribs, and this roughness was inserted in two opposite walls of the square channel.

E. Esmaeilzadeh, *et al.* [15] performed a simulation of 3D rectangular rib channel and achieved maximum heat transfer coefficient in constant heat flux channel. In this research article, the numerical observation of heat transfer with turbulent flowing fluid as in channel was undertaken and also showed the comparison between smooth and rough. Test duct had 570-mm length and 120 mm×120 mm cross section area and three rectangular transverse ribs were arranged in the floor of duct with dimensions of 27 mm×16 mm×120 mm and flowing fluid is applied with Reynolds number along the channel in x-direction with air temperature and pressure 25°C and 100 Kpa, respectively. S.V. Patil and P.V.V. Babu [16] presented a work in a review research article of square channel fitted in circular tube.

R.K. Sahu and B.K. Gandhi [17] numerically observed details of heat transfer increments in rectangular solar air heater with gap in inclined continuous rib. The rectangular channel (2600 × 181 × 31 mm) contains inclined rib at 60° with a gap of 2 mm in flowing fluid direction. In this study, a number of different models were used at Reynolds number in the range of 3500 to 19,000 and their results compared. Aharwal *et al.* [18] worked and obtained heat transfer enhancement of rectangular solar air heater with inclined rib and gap between the roughness.

### 7.1.1 Heat Transfer Enhancement Method

Active method: this method requires the external power source and includes mechanical aids. In this technique, nanosized, high thermal conductivity, metallic powder is used to the base fluid in order to increase the heat transfer rate [19–22].

Passive method: this method does not require any external power source. In this method, various types of inserts can be used such as twisted tapes, wire coil, and vortex rings, and are used to enhance the heat transfer rate without any help from external power. Inserts create the swirl flow in the path of working fluid which disturbs the actual boundary layers which results in an increase in effective surface in the existing systems [23–26].

Compound method: when two or more techniques, *i.e.*, active and passive, are used to enhance the heat transfer rate of any device, which is greater than that produced by any one of those techniques separately. Heat transfer enhancement techniques refer to different methods used to increase the heat transfer rate without affecting the overall performance of the system. Inserts act as an obstacle in the path of fluid flow. So, it enhances the heat transfer rate [27–31].

### 7.1.2 Microchannel Heat Sink Method

With the advancement of miniaturizations of electronic devices, heat dissipation from these miniature electronic devices has become the most difficult task in present design. If heat dissipation is not done properly, it will have a detrimental impact on the life of the device and other electrical equipment, resulting in decreased efficiency. Because of its small size and great thermal efficiency, microchannel is one of the finest options for eliminating heat from electrical equipment. Several scholars have studied the behavior of microchannel heat transfer rates, fluid flow visualizations in microchannels, and their applications both theoretically and empirically. This chapter deals with the review of rigorous behavior of microchannel heat transfer rate by diamond shape geometry numerically.

### 7.1.3 Geometry of the Advanced Heat Sink Channel

Warm management of designing devices, such as electronic chips, integrated circuits (ICs), is fundamental to ensuring their safe and proper operation. Smaller and more remote design of such gears has continually drawn in clients who demand feasible personality of these devices. The extensive reduction in size and increased quantity of semiconductors results in a significant improvement in warmth age per unit volume. It will very certainly reach 1000 W/m$^2$ in the near future. Because standard cooling solutions employing natural convection have limited capability, disseminating a large amount of heat from a confined location need careful planning. Among the new methods for dispersing high heat are small heat pipes, stream impingement cooling, splash cooling, thermoelectric, carbon nanotube, and microchannel. The exploring network of this field has given major focus to microchannel heat sinks. This is a direct outcome of its inherent properties, such as high surface-to-volume proportion and ease of use. Massive exploration activities have been done since the beginning of microchannels to comprehend the key material science of liquid stream and warmth travel via microchannels [32–35].

The examination network has typically studied mathematical change among all of these methods. Changes to the microchannel plan include either a total redesign of the heat move region or specific changes to the stream circle and bay/source complex. One of the proposed changes is the design of a delta/source plenum to encourage consistent coolant appropriation in all channels in the future to reduce stream maldispersion in the microchannel. Warmth passes via a tiny channel heat sink using a single stage air coolant before being grounded and released from the terrible wonder. It should be noted that the microchannel heat sink with single stage air coolant has successfully ensured its business use. Furthermore, reformist work is consistently required to improve the exhibition in order to suit the growing future interest. In terms of heat transfer through a single stage coolant, two important methodologies have been widely drilled in the literature [36–41].

Air cooled internal engines, refrigeration condenser tubes, electric transformers, reciprocating air compressors, semiconductor devices, and automobile radiators are all examples of industrial applications for fins [42, 43].

## 7.2 Objective and Methodology

The primary goal of this research is to statistically explore heat transfer enhancement based on various values such as variations in the Reynolds number Re = 1500–4000, pitch spacing, and friction factor on diamond shape fin geometry inside open microchannel using air as working fluid. Findings can be analyzed and compared to experimental findings based on the parameters listed above.

### 7.2.1 Methodology of the Microchannel Heat Exchanger Rectangular Channel

This section discusses how to use the ANSYS workbench to construct a tiny rectangular heat exchanger with a diamond fin form heat sink. The roughness utilized in the computational fluid dynamics (CFD) created smooth rectangular microchannel. The model was created with CFD ANSYS software since it is very simple to deal with flowing fluid without any intake air or fluid problems [44–47]. The CFD method is a section of the ANSYS workbench 14.0 that specifically uses the microchannel that approaches the working fluid and is also used to acquire all the fluid details such as the heat transfer rate, velocity profile, temperature profile, streamline flow, and shear stresses, among others [48–50].

### 7.2.2 Methodology of Microchannel Diamond Fin Heat Sink

The methodology of the microchannel heat sink with diamond fin shape roughness examined in the ANSYS Fluent and the methodology of the model consisted of diamond fin shape heat sink at $\theta = 45°$ crown angle, as shown in Figure 7.1. The diamond fin shape heat sink has different length and its variations used in the heat sink and the total length of the micro rectangular heat sink is L = 55 mm and test length = 40 mm inlet and exit

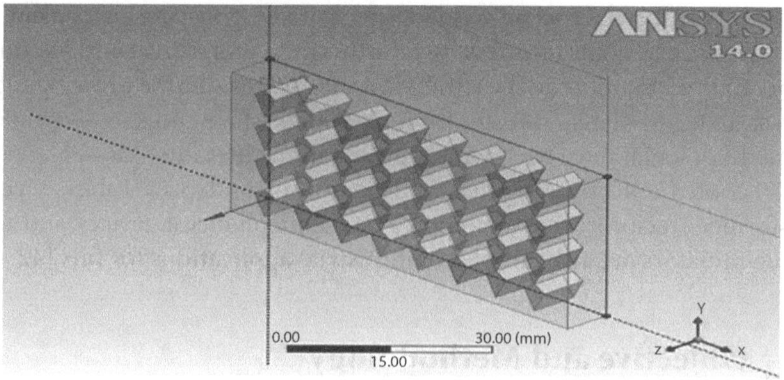


**Figure 7.1** Geometry of the microchannel heat exchanger sink.

length $L_{in}$ = 5 mm and $L_{out}$ = 10 mm, respectively, and specially used as the heat energy depends on the velocity of the working fluid and the velocity was taken from the variation in the Reynolds number Re = 1500, 2000, 2500, 3000, 3500, and 4000 with pitch space between the roughness, P = 3, 4, 5, and 6. In this section, it can be seen that the methodology of the heat sink is used corresponding to the heat transfer, variation in the working fluid, strength of the surface, and flow direction and arrangement. The inserted diamond fin performed a specific value which derives the main part function of heat transfer rate in the heat sink to calculate the heat transfer characteristics, friction factor and thermal performance of the microchannel heat sink. The motive of the microchannel is to transfer the heat energy from one system to required system with the help of working fluid and roughness parts. This complete model is a valuable section which has been carried out with k-e turbulence, and RNG-enhanced wall treatment model is selected by ANSYS Fluent (Figure 7.2).

## 7.3 Parameters of Microchannel Fin Heat Sink

In this chapter, the geometry based on the parameters of microchannel heat sink which are taken for with a different number of inserted diamond fin shape heat sink. All the parameters of the design model are seen in Table 7.1. The inserted parts narrated with diamond fin and its variations in height of the fin of $h_f$ = 1.5 mm, 2 mm, 2.5 mm, and 3 mm and the fin is used with crown angle $\theta_{cf}$ = 45° for transition of air which was converted and heated up by the hot air or cold and conducted in the direction of flowing fluid. The parameter values of the smooth microchannel occupied from the validation

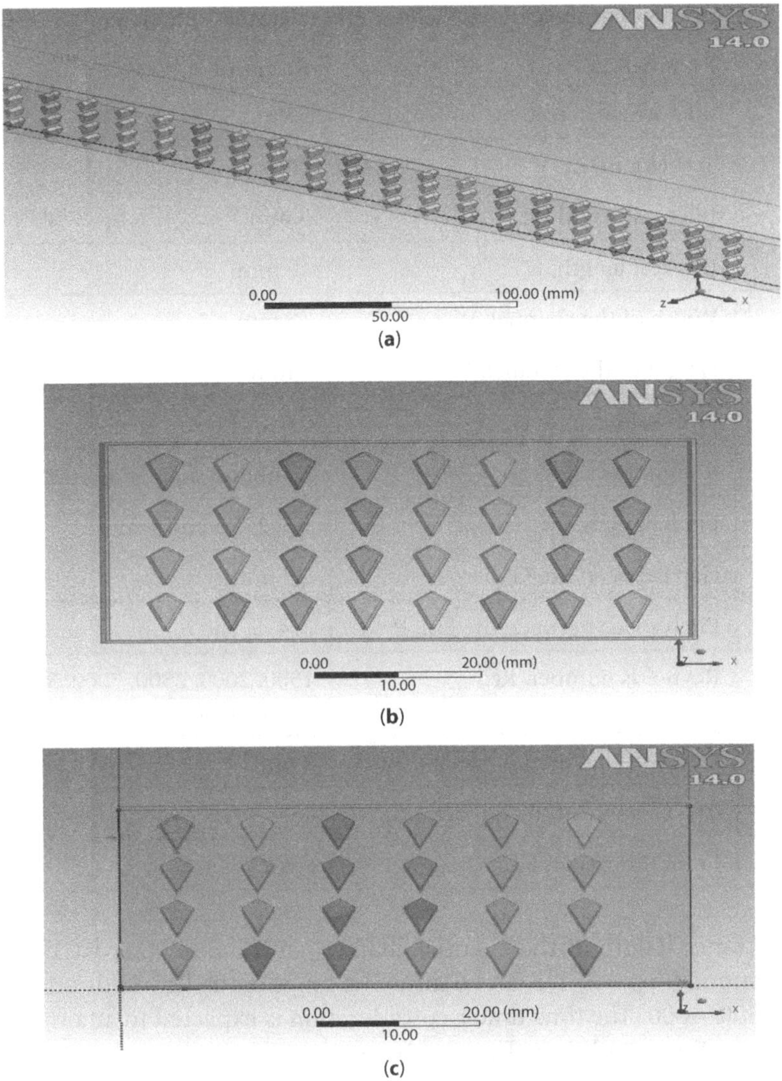


**Figure 7.2** (a–c) Systematic diagram of diamond fin shape heat sink.

research article and inserted roughness parameters are preferred well than the other research article those have been used number of parameters. The designed microchannel heat sink model is effective for heat transfer needs and its solution in all situations and without arrange of maintenance and longer working time. The diamond fin shape roughness was also used to improve heat transfer coefficient, friction factor coefficient, kinetic energy performance and efficiency between the heat sink and diamond fin shape.

**Table 7.1** Details of geometric parameters of the heat sink with diamond shape fin.

| S. no. | Parameters | Range of values |
|---|---|---|
| 1 | Microchannel rectangular length, L | 55 mm |
| 2 | Test length, $L_1$ | 40 mm |
| 3 | Inlet test length, $L_2$ | 5 mm |
| 4 | Exit test length, $L_3$ | 10 mm |
| 5 | Width of the channel, W | 15 mm |
| 6 | Height of the channel, H | 8 mm |
| 7 | Hydraulic diameter, $D_h$ | 0.01043 mm |
| 8 | Roughness used | Diamond shape fin roughness |
| 10 | Fin height, $h_f$ | 1.5, 2, 2.5 and 3 mm |
| 11 | Gap between fin, G | 2.5 mm |
| 12 | Pitch space between the Fin, P | 3, 4, 5, and 6 mm |
| 13 | Reynolds number, Re | 1500, 2000, 2500, 3000, 3500, and 4000 |
| 14 | Depth of the diamond shape, w | 3 mm |
| 15 | Crown angle of the fin, θcf | 45° |
| 16 | Prandtl number, Pr | 0.707 |

The entrance length of the channel changes with the pitch space between the diamond fin p = 3, 4, 5, and 6 mm. The range of the Reynolds number is Re = 1500–4000. The flow under consideration is expected to attain a fully-developed flow condition. The number of parameters are used in the heat sink with the number of variations are shown in Table 7.1.

## 7.4 Governing Equation Used in Microchannel

The governing equation have been developed and maintained for the working flow which is flowing in the microchannel system under three distinct flow condition, namely, steady-laminar, unsteady-laminar, and turbulent flow. In view of the above, the section focuses on conservation equation pertaining to laminar and turbulent flow condition.

### 7.4.1 K-e Model Used in Microchannel Heat Sink

The numerical analysis of the microchannel heat sink with diamond shape fin was generated in the ANSYS Fluent 14.0 with k-e model and the geometry presented the standard, RNG k-e models. It is similar in form to the standard k-e model, but includes the following refinements: the effect of swirl on turbulence is included in the RNG model, enhancing accuracy for swirling flows. Fluctuating velocity fields are the main characteristics of a transition flow. The RNG k-e model was derived using a statistical technique (called renormalization group theory). As the fluctuations in the transported quantities can be scale and high frequency, the variations in velocity fields cause the mixing of transported quantities like momentum and energy, hence the above quantities vary as well. These turbulence models are classified as:

- High Reynolds number turbulence models
- Low Reynolds number turbulence models
- Blending of high and low Reynolds number models

In order to determine the flow characteristics such as velocity, temperature, and pressure field of the flow, the following governing equations needs to be solved:

### 7.4.2 Continuity Equation Applied

The mathematical expression of the principle of conservation of mass applied to an elemental control volume within a fluid under motion is known as continuity equation and is given by:

$$\frac{\partial \vec{u}}{\partial x}+\frac{\partial \vec{v}}{\partial y}+\frac{\partial \vec{w}}{\partial z}=0 \tag{7.1}$$

This is the mass continuity equation for three-dimensional steady flow fluid and it has no breaks in it.

### 7.4.3 Momentum Equation

It is based on law of conservation of moment or on momentum principle for turbulent flow field, which state that the net force acting on a mass acceleration fluid equal to the change in momentum of flow per unit time in the direction.

X-momentum equations:

$$\left(\vec{u}\frac{\partial\vec{u}}{\partial x}+\vec{v}\frac{\partial\vec{u}}{\partial y}+\vec{w}\frac{\partial\vec{u}}{\partial z}\right)=-\frac{1}{\rho}\frac{\partial p}{\partial x}+v\left(\frac{\partial^2\vec{u}}{\partial x^2}+\frac{\partial^2\vec{u}}{\partial y^2}+\frac{\partial^2\vec{u}}{\partial z^2}\right) \tag{7.2}$$

Y-momentum equation:

$$\left(\vec{u}\frac{\partial\vec{v}}{\partial x}+\vec{v}\frac{\partial\vec{v}}{\partial y}+\vec{w}\frac{\partial\vec{v}}{\partial z}\right)=-\frac{1}{\rho}\frac{\partial p}{\partial y}+v\left(\frac{\partial^2\vec{v}}{\partial x^2}+\frac{\partial^2\vec{v}}{\partial y^2}+\frac{\partial^2\vec{v}}{\partial z^2}\right) \tag{7.3}$$

Z –momentum equations:

$$\left(\vec{u}\frac{\partial\vec{v}}{\partial x}+\vec{v}\frac{\partial\vec{v}}{\partial y}+\vec{w}\frac{\partial\vec{v}}{\partial z}\right)=-\frac{1}{\rho}\frac{\partial p}{\partial z}+v\left(\frac{\partial^2\vec{v}}{\partial x^2}+\frac{\partial^2\vec{v}}{\partial y^2}+\frac{\partial^2\vec{v}}{\partial z^2}\right) \tag{7.4}$$

### 7.4.4 Energy Equation

Assuming that the flow is steady and incompressible with constant thermal conductivity no compression work and without heat generation for steady turbulent flow:

In this case no viscous heating,

$$\vec{u}\frac{\partial t}{\partial x}+\vec{v}\frac{\partial t}{\partial y}+\vec{w}\frac{\partial t}{\partial y}=\alpha\left(\frac{\partial^2 t}{\partial x^2}+\frac{\partial^2 t}{\partial y^2}+\frac{\partial^2 t}{\partial y^2}\right) \tag{7.5}$$

### 7.4.5 Assumption for the Microchannel Heat Sink

The following assumptions have been considered during the numerical simulation:

1. Steady flows.
2. Pressure variation in y dir. is zero.
3. Shear force in y dir. is zero.
4. Body force due to gravity has been neglected.
5. Incompressible flow.
6. At the inlet of test section, the flow has been fully developed.

7. The axial heat conduction in the fluid was negligible.
8. The properties of air were constant at atmospheric pressure and temperature.

## 7.5 Material Properties and Boundary Conditions

### 7.5.1 Thermophysical Properties of Copper-Inserted Diamond Fin Heat Sink

Copper material is used for diamond fin shape roughness in the rectangular microchannel heat sink because copper is a good conductor of heat transfer rate and high thermal performance (Table 7.2). Copper materials consume high temperature generated on the surface without any corrosion and tear situation.

**Table 7.2** Thermophysical properties of copper of diamond fin heat sink.

| S. no. | Properties | Value |
|---|---|---|
| 1 | Density, ($\rho$) kg/m$^3$ | 8978 |
| 2 | Specific heat, kJ/kg-k | 381 |
| 3 | Thermal conductivity, K | 387.6 |

**Table 7.3** Thermophysical properties of working fluid air.

| S. no. | Property | Value |
|---|---|---|
| 1 | Density ($\rho$), kg/m$^3$ | 1.225 |
| 2 | Specific heat, kJ/kg-k | 1006.4 |
| 3 | Viscosity ($\mu$), N-s/m$^2$ | 1.7894e-05 |
| 4 | Thermal conductivity, k | 0.707 |

**Table 7.4** Thermophysical properties of aluminum of channel.

| S. no. | Property | Value |
|---|---|---|
| 1 | Density ($\rho$), kg/m$^3$ | 2719 |
| 2 | Specific heat, kJ/kg-k | 871 |
| 3 | Thermal conductivity, K | 202.4 |

### 7.5.2 Boundary Conditions Applied in Diamond Microchannel

The boundary conditions are important sections of the design of the model heat sink with diamond-shaped pin fins, and some conditions applied such as the no-slip condition of the velocity around the solid surfaces in the heat sink, is assumed and the turbulent intensity is set at 5% for all solid walls. The working velocity is taken from the variation in the Reynolds number (Re), ranging from 1500–4000. The boundary condition is limited for the relative roughness of the heat sink, and such condition is applied to the enclosed microchannel as the inlet from where air can enter and the outflow from where effected air can pass out after striking the roughened part in the heat sink; to the inserted parts nominated as circular cutting rings, and then to the wall boundary (Table 7.3 & Table 7.4). The turbulence model is applied with the help of an Ansys workbench tool such as CFD.

### 7.5.3 Mesh Generation of the Microchannel Heat Sink with Diamond Fin Shape

Mesh method is the important part of the designed model in three-dimensional uniform meshing and the tetrahedral and polyhedral meshes format. This type of meshing consists of boundary-based smoothing nodes interior and recovery of the domain boundary. The meshing generation of the heat sink with diamond shape created under the following conditions such as span angle center fine, smooth transition, and virtual topological, and the element size of 0.001 m, as shown in Figure 7.3. Mesh generation is used for accurate or refined results and for maximum thermal performance, more effective quality, validation, automation, and strong mathematical equations of the Microchannel heat sink. Grid test contains important parts in form of number of cells and number of nodes, and both parts create a perfect meshing for the maximum thermal performance. Meshing is automatically generated with each solver within the Ansys workbench environment. As shown in Table 7.5, after 1,816,546 nodes, as shown in Figure 7.3, the variations in the value of heat transfer are negligible and accurate resolution of the working geometric model. There is a lot of meshing and there should be best face size refinement method for the best designed. For grid independence test, a solution adaptive refinement method is used. Thus, to avoid wastages of the computational resources the addition of unnecessary cells was stopped.

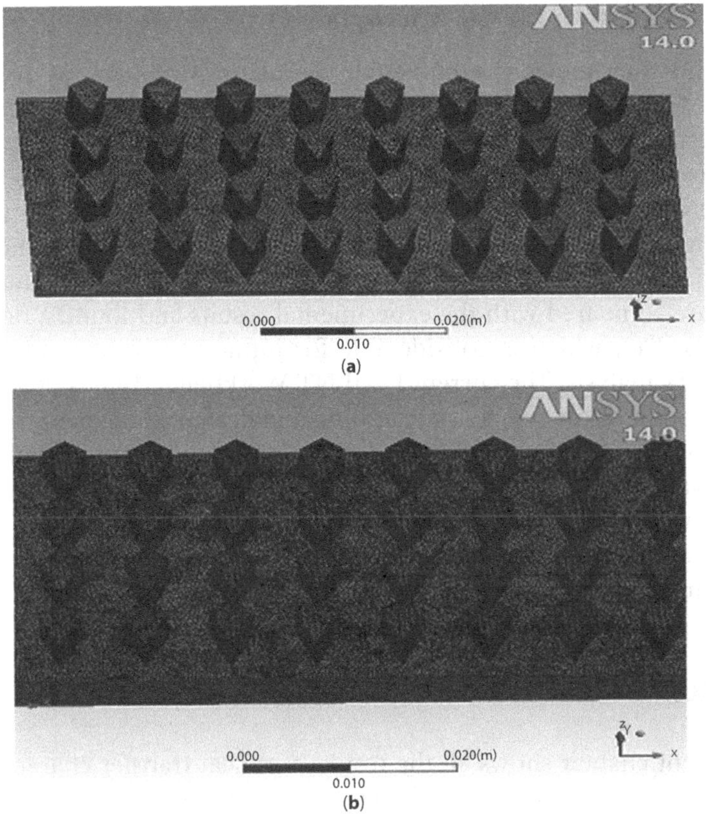


**Figure 7.3** FEA mesh generation of diamond shape heat sink.

**Table 7.5** Grid independence test for diamond shape heat sink.

| Sl. no. | Number of nodes | Number of elements | Nusselt number |
|---|---|---|---|
| 1 | 456,466 | 521,333 | 102.4643 |
| 2 | 813,464 | 1,054,646 | 80.2653 |
| 3 | 1,816,546 | 1,921,646 | 25.8146 |
| 4 | 1,900,256 | 2,164,640 | 25.2646 |
| 5 | 2,221,131 | 2,414,666 | 41.0732 |

### 7.5.4 Validation for the Microplate Fin

This validation shows the numerical simulations which was solved by Sakine Mehrabi, Saeid Kheradmand, Omid iR. Farivar, in their research article entitled "Numerical simulation of thermal and hydraulic performance of a micro plate-pin fin heat sink." The numerical data is used to find the heat transfer from the heat sink where the length is L = 50 mm. The design is validated in the form of Nusselt number, as shown in Figure 7.4, which varies with Reynolds number, Re = 2500–6500. Numerical results are compared with the experimental results and found a deviation of ± (1%–3%), which is considerable for numerical results obtained by using CFD analysis. The current FLUENT Workbench 14.0 is used. When the validation results satisfy the meshing and design, some parameters are taken from articles such as the geometry of the smooth microchannel and several validation efforts that are referenced from scholarly sources, and the user will be able to validate the accuracy of their results. Furthermore, with the assistance of the offered learning modules, the user will be able to gain competence to tackle additional difficulties.

## 7.6 Result and Discussion

The present chapter shows all the results and heat transfer characteristics of the diamond shape heat exchanger sink and also shows the effect of the presented results of the channel with or without inserted roughness. It can be observed that the inserted part as roughness in the microchannel is effective for transferring the thermal energy more than the smooth channel. The heat transfer increment based on the number of variations

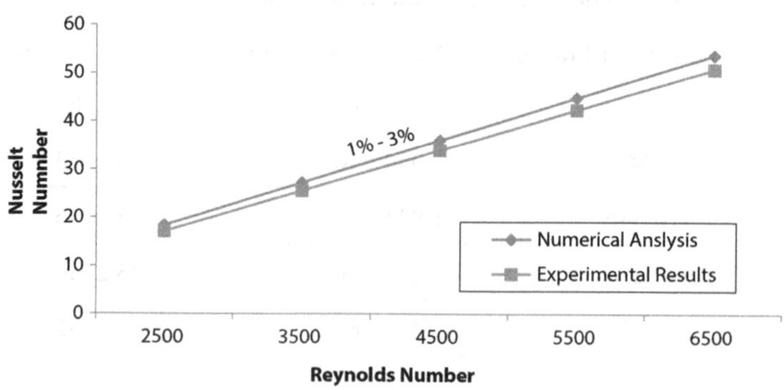


**Figure 7.4** Result validations of Nusselt number with experimental results.

like Reynolds number, Re = 1500–4000, space between the diamond fins roughness, P = 3, 4, 5, and 6 mm. All the results and contour are given in the figures which show the variations in the heat transfer rate, friction factor, velocity contour, temperature contour, and thermal performance of the heat sink. The chapter shows the foundation of heat transfer rate coefficient, friction factor coefficient, and thermal efficiency, and how it can transfer the heat in the heat sink. In ANSYS, the inner microchannel hydraulic diameter was used as a hydraulic diameter, $D_h$ = 0.01043 mm, and the velocity is taken from the base of Reynolds number range. The numerical investigation of the heat sink with diamond fin roughness enclosed the different types of the fluid condition. The fin roughness with number of heights, $H_f$= 1.5, 2, 2.5, and 3 mm, to balance the friction loss in the microchannel heat sink. In the present study, the numerical investigation of the heat sink is generated by the ANSYS Fluent workbench.

### 7.6.1 The Smooth Microchannel and Dittus Boelter Equation

The presented Figure 7.5 shows the comparison between the smooth microchannel, and Dittus-Boelter equation plays an important role for finding the increment of the thermal energy with Reynolds number, Re = 1500–4000. The variations in the Reynolds number in used in the Dittus-Boelter equation and in smooth heat sink for increasing the heat transfer rate characteristics and reduce the energy loss when the inserted fin roughness used in the smooth channel. In this section, the results of the Nusselt number of the smooth microchannel were calculated in the ANSYS Fluent, and these results are compared with the theoretical results of the Dittus-Boelter equation, $Nu = 0.023 \times Re^{0.8} \times Pr^{0.4}$. In the Dittus-Boelter equation,

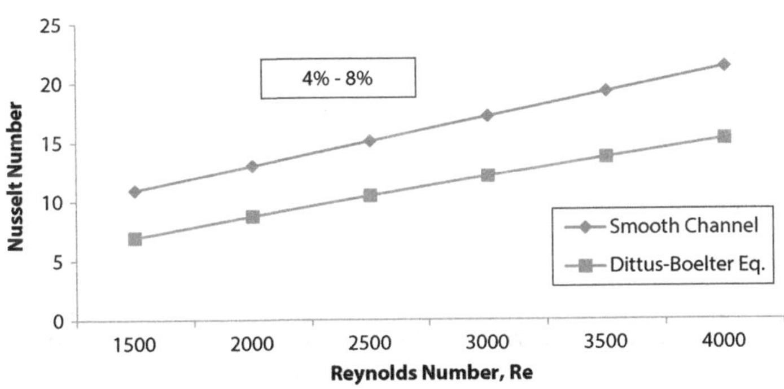


**Figure 7.5** Comparison graph between the smooth microchannel and Dittus-Boelter equation.

the Prandlt number plays an important role in convective effect of the air viscosity and change in the air from laminar to turbulent flow. The comparison between the results of the smooth tube and Dittus-Boelter equation are shown in the given figure below.

The figure shows the comparison between results of the smooth microchannel and results of the Dittus-Boelter equation, and these results show the variation in the Nusselt number with respect to the Reynolds number, Re = 1500, 2000, 2500, 3000, 3500, and 4000. The maximum Nusselt number is Nu = 21.29637373 and minimum Nusselt number Nu = 10.94646465 in smooth microchannel, and in the case of the Dittus-Boelter equation, the maximum Nusselt number is Nu = 15.24556748 and minimum Nusselt I number is Nu = 6.95614472. In the figure, it can be seen that the comparison between the results of the smooth tube and Dittus-Boelter equation are near to each other (4%–8%) so both satisfied the heat transfer rate condition with the use of inserted roughness in the smooth microchannel.

### 7.6.2 Smooth Microchannel and Blasius Friction Equation

Figure 7.6 shows the comparison between the smooth microchannel and Blasius friction equation. The minimum friction of the smooth channel is Fr = 0.002134679 and the Blasius friction factor is Fr = 0.039734896. From these results, it can be observed that smooth heat sink rectangular channel contains the perfect situation in addition of different inserts roughness. The comparison is used for checking the friction factors in the different areas in the microchannel, loss of energy due to roughness, and heat dissipation, and it is also used to maintain the pressure drop. The comparison

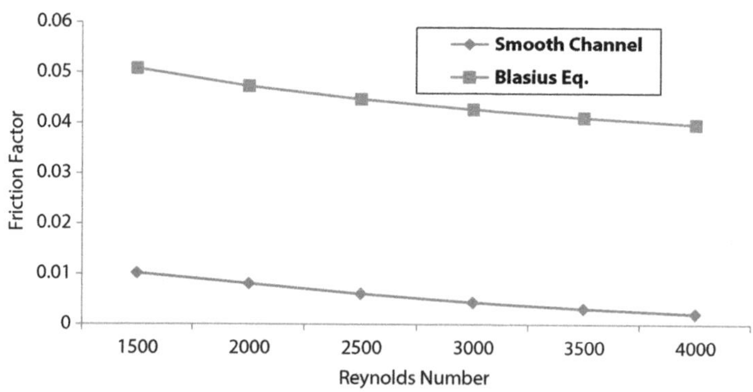


**Figure 7.6** Comparison graph between the variations in friction factor with respect to Blasius equation.

is an improving method to reduce the friction factor and increase the heat transfer enhancement without any error in micro channel. The results are varied with Reynolds number Re = 1500, 2000, 2500, 3000, 3500, and 4000.

### 7.6.3 Results and Effects of Microchannel with Diamond Fin Heat Sink

In this section, all the results and effects of the microchannel roughness of the diamond-shaped fin heat sink are investigated with the help of ANSYS Fluent workbench. Variations of the parameters of the diamond-shaped fins are used to improve the heat transfer characteristic and control the pressure drop and fluid temperature inside the smooth microchannel, the value of which is given in Table 7.1. The heat transfer characteristic of the designed microchannel faced many problems like conduction and convection conditions and how these can be solved. For convection, the solution may depend upon the flow of water, and for radiation, it is considered negligible in ANSYS workbench and is related to the design of the model. In this section, a number of variations are used to explain the results and their effects in the form of velocity contour, temperature contour, Nusselt number, friction factor, and turbulent kinetic energy and efficiency. The space between the roughness is an important part of the microchannel flow because of the cold air that can be transformed into warm air.

### 7.6.4 Results and Effects of the Velocity Contour

The effects of the velocity contour show the variation in the heat transfer rate due to change in the velocity. The velocity in the diamond shape heat sink depends upon the Reynolds number and number of variations used in the inserted roughness to transfer the heat from one region to another region. Figure 7.7 shows the variations in the velocity when the pitch space P = 3 mm is taken between the diamond fin roughness at height of the fin, $H_f$ = 2.5 mm. The velocity contour obtained a maximum velocity 1.5 m/s at the Reynolds number Re = 4000.

Figure 7.8 shows the variations in velocity of the roughened diamond fin shape microchannel with respect to the Reynolds number Re = 1500–4000. The maximum velocity of the working fluid obtained is 2.80 m/s when the pitch space is taken P = 4 mm at $H_f$ = 2.5 mm. The pitch space is used in the roughened heat sink for reducing the friction factor as well as loss of energy in the heat sink.

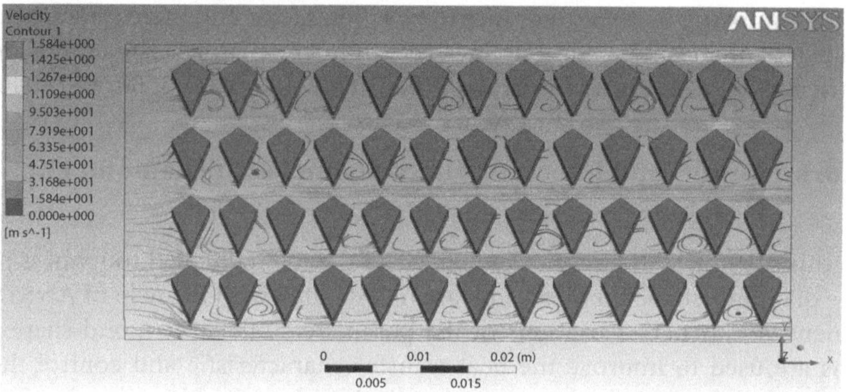


**Figure 7.7** Shows the variation in the velocity with respect to the Reynolds number (Re = 1500–4000) at diamond shape fin with P = 3 mm.

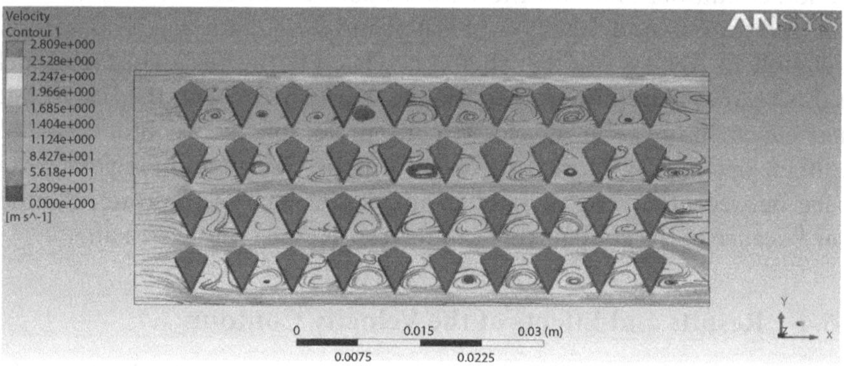


**Figure 7.8** Shows the variation of the velocity with respect to P = 4 mm.

Figure 7.9 shows the variation in velocity contour with respect to the Reynolds number. The maximum velocity of the working fluid in microchannel diamond fin roughness in heat sink obtained is 4.36 m/s at P = 5 mm.

Figure 7.10 shows the variation in velocity of diamond fin shape microchannel with respect to the Reynolds number. The maximum velocity obtained in the tube is 3.98 m/s at P = 6 mm and Re = 4000.

### 7.6.5 Results and Effects of the Temperature Contour

In this section, the temperature contour explains the variation in the temperature in the microchannel with diamond shape fin roughness. Figure 7.11 shows the variation in the temperature contour at P = 3 mm. The maximum temperature obtained is $T_{max}$ = 322.1 K and $T_{min}$ = 300 K.

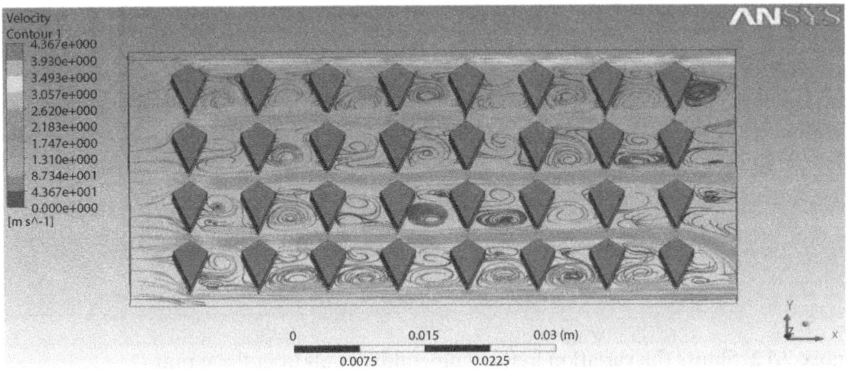


**Figure 7.9** Shows the variation in the velocity contour at Reynolds number Re = 1500–4000 at P = 5 mm.

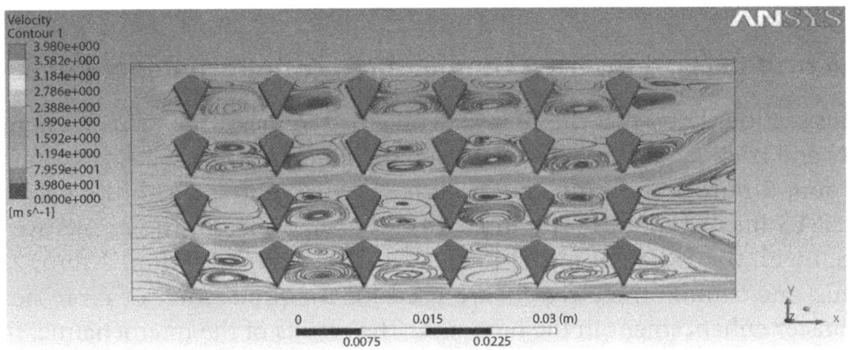


**Figure 7.10** Shows the variation in the velocity at Reynolds number Re = 4000 and P = 6 mm.

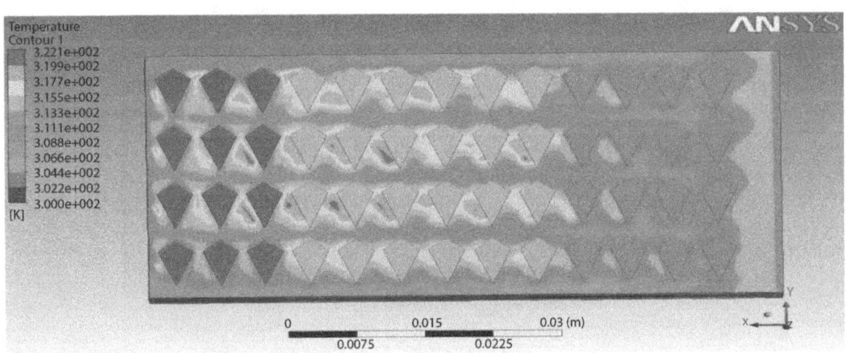


**Figure 7.11** Shows the variation in the temperature contour at P = 3 mm.

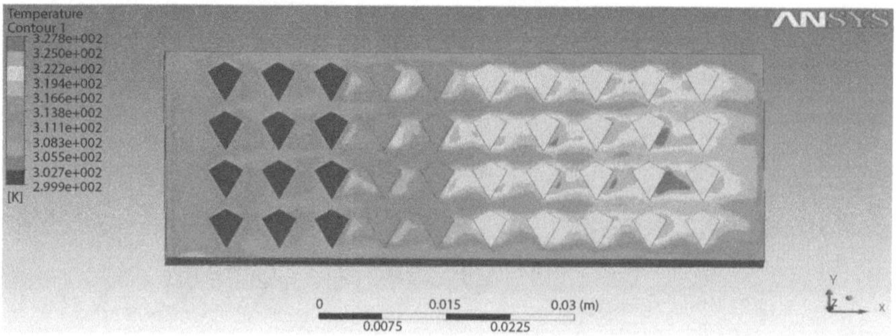


**Figure 7.12** Shows the variation in the temperature contour at P = 4 mm.

Figure 7.12 shows the variation in the temperature contour at P = 4 mm used in the microchannel and the maximum temperature obtained is $T_{max}$ = 327.1 K.

### 7.6.6 Effects of Inserted Diamond Fin with Reynolds Number

This section shows that the all results obtained in the form of Nusselt number and friction factor coefficient when the different pitch space P = 3 mm, 4 mm, 5 mm, and 6 mm were used in the diamond fin with the help of ANSYS fluent (Figures 7.13 and 7.14). The variations in the results were seen with the variation in the Reynolds number $H_f$ = 1.5 mm, 2 mm, 2.5 mm, and 3 mm. These results are used for the production of more heat transfer enhancement in the heat sink. The results of the microchannel fin are specified as the rate of heat transfer by the convection and conduction between the diamond shape fins. These high velocity fluctuations increased

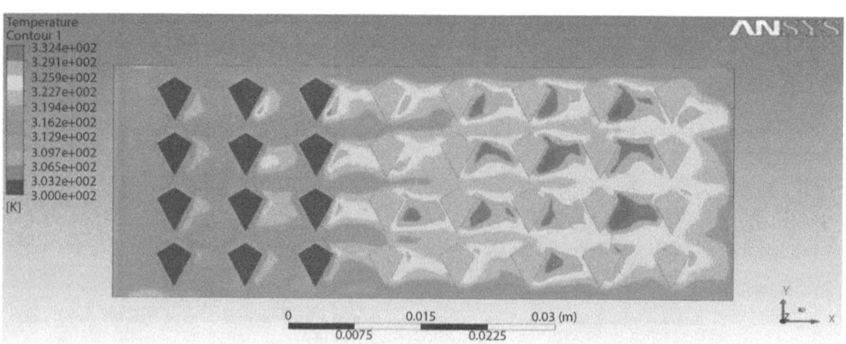


**Figure 7.13** Shows the variation in temperature contour at Re = 4000 and P = 5 mm between the diamond shape rectangular channel.

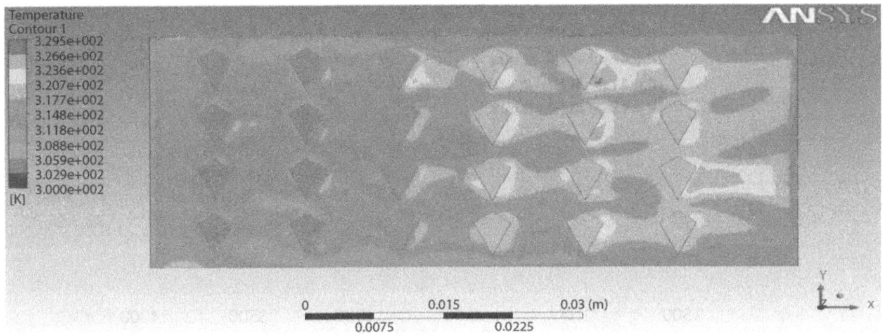


**Figure 7.14** Shows the variation in temperature contour at P = 6 mm.

the change in momentum, thermal energy, and increased the boundary friction as well as pressure drop. The pressure drop was examined by the number of variations in the heat sink and flowing fluid. The turbulent flow in the heat sink is an advantage of increasing the heat characteristics and mass transfer. When the space increased between the roughness, a large turbulent stream function profile and reattachment zone was generated around the solid surface and the cold air gets energy from the conduction and that changed into convection energy and thermal energy of the tube. All the results of roughened heat exchanger tube are shown in Figures 7.15–7.18.

Figure 7.15 shows the variation in the heat transfer rate as Nusselt number Nu of microchannel diamond fin heat sink with respect to the variation in the Reynolds number Re = 1500, 2000, 2500, 3000, 3500, and 4000, and variations in the height of the fin $H_f$ = 1.5 mm, 2 mm, 2.5 mm, and 3 mm at the pitch space between the fin, P = 3 mm. The figure shows maximum Nusselt Nu = 76.36351979 is obtained at $H_f$ = 2.5 mm and Reynolds number Re = 4000, Nusselt number is obtained Nu =55.47366365 at $H_f$ = 1.5 mm,

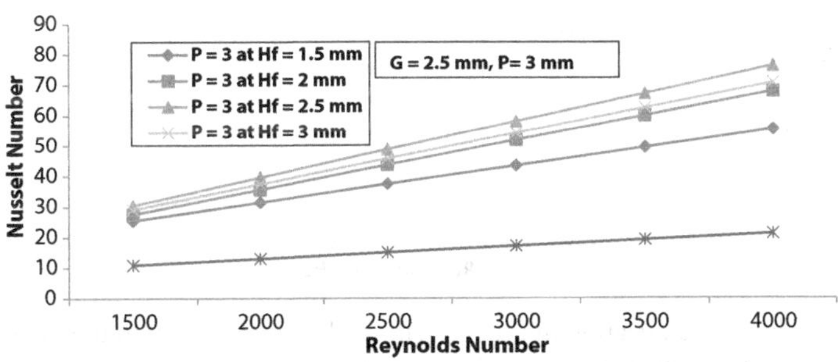


**Figure 7.15** Shows variation in the Nusselt number with respect to the Reynolds number, Re = 1500–4000 at pitch space P = 3 mm.

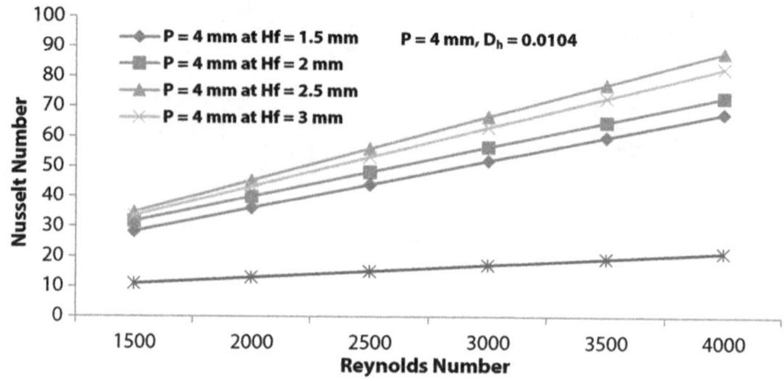


**Figure 7.16** Shows the variation in the Nusselt number with respect to the Reynolds number, Re = 1500–4000 at pitch space P = 4 mm.

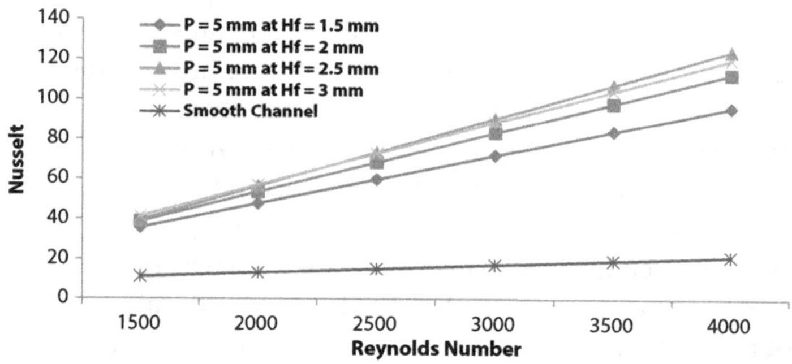


**Figure 7.17** Shows the variations in the Nusselt number of microchannel with diamond shape fin at P = 5 mm.

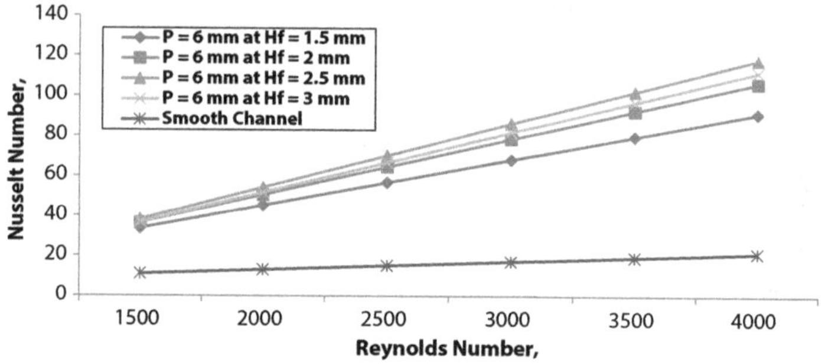


**Figure 7.18** Shows the variation in the Nusselt number with respect to the Reynolds number, Re = 1500–4000 at pitch space between the shape, p = 6 mm.

and the obtained Nusselt number is Nu = 68.04660572 and 70.82537449 at $H_f$ = 2 mm and 3 mm, respectively. In Figure 7.15, the maximum heat transfer rate of the complete model can be observed and it is also compared with smooth tube.

Figure 7.16 shows the variation in the Nusselt number as increasing in the heat transfer rate with increasing Reynolds number 1500–4000, and at some fixed of pitch space, P = 4 mm, with variation of height of the fin $H_f$= 1.5 mm, 2 mm, 2.5 mm, and 3 mm. The maximum Nusselt number Nu = 87.98364947 is obtained at $H_f$= 2.5 mm and obtained other Nusselt numbers, Nu = 67.66600637, 73.10366572, and 82.65621172 at $H_f$= 1.5 mm, 2 mm, and 3 mm, respectively.

Figure 7.17 shows the comparison between the Nusselt number Nu and Reynolds number 1500–4000 at the height of the fin $H_f$= 1.5 mm, 2 mm, 2.5 mm, and 3 mm, at pitch space between the roughness, P = 5 mm. The maximum Nusselt number obtained is Nu = 95.84487348, 112.6342965, 124.1767691, 119.7765006 at $H_f$= 1.5 mm, 2 mm, 2.5 mm, and 3 mm, respectively. All the results found at Re = 4000 and in the figure show that the Nusselt number increases with an increase in the Reynolds number, Re = 1500–4000. In these variations, the maximum heat transfer rate is obtained by the design model.

Figure 7.18 shows the variations in the Nusselt number with respect to the Reynolds number, Re = 1500–4000. The maximum Nusselt number obtained is Nu = 91.12564524, 106.3873662, 118.1448776, and 111.7910075 at $H_f$= 1.5 mm, 2 mm, 2.5 mm, and 3 mm, respectively, with respect to pitch space P = 6 mm. All the results are compared with smooth channel results at Re = 4000.

### 7.6.7 Friction Factor in the Roughened Microchannel Heat Sink

In this section, the friction factor results are examined and their effects on the microchannel heat sink are discussed in Figures 7.15–7.18. The friction results played an important role for defining the performance of the heat sink with increasing heat transfer rate without affecting the material of the microchannel and how long time it can work. When the friction factor increased in the case of roughened heat sink, the loss of energy also increased and required more force for releasing the shear stress in the designed model due to this problem; thermal performance also decreased in the microchannel. So, the number of pitch space P = 3 mm, 4 mm, 5 mm, and 6 mm with the number of height variations of fin, $H_f$= 1.5 mm,

2 mm, 2.5 mm, and 3 mm, respectively, to reduce the capacity of friction factor, pressure drop, and maintain the heat transfer rate in the heat sink.

Figure 7.19 shows the comparison between the friction factor and Reynolds number, Re = 1500–4000. The friction factor increases with a decrease in the Reynolds number, and pressure drop also increased. In the figure, the friction factor varies with the variation in the different roughness shape at P = 3 mm, the maximum friction factor obtained is Fr = 0.008921465 at Re = 4000 and $H_f$= 2.5 mm, and minimum friction factor, Fr = 0.006926465, 0.007313134, and 0.008031646. All the results are compared with the smooth microchannel (Fr = 0.002134679 at Re = 4000). In the figure, it can be seen that friction factor decreases with increasing Reynolds number.

Figure 7.20 shows the variation in the friction factor with respect to the Reynolds number, Re =1500–4000, and minimum friction Fr = 0.007421165, 0.008265649, 0.009221655, 0.008814465, and 0.002134679 are observed for $H_f$= 1.5 mm, 2 mm, 2.5 mm, and 3 mm and smooth microchannel, respectively, at Re = 4000 and P = 4 mm, and $D_h$ = 0.0104.

Figure 7.21 shows the comparison between the friction factors and Reynolds number, Re = 1500–4000. Minimum friction factor, Fr = 0.008046465, 0.008523423, 0.010852363, and 0.009765612 at different heights of diamond fin shape roughness, $H_f$= 1.5 mm, 2 mm, 2.5 mm, and 3 mm, respectively, at Re = 4000 and P = 5 mm, and these results are compared with smooth microchannel. This friction factor is more effective for the maintaining the life of the designed model when produced in the experimental case.

Figure 7.22 shows the variations in friction factors with respect to the Reynolds number, Re = 1500–4000 at P = 6 mm, and varied with diamond fin heights, $H_f$= 1.5 mm, 2 mm, 2.5 mm, and 3 mm. The minimum friction factors observed are Fr = 0.008035646, 0.008365649, 0.009631545, and

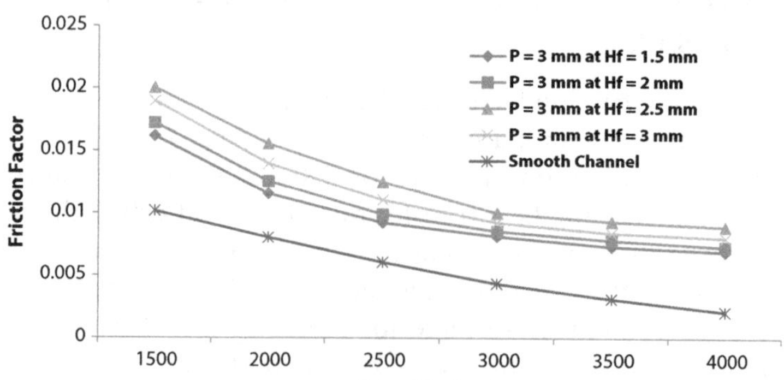


**Figure 7.19** Shows the variation in the friction factor of microchannel diamond fin heat sink at Re = 1500–4000 and P = 3 mm.

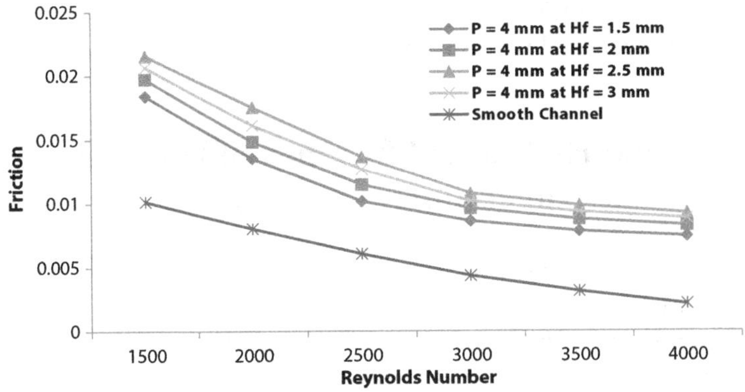


**Figure 7.20** Shows the variations in the friction factor at Re = 4000, P = 4 mm, and diamond fin in microchannel heat sink.

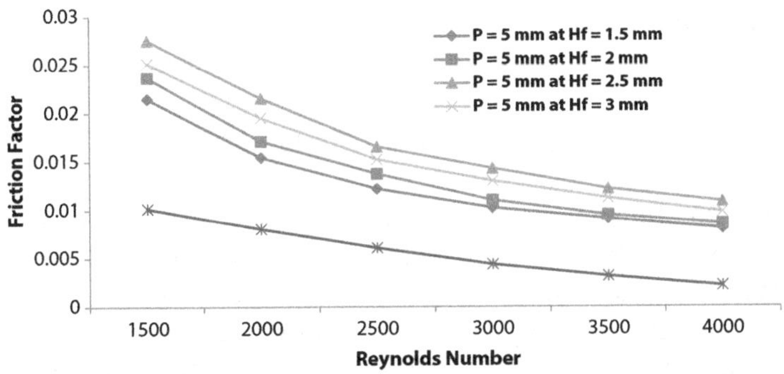


**Figure 7.21** Shows the variation in the friction factor with diamond fin heat sink at P = 5 mm and Re = 4000.

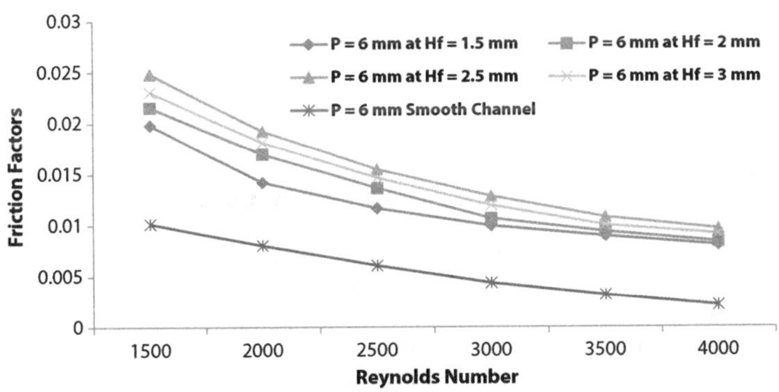


**Figure 7.22** Shows the variations in friction factors at P = 6 mm and Re = 1500–4000.

0.009128441 at $H_f$ = 1.5 mm, 2 mm, 2.5 mm, and 3 mm, respectively. All the results are observed with respect to smooth microchannel.

## 7.7 Thermal Hydraulic Efficiency of Diamond Shape Heat Exchanger Sink

This section presents the efficiency of the microchannel heat sink with diamond fin with roughness. The thermal performance (efficiency) of the model depends upon the heat transfer rate per unit area of the model and loss of energy due to friction. It has been explained that heat transfer rate depends upon the variation in the perforated roughness, variation in flow as turbulent and laminar flow, and most effective variation in Reynolds number. The thermal performance is defined in the form of heat transfer analysis which is calculated by the ANSYS Fluent and it is solved in the form of ratio of thermal conductivity to the heat capacity. The formula for the finding the efficiency is given by the researcher Webb and Eckert.

$$\eta = (Nu/Nus)/(Fr/Frs)^{1/3}$$

Figure 7.23 shows the variation of the efficiency with respect to the Reynolds number, Re = 1500–4000. Thermal performance also validated on the basis of variation in the height of the diamond fin roughness at Re = 4000, P = 5 mm for reducing the friction factor, pressure drop, energy loss, and increasing the capacity of model. The maximum thermal performance, $\eta_{max}$ = 3.409506 is obtained at P = 5 mm and Re = 4000, and $H_f$ = 2.5 mm.

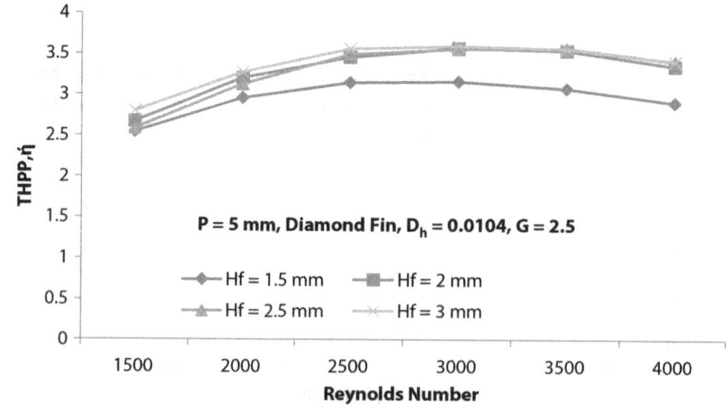


**Figure 7.23** Shows the variation in the efficiency with respect to the Reynolds number.

## 7.8 Conclusion

All results are analyzed using the computational fluid dynamic tool ANSYS workbench software. The heat transfer enhancement based on the various values like variation in the Reynolds number, Re = 1500–4000, pitch space, and $H_f$. On the basis of these variations, the Nusselt number, friction factor, velocity contour, temperature contour, and efficiency of the model was found. This chapter investigates the impact of varying diamond fin heights on heat transfer augmentation within a smooth rectangular microchannel. The study aims to identify the optimal fin height that maximizes heat transfer while minimizing pressure drop. The key findings, including the determination of optimal design parameters, are summarized as follows:

(a) From the results, it is clear that k-ε standard model could provide results with acceptable engineering accuracy for the analysis of flow and heat transfer patterns in heat sink.
(b) The heat transfer increases and friction factor decreases with an increase in the Reynolds number, and the Nusselt number also increased with increasing pitch space between the different height inserted diamond fins.
(c) The highest Nusselt number obtained is Nu = 124.1767691 at pitch space P = 5 mm taken between the diamond shape fin of height Hf = 2.5 mm, and at Re = 4000.
(d) The lowest friction factor Fr = 0.006926465 is obtained at P = 3 mm, and at Hf = 1.5 mm and Re = 4000.
(e) The heat transfer rate maximum increased 5.830887958 times as compared to the heat transfer rate of smooth microchannel.
(f) The maximum efficiency is obtained, $\eta_{max}$ = 3.409505726 at P = 5 mm, Re = 4000, and $H_f$ = 2.5 mm.

## References

1. Bhandari, P. and Prajapati, Y.K., Thermal performance of open microchannel heat sink with variable pin fin height. *Int. J. Therm. Sci.*, 159, 106609, January 2021, https://doi.org/10.1016/j.ijthermalsci.2020.106609.
2. Qomi, M.E., Sheikhzadeh, G.A., Fattahi, A., Heat transfer enhancement in a microchannel using a pulsating MHD hybrid nanofluid flow, https://doi.org/10.1080/15567036.2020.1834031.

3. Bhandari, P. and Prajapati, Y.K., Influence of fin height on heat transfer and fluid flow characteristics of rectangular microchannel heat sink April 2019. *Int. J. Heat Mass Transf.*, 137, 1, 1041–1052, DOI:10.1016/j.ijheatmasstransfer.2019.04.012.
4. Mehrabi, S., Kheradmand, S., Farivar, O.R., Numerical simulation of thermal and hydraulic performance of a micro plate - pin fin heat sink, IIETA, 2018, DOI: https://doi.org/10.18280/ijht.360315.
5. Kumar, R. *et al.*, Investigation of heat transfer and friction factor in triangular duct solar air heater due to forward facing chamfered rectangular ribs. *Renew. Energy*, 824–835, 2018.
6. Bhandari, P. and Prajapati, Y.K., Flow boiling instabilities in microchannels and their promising solutions- A review. *Exp. Therm Fluid Sci.*, 88, July 2017, DOI:10.1016/j.expthermflusci.2017.07.014.
7. Sharma, S.K. *et al.*, Expt. And numerical investigation of forced convective heat transfer in solar air heater with thin ribs. *Sol. Energy*, 147, 277–291, 2017.
8. Varshney, L. *et al.*, Performances predictions for the solar air heater having a rectangular sectional tapered rib roughness using CFD. *Therm. Sci. Eng. Prog.*, 4, 122–132, 2017.
9. Pitak, P. *et al.*, Turbulent heat transfer and pressure loss in a square channel with discrete broken V-rib turbulators. *J. Hydrodyn.*, 28, 275–283, 2016.
10. Narendran, G., Gnanasekaran, N., Perumal, D.A., A Review on Recent Advances in Microchannel Heat Sink Configurations. *Recent Pat. Mech. Eng.*, 11, 3, 2018.
11. Montelpare, S., An experimental IR thermographic method for the evaluation of the heat transfer coefficient of liquid-cooled short pin fins sink arranged in line. March 2006. *Exp. Therm Fluid Sci.*, 30, 4, 381–391, DOI:10.1016/j.expthermflusci.2005.09.004.
12. Yildizeli, A. and Cadirci, S., objective optimization of a micro-channel heat sink through genetic algorithm. *Int. J. Heat Mass Transf.*, 146, 118847, January 2020, https://doi.org/10.1016/j.Ijheatmasstransfer.2019.118847.
13. Tariq, A., Panigrahi, P.K., Muralidhar, K., Flow And Heat Transfer in the wake of a surface-mounted rib with a slit. *Exp. Fluids*, 37, 701–719, 2004.
14. Kumar, A., Saravana, R., Prabhahar, S., Numerical investigation of heat transfer characteristics in a square duct with internal ribs. *IJIRSET*, 3, 1, 2014.
15. Esmaeilzadeh, E., Alamgholilou, A., Mirzaie, H., Khoshnevis, A., Numerical investigation Of heat transfer enhancement of rectangular ribs with constant heat flux located in the floor Of A 3d duct Flow. *Proceedings Of The 12th Asian Congress Of fluid mechanics*, 2008.
16. Patil, S.V. and Vijay Babu, P.V., Heat Transfer Augmentation in a circular tube and square duct fitted with swirl flow generators: a Review. *Int. J. Chem. Eng. Appl.*, 2, 5, 2011.

17. Sahu, R.K. and Gandhi, B.K., Numerical simulation of heat transfer enhancement due to a gap in an inclined continuous rib arrangement in a solar air heater duct. *Int. J. Adv. Mech. Eng.*, 4, 6, 687–693, 2014, ISSN 2250-3234.
18. Aharwal, K.R., Gandhi, B.K., Saini, J.S., Experimental investigation on heat transfer enhancement due to a gap in an inclined continuous rib arrangement in a rectangular duct of solar air heater. *Renew. Energy*, 33, 585e96, 2008.
19. Kundu, A., Kumar, A., Dutt, N., Meena, C.S., Singh, V.P., Introduction to Thermal Energy Resources and their Smart Applications, in: *Thermal Energy Systems: Design, Computational Techniques and Applications*, A. Kumar, V.P. Singh, C.S. Meena, N. Dutt, (Eds.), pp. 1–15, CRC Press, Boca Raton, FL, USA, 2023, Chapter 1; http://dx.doi.org/10.1201/9781003395768-1.
20. Kundu, A., Kumar, A., Dutt, N., Singh, V.P., Meena, C.S., Modelling and Simulation of Thermal Energy System for Design Optimization, in: *Thermal Energy Systems: Design, Computational Techniques and Applications*, A. Kumar, V.P. Singh, C.S. Meena, N. Dutt, (Eds.), pp. 103–140, CRC Press, Boca Raton, FL, USA, 2023, Chapter 7; http://dx.doi.org/10.1201/9781003395768-7.
21. Dewangan, A.K., Moinuddin, S.Q., Cheepu, M., Sajjan, S.K., Kumar, A., Thermal Energy Storage: Opportunities, Challenges and Future Scope, in: *Thermal Energy Systems: Design, Computational Techniques and Applications*, A. Kumar, V.P. Singh, C.S. Meena, N. Dutt, (Eds.), pp. 17–28, CRC Press, Boca Raton, FL, USA, 2023, Chapter 2; http://dx.doi.org/10.1201/9781003395768-2.
22. Dutt, N., Hedau, A.J., Kumar, A., Awasthi, M.K., Singh, V.P., Dwivedi, G., Thermo-hydraulic performance of solar air heater having discrete D-shaped ribs as artificial roughness. *Environ. Sci. Pollut. Res.*, 2023, https://doi.org/10.1007/s11356-023-28247-9.
23. Kr. Kushwaha, P., Kr. Sharma, N., Kumar, A., Meena, C.S., Recent Advancements in Augmentation of Solar Water Heaters Using Nanocomposites with PCM: Past, Present, and Future. *Buildings*, 13, 79, 2023, https://doi.org/10.3390/buildings13010079.
24. Singh, V.P., Jain, S., Kumar, A., Mishra, S., Sharma, N.K., Heat transfer and friction factor correlations development for double pass solar air heater artificially roughened with perforated multi-V ribs. *Case Stud. Therm. Eng.*, 39, 102461, 2022. ISSN 2214-157X, https://doi.org/10.1016/j.csite.2022.102461.
25. Singh, V.P., Jain, S., Karn, A., Kumar, A., Dwivedi, G., Meena, C.S., Recent Developments and Advancements in Solar Air Heaters: A Detailed Review. *Sustainability*, 14, 12149, 2022, https://doi.org/10.3390/su141912149.
26. Singh, V.P., Jain, S., Karn, A., Kumar, A., Experimental Assessment of Variation in Open Area Ratio on Thermohydraulic Performance of Parallel Flow Solar Air Heater. *Arabian J. Sci. Eng.*, 2022, https://doi.org/10.1007/s13369-022-07525-7.
27. Meena, C.S., Prajapati, A.N., Kumar, A., Kumar, M., Utilization of Solar Energy for Water Heating Application to Improve Building Energy Efficiency:

An Experimental Study. *Buildings*, 12, 2166, 2022, https://doi.org/10.3390/buildings12122166.
28. Meena, C.S., Kumar, A., Roy, S., Cannavale, A., Ghosh, A., Review on Boiling Heat Transfer Enhancement Techniques. *Energies*, 15, 5759, 2022, https://doi.org/ 10.3390/en15155759.
29. Singh, V.P., Jain, S., Kumar, A., Establishment of Correlations for the Thermo-Hydraulic Parameters Due to Perforation in a Multi-V Rib Roughened Single Pass Solar Air Heater. *Exp. Heat Transf.*, 2022, DOI: 10.1080/08916152.2022.2064940.
30. Meena, C.S., Kumar, A., Jain, S., Ur Rehman, A., Mishra, S., Innovation in Green Building Sector for Sustainable Future. *Energies*, 15, 6631, 2022, https://doi.org/10.3390/en15186631.
31. Singh, V.P., Jain, S., Karn, A., Kumar, A., Dwivedi, G., Meena, C.S., Cozzolino, R., Mathematical Modeling of Efficiency Evaluation of Double-Pass Parallel Flow Solar Air Heater. *Sustainability*, 14, 10535, 2022, https://doi.org/10.3390/su141710535.
32. Pant, G., Meena, C.S., Saxena, A., Kumar, A., Singh, V.P., Dutt, N., Study the Temperature Variation in Alternate Coils of Insulated Condenser Cum Storage Tank: Experimental Study, in: *Advances in Fluid and Thermal Engineering. FLAME 2022.* Lecture Notes in Mechanical Engineering, B.S. Sikarwar, S.K. Sharma, A. Jain, K.M. Singh, (Eds.), Springer, Singapore, 2023, https://doi.org/10.1007/978-981-99-2382-3_52.
33. Prajapati, Y.K., Gupta, P.K., Kumar, A., Free Surface Undulation and Air Entrainment in a Rectangular Tank. *Int. J. Appl. Eng. Res.*, 6, 4, 409–419, 2011.
34. Pathak, A., Kumar, A., Kumar, A., Kumar, A., Francis Luther King, M., Application of Laser Technology in the Mechanical and Machine Manufacturing Industry, in: *Laser-based Technologies for Sustainable Manufacturing*, A. Kumar, A. Kumar, A. Kumar (Eds.), pp. 107–155, CRC Press, Boca Raton, FL, USA, 2023, Chapter 06; DOI: 10.1201/9781003402398-6.
35. Saxena, A., Prajapati, A.N., Pant, G., Meena, C.S., Kumar, A., Singh, V.P., Water Consumption Optimization of Hybrid Heat Pump Water Heating System, in: *Recent Advances in Mechanical Engineering. FLAME 2022.* Lecture Notes in Mechanical Engineering, A.K. Shukla, B.P. Sharma, A. Arabkoohsar, P. Kumar, (Eds.), Springer, Singapore, 2023, https://doi.org/10.1007/978-981-99-1894-2_61.
36. Kumar, A., Singh, V.P., Meena, C.S., Dutt, N., *Thermal Energy Systems: Design, Computational Techniques and Applications*, Taylor & Francis (CRC Press), 2023, https://doi.org/10.1201/9781003395768.
37. Kumar, A., Kumar, A., Kumar, A., *Laser Based Technologies for Sustainable Manufacturing*, Taylor & Francis (CRC Press), 2023, ISBN: 9781003402398, https://doi.org/10.1201/9781003402398.
38. Kumar, R., Prasad, A., Kumar, A., *Sustainable Smart Manufacturing Processes in Industry 4.0*, Taylor & Francis (CRC Press), 2023, ISBN: 9781003436072, https://doi.org/10.1201/9781003436072.

39. Tomar, P. and Gope, P.C., Impact Strength Enhancement of PMMA base Denture Material by Fibre Addition. *IJMMST*, 2023, DOI: https://doi.org/10.56896/IJMMST.2023.2.1.005.
40. Sharma, V.K., Kumar, A., Gupta, M., Kumar, V., Sharma, D.K., Sharma, S., *Additive Manufacturing in Industry 4.0: Methods, Techniques, Modeling, and Nano Aspects*, Taylor & Francis (CRC Press), 2022, https://doi.org/10.1201/9781003360001.
41. Verma, V., Meena, C.S., Thangavel, S., Kumar, A., Choudhary, T., Dwivedi, G., Ground and solar assisted heat pump systems for space heating and cooling applications in the northern region of India – A study on energy and CO2 saving potential. *Sustain. Energy Technol. Assess.*, 59, 103405, 2023, ISSN 2213-1388, https://doi.org/10.1016/j.seta.2023.103405.
42. Tomar, P., Gope, P.C., Chauhan, S., Influence of fiber Reinforcement on properties of PMMA. *Int. J. Mech. Therm. Eng.*, 4, 1, 23–27, 2023.
43. Awasthi, M.K., Dutt, N., Kumar, A., Kumar, S., Electrohydrodynamic capillary instability of Rivlin–Ericksen viscoelastic fluid film with mass and heat transfer. *Heat Transfer*, 1–19, 2023, doi: 10.1002/htj.22944.
44. Tomar, P. and Gope, P.C., Effect of glass fiber and nylon fiber reinforcement on the mechanical and thermal properties of styrene butadiene rubber mixed PMMA denture base material. *J. Mech. Behav. Biomed. Mater.*, 150, 106308, February 2024.
45. Dutt, N., Hedau, A., Awasthi, M.K., Kumar, A., Meena, C.S., Thermo-hydraulic performance investigation of solar air heater duct having staggered D-shaped ribs: Numerical approach. *Heat Transfer*, 53, 3, 1501–1531, 2024, https://doi.org/10.1002/htj.22998.
46. Tomar, P. and Gope, P.C., Enhancement of PMMA base Denture Material: A Review. *Int. J. Mater. Manuf. Sustain. Technol.*, 2023, DOI: https://doi.org/10.56896/IJMMST.2023.2.1.003.
47. *Highly Efficient Thermal Renewable Energy Systems: Design, Design, Optimization and Applications*, V. Verma, S. Thangavel, N. Dutt, A. Kumar, R. Weerasinghe (Eds.),CRC Press, Boca Raton, FL, USA, 2024, ISBN 9781032595641.
48. Verma, V., Thangavel, S., Dutt, N., Kumar, A., Weerasinghe, R., Recent Development of Thermal Energy Storage: Solar, Geothermal and Hydrogen Energy, in: *Highly Efficient Thermal Renewable Energy Systems: Design, Design, Optimization and Applications*, V. Verma, S. Thangavel, N. Dutt, A. Kumar, R. Weerasinghe (Eds.), CRC Press, Boca Raton, FL, USA, 2024, Chapter 01; ISBN 9781032595641.
49. Ahmadizadeh, M., Heidari, M., Thangavel, S., Naamani, E.A., Khashehchi, M., Verma, V., Kumar, A., Technological Advancements in Sustainable and Renewable Solar Energy Systems, in: *Highly Efficient Thermal Renewable Energy Systems: Design, Design, Optimization and Applications*, V. Verma, S. Thangavel, N. Dutt, A. Kumar, R. Weerasinghe (Eds.), CRC Press, Boca Raton, FL, USA, 2024, Chapter 02; ISBN 9781032595641.

50. Khashehchi, M., Thangavel, S., Rahmanivahid, P., Heidari, M., Moazzeni, T., Verma, V., Kumar, A., Solar Desalination Techniques: Challenges and Opportunities, in: *Highly Efficient Thermal Renewable Energy Systems: Design, Design, Optimization and Applications*, V. Verma, S. Thangavel, N. Dutt, A. Kumar, R. Weerasinghe (Eds.), CRC Press, Boca Raton, FL, USA, 2024, Chapter 19; ISBN 9781032595641.

8

# Particle Swarm Optimization Technique for Determining Optimal Process Parameters for Counter Flow Double Pipe Heat Exchanger

**Sridharan M.**

*Department of Mechanical Engineering, SRM Institute of Science and Technology, Tiruchirappalli, India*

***Abstract***

In this study, the particle swarm optimization (PSO) is used to find the optimal operating parameters for maximizing the heat transfer rate in the double pipe heat exchanger (DPHE). For this, the real-time experiments are conducted with the counter-flow double pipe heat exchanger rig. Empirical models for two different heat transfer rates are described for exploring optimized operating parameters in the heat transfer operation. The mass flow rates of two fluids (hot and cold) at different temperatures are observed as constraints for both the theoretical and experimental approach. It is also observed that the heat transfer rate obtained by the PSO predicted operating parameters are highly comparable than the results of initial experiments.

***Keywords*:** PSO, heat exchanger, heat transfer rate, optimal and parameters

## Nomenclature

$T_1$ - Hot water inlet temperature (°C)
$t_1$ - Cold water inlet temperature (°C)
$T_2$ - Hot water outlet temperature (°C)
$t_2$ - Cold water outlet temperature (°C)

*Email*: sridhudhiya@gmail.com; ORCID: https://orcid.org/0000-0003-0792-4161

Ashwani Kumar, Nitesh Dutt and Mukesh Kumar Awasthi (eds.) Heat Transfer Enhancement Techniques: Thermal Performance, Optimization and Applications, (193–220) © 2025 Scrivener Publishing LLC

$m_c$ - Mass flow rate of cold fluid (kg/min)
$C_{pc}$ - Specific heat capacity of cold fluid (J/kg-K)
$Q_c$ - Heat transfer rate of cold fluid (W)
$m_h$ - Mass flow rate of hot fluid (kg/min)
$C_{ph}$ - Specific heat capacity of hot fluid (J/kg-K)
$Q_h$ - Heat transfer rate of hot fluid (W)
$\Delta T$- Hot fluid temperature difference (°C)
$\Delta t$- Cold fluid temperature difference (°C)

## Abbreviations

TS – Temperature sensor
PSO – Particle swarm optimization
CF-DPHE – Counter flow double pipe heat exchanger
SOTO – Single objective type optimization
MOTO – Multi objective type optimization
NTOT - Non-traditional optimization technique
r ( ) – Random number
v ( ) - Velocity of the particle
p ( ) – Position of the particle

## 8.1 Introduction

The heat exchanger is a device in which the process of heat exchange between two different fluids with temperature difference takes place. Such a device with higher effectiveness is always a demand among the industries. The higher effectiveness delivered by a heat exchanger is directly proportional to its rate of heat transfer. The heat transfer rate includes two sub types such as (i) hot fluid side heat transfer rate ($Q_h$), and (ii) cold fluid side heat transfer ($Q_c$).

Both the $Q_h$ and $Q_c$ are the direct dependent variables of eight mass parameters such as

i. Mass flow rate of hot fluid ($m_h$)
ii. Hot fluid inlet temperature ($T_1$)
iii. Specific heat capacity of the hot fluid ($C_{ph}$)
iv. Hot fluid outlet temperature ($T_2$)
v. Mass flow rate of cold fluid ($m_c$)

vi. Cold fluid inlet temperature ($t_1$)
vii. Specific heat capacity of the cold fluid ($C_{pc}$)
viii. Cold fluid outlet temperature ($t_2$)

Among eight parameters, four parameters ($T_1$, $t_1$, $C_{ph}$, $C_{pc}$) remain constant throughout the experimental phase. The remaining four parameters ($m_h$, $m_c$, $T_2$, $t_2$) are transient variables. The rate of heat transfer (both hot and cold) mainly depends on the above mentioned four parameters such as $m_h$, $m_c$, $T_2$, and $t_2$.

Many researchers involved themselves in identifying the optimum process parameters for heat exchangers. Such research works are mainly classified into two types: (i) single objective type optimization [SOTO] and (ii) multi objective type optimization [MOTO]. Among such classifications, majority of researchers concentrated with the MOTO, the process in which more than one objective function is optimized. Table 8.1 presents the classification of different types of objective function considered by researchers References [2] to [11]. Table 8.2 presents the highlighted review and its inference from the SOTO and MOTO. In this work, heat exchange experiments were conducted on the double pipe heat exchanger (DPHE) assembly. The objectives of this work are:

i. To study the effects of thermodynamic parameters ($T_1$, $t_1$, $T_2$, $t_2$) on the performance of the DPHE through real-time experiment.
ii. To study the effects of flow parameters ($m_h$, $m_c$).
iii. To implement a particle swarm optimization (PSO)-based optimization technique that could give optimum parameters to maximize the heat transfer rate in DPHE.

The proposed PSO-based optimization technique is capable of replacing the tedious process of operating parameters selection by conventional trial and error method.

### 8.1.1 Need for Optimization in Heat Exchangers

The optimization process involves finding an alternative with the highest achievable performance under the given constraints, by maximizing desired components and minimizing undesired ones. In comparison, maximization means an attempt to attain the maximum or most top result without regard to expense or cost. In this regard, heat exchanger performance depends on many desirable and undesirable factors. Numerous single and multi-objective optimization studies are conducted to maximize

**Table 8.1** Major objective function considered in the process optimization of heat exchangers.

| **Heat exchanger objective function** | | | |
|---|---|---|---|
| **Geometry based** | **Flow characteristic based** | **Performance based** | **Cost based** |
| 1. Total heat exchanger weight<br>2. Reduction of length<br>3. Reduction of shell diameter<br>4. Reduction of tube diameter<br>5. Total heat transfer area<br>6. Overall size of the heat exchanger<br>7. Thickness, width, and pitch of tapes<br>8. Channel geometry | 1. Mass flow rate<br>2. Volumetric flow rate<br>3. Pressure drop<br>4. Velocity field heat flow<br>5. Hydraulic performance<br>6. Nusselt number<br>7. Colburn factor<br>8. Reynolds number<br>9. Geometric and flow parameters | 1. Equivalent hot water output<br>2. Specific energy consumption<br>3. Energy loss and efficiency<br>4. Design and input process parameters<br>5. Overall heat-transfer coefficients<br>6. Heat exchange rate<br>7. Efficiency of the heat exchanger<br>8. Entropy generation number<br>9. Thermal performance.<br>10. Heat exchanger effectiveness<br>11. Modified entropy number<br>12. Rib effect coefficient<br>13. Enthalpy balances | 1. Capital investment<br>2. Operating cost<br>3. Total annual cost<br>4. Maintenance cost<br>5. Cost function<br>6. Specific material cost<br>7. Annual total amount of produced energy<br>8. Ecological function cost<br>9. Energy costs<br>10. Maximum field synergy number |

**Table 8.2** Highlighted literature works based on the PSO in the field of heat exchangers.

| Author | Type of heat exchanger | Objective function | Objective type | Method/ algorithm | Inference |
|---|---|---|---|---|---|
| Viviana Cocco Mariani *et al.* [1] | Shell and tube heat exchanger | Minimization of total capital cost | Single | Quantum particle swarm optimization (QPSO) | A new PSO approach integrated with Zaslavskii chaotic map sequence optimization is applied to two cases of shell and tube heat exchanger experiments on the minimization.<br>As a result of which total cost decreased by 30% and 27% for case-1 and case-2. |
| Patel *et al.* [2] | Shell and tube heat exchanger | Minimization of total capital cost | Single | Particle swarm optimization (PSO) and genetic algorithm (GA) | A new PSO approach is applied to four cases of shell and tube heat exchanger experiments on the minimization.<br>As a result of which total cost decreased by 30% and 5.7%, 5.1%, 2.86%, and 3.77%. |
| Guo *et al.* [3] | Double pipe heat exchanger | Maximizing the thermal efficiency of micro-finned tubes | Single | Particle swarm algorithm | The PSO results indicate that the optimum microfin height increases with increasing the Reynolds number. The opposite trend is observed for the optimal helix angle. |

*(Continued)*

**Table 8.2** Highlighted literature works based on the PSO in the field of heat exchangers. (*Continued*)

| Author | Type of heat exchanger | Objective function | Objective type | Method/ algorithm | Inference |
|---|---|---|---|---|---|
| Aline P. Silva *et al.* [4] | Heat exchanger network (HEN) | Minimization of total capital cost (cost of utilities and the cost of the new equipment) | Multi | Particle swarm optimization | Significant reduction in the total cost is observed. The results obtained using PSO are promising and more consistent. |
| Sadeghzadeh *et al.* [5] | Shell and tube heat exchanger | Minimization of total capital cost | Multi | Genetic algorithms and particle swarm algorithms | A new PSO approach integrated is applied to a shell and tube heat exchanger rig. As a result of which total cost decreased by 22.25%. |
| Hossein Zarea *et al.* [6] | Plate fin heat exchanger | Minimization of the entropy generation units | Multi | Particle swarm optimization (PSO) | Preliminarily PFHE design and effectiveness was improved by minimizing the entropy generation units. The optimal results obtained minimize entropy generation units by 25%. |

(*Continued*)

**Table 8.2** Highlighted literature works based on the PSO in the field of heat exchangers. (*Continued*)

| Author | Type of heat exchanger | Objective function | Objective type | Method/ algorithm | Inference |
|---|---|---|---|---|---|
| Rao *et al.* [7] | Cross flow plate-fin heat exchanger | Number of entropy generation units, total volume, and total annual cost | Multi | Particle Swarm Optimization | Performance process parameter (17 numbers) based optimization analysis is also carried out to demonstrate the effect of heat exchanger dimensions on the optimum solution.<br>Significant improvement in the performance of cross flow heat exchanger is obtained. |
| Jyothiprakash *et al.* [8] | Cross-flow heat exchanger | Maximizing effectiveness and minimizing entropy generation units | Multi | Genetic algorithm (GA) and particle swarm optimization (PSO) | The results obtained from the GA and the PSO methods are compared. Both the optimization techniques reflect similar optimal results for generation units and effectiveness. Among which PSO takes lesser time for execution. |
| Mohammad Hemmat Esfe *et al.* [9] | Heat exchanger | Minimum pressure drop and heat transfer coefficient | Multi | particle swarm optimization algorithm and response surface method | The results obtained from the GA and the RSM methods are compared. Among which the results obtain by the PSO is more optimal. |

(*Continued*)

**Table 8.2** Highlighted literature works based on the PSO in the field of heat exchangers. *(Continued)*

| **Author** | **Type of heat exchanger** | **Objective function** | **Objective type** | **Method/ algorithm** | **Inference** |
|---|---|---|---|---|---|
| Hossein Zarea *et al.* [10] | Plate-fin heat exchanger | Maximum effectiveness and the minimum total annual cost | Multi | Bees algorithm hybrid with particle swarm optimization (BAHPSO) | The results obtained from the bees-PSO hybrid algorithm produces more optimal output than conventional PSO. |
| Sridharan *et al.* | Double pipe heat exchanger | Maximum heat transfer rate (hot and cold side) | Multi | PSO | Current study |

desirable and minimize undesirable ones. Such desirable and undesirable constraints are mentioned below [3–10, 15, 16, 18, 19].

- ✓ To maximize the heat transfer rate
- ✓ To maximize overall heat transfer coefficients
- ✓ To maximize the heat flow rate
- ✓ To maximize effectiveness
- ✓ To maximize tube side Nusselt number
- ✓ To maximize the performance rate
- ✓ To minimize total annual cost
- ✓ To minimize pressure drop
- ✓ To minimize flow non-uniformity
- ✓ To minimize entropy generation
- ✓ To minimize energy loss
- ✓ To minimize the total heat transfer area
- ✓ To minimize energy consumption
- ✓ To minimize material cost

## 8.2 Experimental Setup

Tables 8.3 and 8.4 present the detailed technical specifications of a DPHE considered under this study. Figure 8.1 shows the 2D schematic layout of the DPHE. Hot fluid from the reservoir (left end) flow through the inner tube (from the left to the right) and the cold fluid flow through the outer pipe (from the right to the left). With the help of control valves (V1–V3 remains opened and V2–V4 remains closed), counter flow mode of heat exchange takes place between the hot fluid and cold fluid. Temperatures of hot fluid at inlet and outlet are measured using temperature sensors TS1 and TS2. Similarly, the inlet and outlet temperatures of cold fluid are measured using temperature sensors (TS3 and TS4). In this experiment, inlet of both the hot (65°C) and cold fluid (32°C) are maintained as constant. Hot fluid at an inlet temperature of 65°C ± 0.1°C ($T_1$) enters into the core tube through valve V1 from the hot fluid reservoir. At the end of the core tube, the hot fluid exits with the temperature of $T_2$ measured using TS2. The cold fluid of temperature 32°C ± 0.1°C ($t_1$) enters through valve V3 (valves V2 and V4 are closed) into the outer tube resulted in counter-current flow and exit through valve V5 with temperature $t_2$ measured using TS4. By varying the mass flow rate of hot fluid and cold fluid between 1 kg/min ± 1.42 % to 4 kg/min ± 1.42 %, 15 numbers of experiments were conducted. Table 8.4 presents the responses obtained from the real-time counter flow experimental setup.

**Table 8.3** Specifications of the CF-DPHE.

| Parameter | Value |
|---|---|
| The core tube material<br>Core length of the copper tube (m) | Copper<br>1 |
| Heat transport fluid | Water |
| Inner copper pipe diameter (m) | 0.0127 |
| Outer copper pipe diameter (m) | 0.0254 |
| Volume of hot and cold fluid reservoir (L) | 50 |
| Hot fluid inlet temperature (°C) | 65 |
| Cold fluid inlet temperature (°C) | 32 |
| Mass flow rate of hot fluid (kg/min) | 1 to 4 |
| Mass flow rate of cold fluid (kg/min) | 1 to 4 |
| Heat exchanger surface area ($m^2$) | 1 |
| Weight of the heat exchanger (kg) | 50 |

**Table 8.4** Specifications of the measuring instruments.

| Parameter | Range |
|---|---|
| Thermometer | |
| Manufacturer<br>Type of thermometer<br>Number of thermometers<br>Range of thermometer (°C)<br>Accuracy of thermometer (°C)<br>Temperature resolution (°C)<br>Survey cycle time (s) | Mextech Technologies<br>PM-10 digital thermometer<br>04<br>–50 to 110<br>± 1<br>0.1<br>10 |
| Heating Element | |
| Manufacturer<br>Type of heating element<br>Number of heating element<br>Power of heating element (W)<br>Time taken to reach 65°C in (min)<br>Weight of heating element (g) | Bajaj Inc<br>Immersion type<br>1<br>1500<br>55<br>540 |

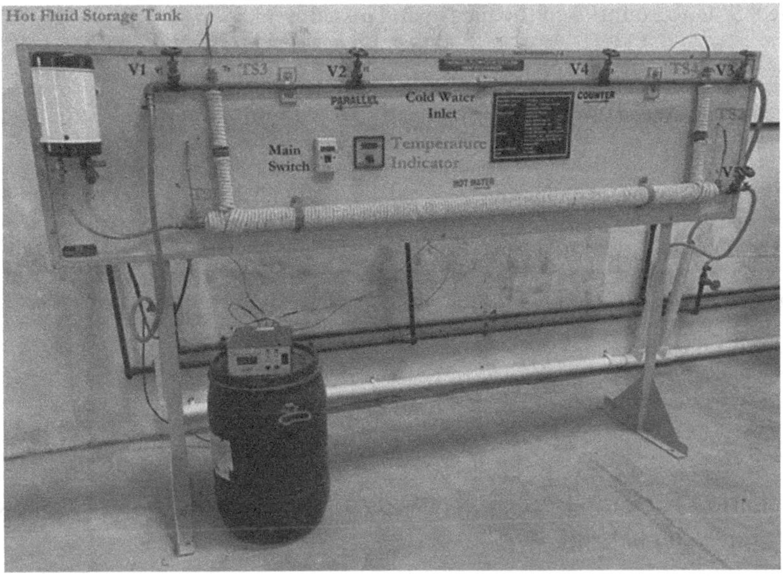


**Figure 8.1** The actual CF-DPHE experimental setup.

### 8.2.1 Data Reduction

This study mainly focuses on predicting the final exit temperatures ($T_2$ and $t_2$) and the rate of heat transfer between the hot and the cold fluids. The detailed formula for calculating the performance of counter flow double pipe heat exchanger (CF-DPHE) is by Sridharan [1]. Equation 8.1 presents the mathematical equation used in calculating the hot fluid side heat transfer rate [13, 14],

$$Q_h = m_h \times C_{ph} \times (T_1 - T_2) \tag{8.1}$$

Equation 8.1 can also be expressed in terms of heat capacity rate ($C_h$) using Equation 8.2,

$$Q_h = C_h \times (T_1 - T_2) \tag{8.2}$$

The mathematical equation used in calculating the cold fluid side heat transfer rate is given by Equation 8.3,

$$Q_c = m_h * C_{ph} * (t_2 - t_1) \tag{8.3}$$

**Table 8.5** Uncertainties in the measuring instrument.

| Measured parameter | Value or limit | Error |
|---|---|---|
| Inner and outer diameter of the core copper tube | Di = 11 mm<br>Do = 25.74 mm | ± 0.006 |
| Length of the core copper tube | L = 1000 mm | ± 0.252 |
| Hot and cold fluid temperatures | -50°C to 110°C | ± 0.1°C |
| Hot and cold fluid mass flow rate | 2.65 to 26.5 L/min | ± 4 % |
| Hot fluid heat transfer rate | 700 W – 2900 W | ± 1.7% |
| Cold fluid heat transfer rate | 600 W – 2100 W | ± 1.5% |

Equation 8.3 can also be expressed in terms of cold capacity rate ($C_c$) using Equation 8.4,

$$Q_c = C_c \times (t_2 - t_1) \tag{8.4}$$

Table 8.5 presents uncertainties values related to the instruments used in this study.

## 8.3 Mathematical Model

### 8.3.1 Formulation of Objective Function

The objective of this work is to improve the performance of the DPHE. Such performance improvement is directly proportional to the maximization of two different heat transfer rates ($Q_h$ and $Q_c$). $Q_h$ and $Q_c$ are the direct dependent variables of eight mass parameters such as mass flow rate of hot fluid ($m_h$), hot fluid inlet temperature ($T_1$), specific heat capacity of the hot fluid ($C_{ph}$), hot fluid outlet temperature ($T_2$), mass flow rate of cold fluid ($m_c$), cold fluid inlet temperature ($t_1$), and specific heat capacity of the cold fluid ($C_{pc}$). The dependent and performance deciding variables $Q_h$ and $Q_c$ are expressed using Equations 8.5 and 8.7. Equations 8.6 and 8.8 are the Minitab software-generated linear regression equations. Datasets presented in Table 8.7 by Reference [1] are used to obtain such linear equations.

$$Q_h = f(m_h, C_{ph}, T_1, T_2) \tag{8.5}$$

$$Q_h = 6864 + 669 \times m_h - 126 \times T_2 \tag{8.6}$$

$$Q_c = f(m_c, C_{pc}, t_1, t_2) \tag{8.7}$$

$$Q_c = -4209 + 476 \times m_c + 106 \times t_2 \tag{8.8}$$

### 8.3.2 Process Parameters

Equations 8.5 and 8.7 consist of eight different process parameters. The parameters such as $m_h$, $m_c$, $T_2$, and $t_2$ dominantly decide the rate of heat exchange inside the DPHE [1]. The first and second parameters constitute flow parameter, while the last two are thermal parameters. Therefore, the effects of $m_h$, $m_c$, $T_2$, and $t_2$ alone are considered in this work.

#### *8.3.2.1 Hot Fluid Outlet Temperature*

Table 8.6 shows the lower bound and upper bound value of DPHE process parameters and constant values. The range of $T_2$ is mathematically represented using Equation 8.9 below,

$$T_{2-min} \leq T_i \leq T_{2-max} \tag{8.9}$$

**Table 8.6** Boundary values of the CF-DPHE process parameters.

| **Parameters** | **Values** |
|---|---|
| $m_c$ min | 1 kg/min |
| $m_c$ max | 4 kg/min |
| $m_h$ min | 1 kg/min |
| $m_h$ max | 4 kg/min |
| Cpc min | 4178 J/kg-K |
| Cph max | 4183 J/kg-K |
| $T_2$ min | 48.3°C |
| $T_2$ max | 60°C |
| $t_2$ min | 36°C |
| $t_2$ max | 45.8°C |

**Table 8.7** Experimental input and output values of the CF-DPHE by Reference [1].

| Sl no. | mh (kg/min) | mc (kg/min) | T1 (°C) | Cph (J/kg-K) | t1 (°C) | Cpc (J/kg-K) | T2 (°C) | t2 (°C) | ΔT (°C) | Δt (°C) | Ch | Qh (W) | Cc | Qc (W) |
|---|---|---|---|---|---|---|---|---|---|---|---|---|---|---|
| **1** | 1 | 1 | 65 | 4183 | 32 | 4178 | 54 | 44 | 11 | 12 | 4183 | 767 | 4178 | 836 |
| **2** | 1 | 2 | 65 | 4183 | 32 | 4178 | 48.3 | 36.5 | 16.7 | 4.5 | 4183 | 1164 | 8356 | 627 |
| **3** | 1 | 3 | 65 | 4183 | 32 | 4178 | 50 | 37.9 | 15 | 5.9 | 4183 | 1046 | 12534 | 1233 |
| **4** | 1 | 4 | 65 | 4183 | 32 | 4178 | 53.8 | 39.3 | 11.2 | 7.3 | 4183 | 781 | 16712 | 2033 |
| **5** | 2 | 1 | 65 | 4183 | 32 | 4178 | 56.5 | 45.8 | 8.5 | 13.8 | 8366 | 1185 | 4178 | 961 |
| **6** | 2 | 2 | 65 | 4183 | 32 | 4178 | 56 | 38 | 9 | 6 | 8366 | 1255 | 8356 | 836 |
| **7** | 2 | 3 | 65 | 4183 | 32 | 4178 | 55.2 | 36 | 9.8 | 4 | 8366 | 1366 | 12534 | 836 |
| **8** | 2 | 4 | 65 | 4183 | 32 | 4178 | 53.4 | 37.3 | 11.6 | 5.3 | 8366 | 1617 | 16712 | 1476 |
| **9** | 3 | 1 | 65 | 4183 | 32 | 4178 | 56.3 | 43.9 | 8.7 | 11.9 | 12549 | 1820 | 4178 | 829 |
| **10** | 3 | 2 | 65 | 4183 | 32 | 4178 | 60 | 40 | 5 | 8 | 12549 | 1046 | 8356 | 1114 |
| **11** | 3 | 3 | 65 | 4183 | 32 | 4178 | 57.6 | 39.1 | 7.4 | 7.1 | 12549 | 1548 | 12534 | 1483 |
| **12** | 3 | 4 | 65 | 4183 | 32 | 4178 | 54.7 | 36.8 | 10.3 | 4.8 | 12549 | 2154 | 16712 | 1337 |
| **13** | 4 | 1 | 65 | 4183 | 32 | 4178 | 54.8 | 41.4 | 10.2 | 9.4 | 16732 | 2844 | 4178 | 655 |
| **14** | 4 | 2 | 65 | 4183 | 32 | 4178 | 56.5 | 39.8 | 8.5 | 7.8 | 16732 | 2370 | 8356 | 1086 |
| **15** | 4 | 3 | 65 | 4183 | 32 | 4178 | 58.5 | 40 | 6.5 | 8 | 16732 | 1813 | 12534 | 1671 |

Where, i = 1, 2, 3, 4.....

#### 8.3.2.2 *Mass Flow Rate of Hot Fluid*

Equation 8.10 presents the hot fluid mass flow rate range,

$$m_{h-min} \leq m_{hi} \leq m_{h-max} \tag{8.10}$$

Where, hi = 1,2,3,4.....

#### 8.3.2.3 *Cold Fluid Outlet Temperature*

Equation 8.11 presents the cold fluid outlet temperature range,

$$t_{2-min} \leq t_i \leq t_{2-max} \tag{8.11}$$

Where, i = 1, 2, 3, 4.....

#### 8.3.2.4 *Mass Flow Rate of Cold Fluid*

Equation 8.12 presents the cold fluid mass flow rate range,

$$m_{c-min} \leq m_{ci} \leq m_{c-max} \tag{8.12}$$

Where, ci=1,2,3,4.....

## 8.4 Implementation of Multi-Objective Type Optimization Technique [MOTOT]

The objective of this work is to determine the optimal parameters and to maximize the heat transfer rate using MOTOT. The experimental results reveal that the product of mass flow rate and the specific heat capacity of the fluid play a vital role in the rate of heat transfer. Experimental trial with higher (m x $C_p$) contributes to the maximum rate of heat transfer while the experiments with lower (m x $C_p$) contributes to the minimum rate of heat transfer. In this stage, an optimization technique is required to identify optimal flow parameter which contributes to the effective heat transfer rate. Different optimization schemes and techniques are available to some mathematical models for heat transfer problems. Among which, the technique PSO is implemented to identify optimal operating parameters in this work.

### 8.4.1 Particle Swarm Optimization

Russell and James first formulated PSO in the middle of 1990's [12]. Particle swarm optimization is among one of the direct search and non-traditional optimization techniques (NTOT). Such NTOT evolved from social learning. Particle swarm optimization is based on the intelligence of swarm where all the swarms in the same group could reach the available food more comfortably. The swarm intelligence is based on the working principle of psychological and social behavior of the swarm. In PSO, each and every swarm is considered as particle and they are allowed to move in multiple directions. Each swarm particle has its very own velocity and position. Swarm (particle) in a population mode communicate each other in order to reach good position. In a maximization-based optimization problem, "best" indicated the particle position with large objective junction value. All the swarm have to experience the following conditions to reach the best position.

#### *8.4.1.1 Condition 1*

Global best [g-best]: in these criteria, the new best position is formed based on random particle in the population.

#### *8.4.1.2 Condition 2*

Local best [l-best]: in these criteria, the suitable particle position is identified by the same particle itself.

Equation 8.13 presents the mathematical relations for calculating the velocity of the particle by Reference [12],

$$v[\,] = W_{vi} r_{1(\,)} \times c_1 (p_i - x_i) + r_{2(\,)} \times c_2 (g_i - x_i) \tag{8.13}$$

Where, random numbers ($r_1$, $r_2 \sim u\ (0, 1)$); $p_i$ – pbest; $g_i$ – global best; $x_i$ – present; $C_1$ & $C_2$ - constants.

Equation 8.14 presents the mathematical relations for calculating the particle position by Reference [12],

$$p[\,] = v[\,] + x_i \tag{8.14}$$

v[ ] is the velocity of the particle, p[ ] is the new particle position, and r ( ) is the random number. The range of random number varies between 0 and 1, and $c_1$ and $c_2$ are the factors for learning. In general, learning factors varies between a minimum of 1 and to a maximum of 4.

### 8.4.2 Merits of PSO

- ✓ The PSO is capable of acquiring at the best result in a faster way.
- ✓ Computationally efficient when compared with heuristic optimization schemes.
- ✓ PSO is simpler, robust, and converges fast.
- ✓ PSO algorithm also reduces costs and time related to the real-time experimental activities.

### 8.4.3 Algorithm

STEP-1 The process of PSO starts with the creation of uniformly distributed particles population.

STEP-2 Then the particle position is estimated based on objective function.

STEP-3 Current particle position is updated when the current is better than the past position.

STEP-4 Particle best [$P_{best}$] is determined based on previous best positions of the particles.

STEP-5 v [ ] is updated.

STEP-6 Particles are moved to their next positions.

STEP-7 Repeat S-2 until the stopping conditions are satisfied.

STEP-8 End.

### 8.4.4 Parameters of PSO

For the experiment number 1, iterations were carried out with populations [100]. The point of convergence near the optimal result is obtained in the 40th iterations. The remaining values obtained after first to 100th are equal with the 40th iteration, for further improvement and verification, iteration numbers are increased up to 1000. But each and every time of the PSO program execution, the point of convergence will vary. This is mainly due to variations in the selection of random numbers. Bharathi Raja *et al.* [12] suggested 100 population and 1000 iterations after side for all the PSO program runs. In this study, the PSO program is allowed to run for 1000 iterations with 100 population size. Table 8.6 presents the boundary parameters considered for executing PSO for optimizing the performance process parameters of DPHE.

### 8.4.5 Numerical Illustration of PSO

The procedure is explained with respect to the particle in iteration 1. The remaining particles are executed by this same way.

#### 8.4.5.1 *Calculation of $m_h$*

The hot fluid mass flow rate is calculated randomly within the limits using Equation 8.15,

$$m_h = m_{h\,min} + (m_{h\,max} - m_{h\,min}) * r(\ ) \tag{8.15}$$

#### 8.4.5.2 *Calculation of $m_c$*

The cold fluid mass flow rate is calculated randomly within the limits using Equation 8.16,

$$m_c = m_{c\,min} + (m_{c\,max} - m_{c\,min}) * r(\ ) \tag{8.16}$$

#### 8.4.5.3 *Calculation of $T_2$*

The hot fluid outlet temperature is calculated randomly within the limits using Equation 8.17,

$$T_2 = T_{2\,min} + (T_{2\,max} - T_{2\,min}) * r(\ ) \tag{8.17}$$

#### 8.4.5.4 *Calculation of $t_2$*

The cold fluid mass flow rate is calculated randomly within the limits using Equation 8.18,

$$t_2 = t_{2\,min} + (t_{2\,max} - t_{2\,min}) * r(\ ) \tag{8.18}$$

**Table 8.8** PSO algorithm executed results.

| Sl no. | mh (kg/min) | mc (kg/min) | T1 (°C) | t1 (°C) | T2 (°C) | t2 (°C) | ΔT (°C) | Δt (°C) | Qh (W) | Qc (W) |
|---|---|---|---|---|---|---|---|---|---|---|
| **1** | 1.05 | 1 | 65 | 32 | 54.2 | 44.6 | 10.8 | 12.6 | 790.58 | 877.38 |
| **2** | 1.14 | 2.2 | 65 | 32 | 49 | 37.2 | 16 | 5.2 | 1271.63 | 796.61 |
| **3** | 1.32 | 3.18 | 65 | 32 | 50.3 | 37.2 | 14.7 | 5.2 | 1352.78 | 1151.46 |
| **4** | 1 | 4.2 | 65 | 32 | 53.4 | 38.9 | 11.6 | 6.9 | 808.71 | 2017.97 |
| **5** | 2 | 1.6 | 65 | 32 | 56 | 45 | 9 | 13 | 1254.90 | 1448.37 |
| **6** | 2.1 | 2.1 | 65 | 32 | 56.9 | 37.8 | 8.1 | 5.8 | 1185.88 | 848.13 |
| **7** | 2.3 | 3.05 | 65 | 32 | 54.8 | 36.5 | 10.2 | 4.5 | 1635.55 | 955.72 |
| **8** | 2.16 | 4.6 | 65 | 32 | 53.5 | 38 | 11.5 | 6 | 1731.76 | 1921.88 |
| **9** | 3.45 | 1.22 | 65 | 32 | 55 | 43.2 | 10 | 11.2 | 2405.23 | 951.47 |
| **10** | 3.08 | 2.16 | 65 | 32 | 59.6 | 39.9 | 5.4 | 7.9 | 1159.53 | 1188.22 |
| **11** | 3 | 3.52 | 65 | 32 | 57.9 | 39.6 | 7.1 | 7.6 | 1484.97 | 1862.83 |
| **12** | 3.3 | 4 | 65 | 32 | 54.2 | 36 | 10.8 | 4 | 2484.70 | 1114.13 |
| **13** | 4 | 1.03 | 65 | 32 | 55 | 41 | 10 | 9 | 2788.67 | 645.50 |
| **14** | 4.4 | 2.19 | 65 | 32 | 56.3 | 39.1 | 8.7 | 7.1 | 2668.75 | 1082.73 |
| **15** | 4.28 | 3.65 | 65 | 32 | 58.5 | 40 | 6.5 | 8 | 1939.52 | 2033.29 |

#### 8.4.5.5 *Calculation of Objective Function*

The optimized process parameters values are then substituted in the heat rate equations (Equations 8.6 and 8.8).

#### 8.4.5.6 *Calculation of Particle Best Value*

The maximum heat transfer rate for each initial solution is considered as the particle best value.

#### 8.4.5.7 *Calculation of Global Best Value*

The maximum heat transfer rate for the whole iteration is considered as the global best value.

### 8.4.6 Computational Result of PSO

Table 8.8 presents the PSO-obtained heat transfer rate and its corresponding parameters. The heat transfer rate obtained from the real-time experiment is maintained constant and 100 population sizes with 1000 iterations were executed to obtain optimized operating parameters.

## 8.5 Confirmation Experiments

The optimum process parameters obtained from the PSO are validated by repeating experiments. Table 8.9 presents the validation experiments results.

## 8.6 Results and Discussion

In this study, the PSO-based process parameter optimization of the CF-DPHE is tested. These initial real time experiments are conducted with the CP-DPHE rig at the heat and mass transfer laboratory, Saranathan college of Engineering. It is followed by the design of the PSO algorithm. For the 100 uniformly distributed population and 1000 iterations are executed for obtaining optimal solution. In order to validate such optimal results, confirmation experiments are conducted. Results obtained using such validation experiments are in line with the PSO algorithm predicted optimal results. This section presents the discussions related to the process parameter optimization of the CF-DPHE using the PSO algorithm.

**Table 8.9** The results of experimental validation.

| Sl no. | mh (kg/min) | mc (kg/min) | T1 (°C) | t1 (°C) | T2 (°C) | t2 (°C) | ΔT (°C) | Δt (°C) | Qh (W) | Qc (W) |
|---|---|---|---|---|---|---|---|---|---|---|
| **1** | 1.05 | 1 | 65 | 32 | 54 | 44.8 | 11 | 12.8 | 805.23 | 891.31 |
| **2** | 1.14 | 2.2 | 65 | 32 | 48 | 37.6 | 17 | 5.6 | 1351.11 | 857.88 |
| **3** | 1.32 | 3.18 | 65 | 32 | 50.6 | 37 | 14.4 | 5 | 1325.17 | 1107.17 |
| **4** | 1 | 4.2 | 65 | 32 | 53.1 | 38.2 | 11.9 | 6.2 | 829.63 | 1813.25 |
| **5** | 2 | 1.6 | 65 | 32 | 56.4 | 44.8 | 8.6 | 12.8 | 1199.13 | 1426.09 |
| **6** | 2.1 | 2.1 | 65 | 32 | 56.4 | 39.2 | 8.6 | 7.2 | 1259.08 | 1052.86 |
| **7** | 2.3 | 3.05 | 65 | 32 | 55 | 37 | 10 | 5 | 1603.48 | 1061.91 |
| **8** | 2.16 | 4.6 | 65 | 32 | 53.6 | 38.5 | 11.4 | 6.5 | 1716.70 | 2082.04 |
| **9** | 3.45 | 1.22 | 65 | 32 | 55.2 | 43 | 9.8 | 11 | 2357.12 | 934.48 |
| **10** | 3.08 | 2.16 | 65 | 32 | 59.2 | 40.5 | 5.8 | 8.5 | 1245.42 | 1278.47 |
| **11** | 3 | 3.52 | 65 | 32 | 58 | 39.9 | 7 | 7.9 | 1464.05 | 1936.36 |
| **12** | 3.3 | 4 | 65 | 32 | 54.2 | 36.6 | 10.8 | 4.6 | 2484.70 | 1281.25 |
| **13** | 4 | 1.03 | 65 | 32 | 54.7 | 40.9 | 10.3 | 8.9 | 2872.33 | 638.33 |
| **14** | 4.4 | 2.19 | 65 | 32 | 56.8 | 38 | 8.2 | 6 | 2515.38 | 914.98 |
| **15** | 4.28 | 3.65 | 65 | 32 | 58.1 | 40 | 6.9 | 8 | 2058.87 | 2033.29 |

### 8.6.1 Initial Experiments

Initially, 15 set of experiments are designed by trial and error method. Such experiments are specifically different in terms of mass flow rate (1 to 4 kg/min). Figures 8.2 to 8.5 present the variation in between (mh, mc,T2, t2, Qh, Qc). Table 8.7 presents the response obtained from the initial trial and error experiment. From Figure 8.4, it is observed that experiment-13 with mh = 4, T1 = 65°C, and T2 = 54.8°C resulted in the maximum rate of hot side heat transfer of 2844 W. This is mainly due to maximum product value of the hot fluid mass flow rate and its heat carrying capacity ($C_{ph}$). From Figure 8.5, it is also observed that experiment-4 with mc = 4, t1 = 32°C, and t2 = 39.3°C resulted in the maximum cold fluid side heat transfer rate of 2033 W. This is mainly due to maximum product value of cold fluid mass flow rate and its heat carrying capacity ($C_{pc}$).

### 8.6.2 The PSO Analysis

The results of trial and error experimental analysis are considered for designing the PSO algorithm. Such algorithm is coded using MATLAB R2016a software. Each and every experiment is allowed to proceed with the PSO steps listed in Section 8.4.3. Figures 8.2 to 8.5 show the relationship

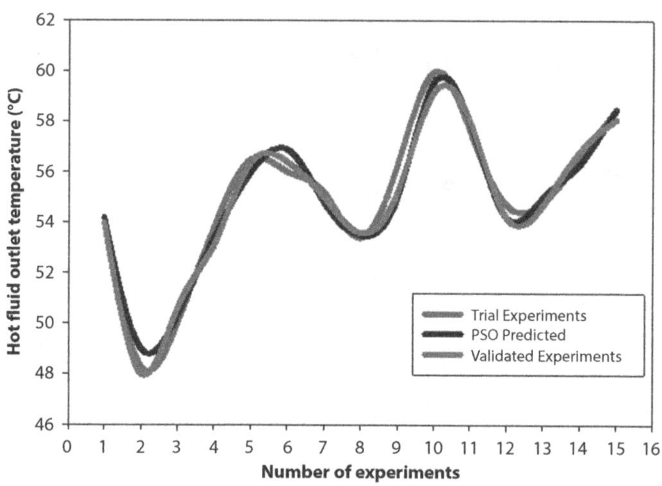


**Figure 8.2** Variations in the hot fluid outlet temperature with respect to number of experiments (with different combinations of mass flow rate).

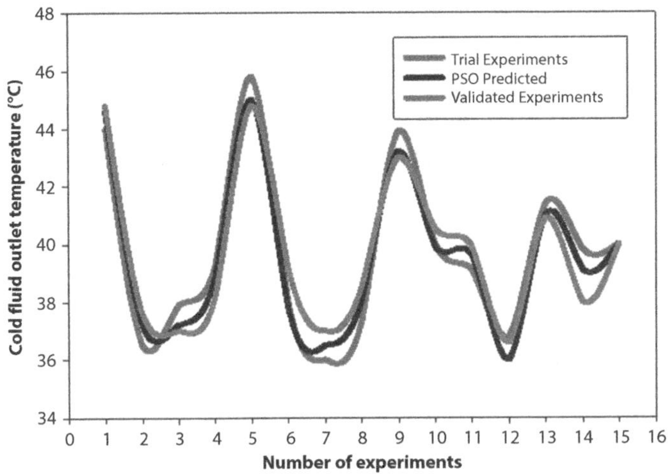


**Figure 8.3** Variations in the cold fluid outlet temperature with respect to number of experiments (with different combinations of mass flow rate).

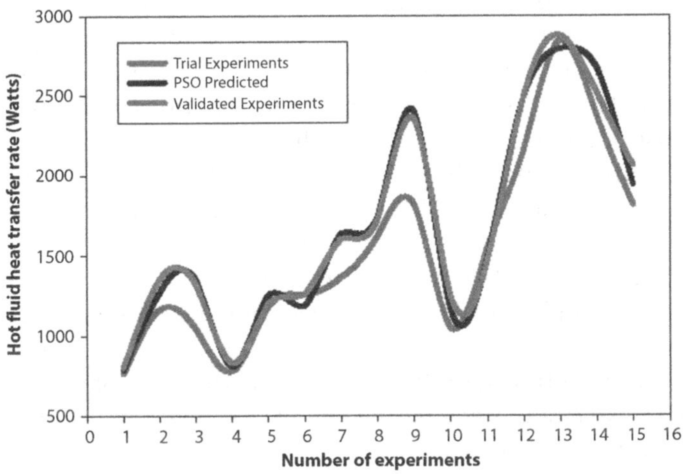


**Figure 8.4** Variations in the hot fluid side heat transfer rate with respect to number of experiments (with different combinations of mass flow rate).

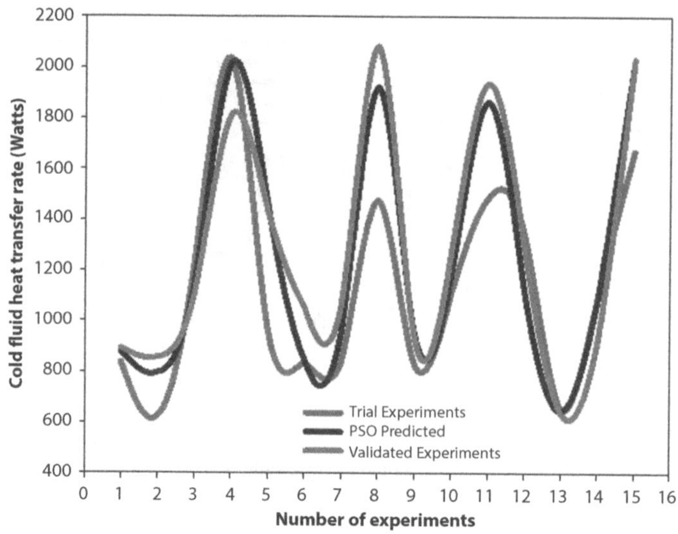


**Figure 8.5** Variations in the cold fluid side heat transfer rate with respect to number of experiments (with different combinations of mass flow rate).

between the mass flow rate, T2, t2, Qh, and Qc. For the same input parameters, the heat transfer rate obtained by PSO and validation experiment is the same. At the same time, the cold fluid heat transfer rate obtained by the PSO and validation experiments has minor deviations. Table 8.8 presents the response obtained from the PSO technique.

From Figure 8.4 and Table 8.8, it is observed that experiment-13 with mh = 4, T1 = 65°C, and T2 = 55°C resulted in the maximum rate of hot side heat transfer of 2788.67 W. The value of maximum hot side heat transfer rate predicted by the PSO shows a deviation of 1.9% with the real time experiments. From Figure 8.5 and Table 8.8, it is also observed that experiment-4 with mc = 3.65, t1 = 32°C, and t2 = 40°C resulted in the maximum cold fluid side heat transfer rate of 2033.29 W. The value of maximum cold side heat transfer rate predicted by the PSO shows a deviation of 0.01% with the real-time experiments.

### 8.6.3 The Validation Experiments

The results of the PSO analysis are then considered for validating the optimal result prediction accuracy of the proposed algorithm. From Figure 8.4 and Table 8.8, it is observed that experiment-13 with mh = 4, T1 = 65°C, and T2 = 54.7°Cq resulted in the maximum rate of hot side heat transfer of 2872.33 W. The value of maximum hot side heat transfer rate obtained by the validation experiment shows a deviation of only 0.98% with the PSO obtained results. From Figure 8.5 and Table 8.8, it is also observed that experiment-8 with mc = 4.6, t1 = 32°C, and t2 = 38.5°C resulted in the maximum rate of hot side heat transfer of 2082.04 W. The value of maximum cold side heat transfer rate obtained by the validation experiment shows a deviation of only 2.34% with the PSO-obtained results.

As a whole, for the CF-DPHE under this study, the PSO suggested optimal input parameters to attain the maximum heat transfer rates (Qh and Qc), which are: Mh =4, T1 = 65°C, mc = 4.6, and t1=32°C.

## 8.7 Conclusions

In general, the desired heat transfer rate in a DPHE can be achieved by trial and error method, but such method is a time consuming one. There is a need to design a model that is capable of finding the optimum process parameters for maximizing the heat transfer rate. The following points were observed as conclusion from the present PSO base process optimization work. Particle swarm optimization provides an efficient and systematic methodology for process parameter optimization:

- ✓ Validation experiments confirm the efficiency of the PSO approach in determining the optimal operating process parameters.
- ✓ The proposed PSO technique is capable of obtaining a global optimum solution in less than 4 s.
- ✓ The PSO algorithm is problem-independent and completely generalized, so that it can be instantly modified to any thermal parameter optimization.
- ✓ Both heat transfer rate equation developed can be used for obtaining optimal operating process parameters.
- ✓ Heat transfer rate can be improved further by the proposed approach instead of a thumb-rule judgement.
- ✓ The product (m x Cp) has a greater influence on the heat transfer rate.
- ✓ Use of higher hot and cold fluid mass flow rate is recommended to obtain maximum heat transfer rate.

## References


1. Sridharan, M., Predicting Performance of Double-Pipe Parallel- and Counter-Flow Heat Exchanger Using Fuzzy Logic. *ASME J. Therm. Sci. Eng. Appl.*, 12, 3, 031006, 2020, 10.1115/1.4044696.
2. Mariani, V.C., Duck, A.R.K., Guerra, F.A., dos Santos Coelho, L., Rao, R.V., A chaotic quantum-behaved particle swarm approach applied to optimization of heat exchangers. *Appl. Therm. Eng.*, 42, 119–128, 2012.
3. Patel, V.K. and Rao, R.V., Design optimization of shell-and-tube heat exchanger using particle swarm optimization technique. *Appl. Therm. Eng.*, 30, 1417–1425, 2010.
4. Guo, D., Liu, M., Wang, J., Xie, L., Optimization in Plate-fin Safety Structure of Heat Exchanger Using Genetic and Monte Carlo Algorithm. *Appl. Therm. Eng.*, 2014.
5. Silva, A.P., Ravagnani, M.A.S.S., Biscaia Jr., E.C., Particle Swarm Optimisation Applied in Retrofit of Heat Exchanger Networks.
6. Sadeghzadeh, H., Ehyaei, M.A., Rosen, M.A., Techno-economic optimization of a shell and tube heat exchanger by genetic and particle swarm algorithms. *Energy Convers. Manage.*, 93, 84–91, 2015.
7. Zarea, H., Kashkooli, F.M., Soltani, M., Rezaeian, M., A novel single and multi-objective optimization approach based on Bees Algorithm Hybrid with Particle Swarm Optimization (BAHPSO): Application to thermal-economic design of plate fin heat exchangers. *Int. J. Therm. Sci.*, 129, 552–564, 2018.
8. Rao, R.V. and Patel, V.K., Thermodynamic optimization of cross flow plate-fin heat exchanger susing a particle swarm optimization algorithm. *Int. J. Therm. Sci.*, 49, 1712–1721, 2010.
9. Jyothiprakash, K.H., Harshith, J., Sharan, A., Seetharamu, K.N., Krishnegowda, Y.T., Thermodynamic Optimization of Three-Fluid Cross-Flow Heat Exchanger Using GA and PSO Heuristics. *Therm. Sci. Eng. Prog.*, 11, 289–301, 2019.
10. Esfe, M.H., Mahian, O., Hajmohammad, M.H., Wongwises, S., Design of a heat exchanger working with organic nanofluids using multi-objective particle swarm optimization algorithm and response surface method. *Int. J. Heat Mass Transfer*, 119, 922–930, 2018.
11. Zarea, H., Kashkooli, F.M., Soltani, M., Rezaeian, M., A novel single and multi-objective optimization approach based on Bees Algorithm Hybrid with Particle Swarm Optimization (BAHPSO): Application to thermal-economic design of plate fin heat exchangers. *Int. J. Therm. Sci.*, 129, 552–564, 2018.
12. Bharathi Raja, S. and Baskar, N., Particle Swarm Optimization Technique for determining optimal machining parameters of different work piece materials in turning operation. *Int. J. Adv. Manuf. Technol.*, 54, 445–463, 2011.

13. Sridharan, M., Application of fuzzy logic expert system in predicting cold and hot fluid outlet temperature of counter-flow double-pipe heat exchanger, in: *Advanced Analytic and Control Techniques for Thermal Systems with Heat Exchangers*, pp. 307–323, Academic Press, 2020, ISBN 978-0-12-819422-5.
14. Sridharan, M., Applications of artificial intelligence techniques in heat exchanger systems, in: *Advanced Analytic and Control Techniques for Thermal Systems with Heat Exchangers*, pp. 325–334, Academic Press, 2020, ISBN 978-0-12-819422-5.
15. Sridharan, M., Verification and Validation of Solar Photovoltaic Thermal Water Collectors Performance Using Fuzzy Logic. *J. Verif. Valid. Uncertain. Quantif.*, 12, 3, 1–8, 2019.
16. Sridharan, M., Jayaprakash, G., Chandrasekar, M., Vigneshwar, P., Paramaguru, S., Amarnath, K., Prediction of Solar Photovoltaic/ Thermal Collector Power Output Using Fuzzy Logic. *J. Sol. Energy Eng.*, 140, 6, 1–6, 2018.
17. Sridharan, M., Application of Generalized Regression Neural Network in Predicting the Performance of Natural Convection Solar Dryer. *J. Sol. Energy Eng.*, 142, 3, 1–8, 2020.
18. Sridharan, M., Application of Generalized Regression Neural Network in Predicting the Performance of Solar Photovoltaic Thermal Water Collector. *Ann. Data Sci.*, 2020, https://doi.org/10.1007/s40745-020-00273-1.
19. Shribalaji Shenbagaraj, S.M., Application of Generalized Regression Neural Network in Predicting the Thermal Performance of Solar Flat Plate Collector Systems. *J. Therm. Sci. Eng. Appl.*, 13, 2, 1–11, 2020.

# 9

# Application of Geothermal Energy-Based Earth-Air Heat Exchanger in Sustainable Buildings

**Arijit Kundu**

*Department of Mechanical Engineering, Jalpaiguri Government Engineering College, Jalpaiguri, West Bengal, India*

***Abstract***

The global pursuit of sustainable and environmentally responsible building requirements has accelerated innovations in energy-efficient technologies. This chapter discuss the augmentation of solar, geothermal, and Earth-air heat exchanger systems within the context of sustainable building designs and operations. By utilizing the renewable energy sources and optimizing the design of heat exchanger, it is aimed to reduce the environmental impact of buildings without hampering the occupant comfort. This chapter also provides a comprehensive overview of the solar energy harnessing process through photovoltaic panels and solar thermal collectors, and the geothermal systems that are taped into the Earth's stable temperature to provide heating and cooling. The effect of integration of these systems within the building's HVAC systems is discussed in detail.

***Keywords:*** Sustainable buildings, energy-efficient technologies, solar energy, geothermal energy, earth-air heat exchanger, renewable energy, HVAC systems, occupant comfort

## 9.1 Introduction to Sustainable Building

To secure future well-being, we must cut our ecological footprint or environmental impact, in the built environment by two thirds in order to reach a sustainable and globally fair level. The only way to do this is through

*Email:* arijit.kundu@me.jgec.ac.in; ORCID: 0000-0003-0719-5907

Ashwani Kumar, Nitesh Dutt and Mukesh Kumar Awasthi (eds.) Heat Transfer Enhancement Techniques: Thermal Performance, Optimization and Applications, (221–232) © 2025 Scrivener Publishing LLC

sustainable design. Decision-makers frequently believe that conservation of energy, water, and material resources is the primary focus of sustainable design. With an emphasis on the environment, sustainable design combines the cutting-edge technologies of today with the low-tech, natural conditioning solutions of the past (natural ventilation, solar heat, and daylight) to create an integrated, "intelligent" system that promotes individual control while negotiating for resource consciousness and environmental quality [1]. Buildings that are smarter and more efficient are desperately needed. The pinnacle of the intelligent building concept is integration between the building and its end users. It takes skill to "connect" a building and its end user. The study and development of user-connectivity—the amalgamation of usability and user interface—continues. People can engage with one another and the building using information and communication technology (Clements-Croome 1997). Building automation has emerged as a critical component for achieving desired comfort levels while consuming the least amount of energy. From basic thermostatic controls, it developed into a specialist specialty employing the most recent advances in data-communication and control algorithm technologies. To maximize building comfort, vital information about the condition and functionality of the equipment is collected. Additional optimization seeks to lower energy usage without sacrificing indoor comfort. When analyzing the thermal exchanges that occur between a building and its surroundings and the energy consumption that results, it is crucial to take human behavior into account [2]. Users' contentment and the degree of comfort they experience are directly correlated with their ability to make decisions and manage their environment. Up to 20% of current energy consumption might be saved through dynamic online steering of building management and individual comfort control [3]. It is necessary to further integrate users with the existing systems. The idea of intelligent software agents (ISA) is an excellent way to realize further building system optimization and integration. Software agents provide an extremely versatile idea for the integration of optimization at various levels because of their modular structure, independent operation, and communication capabilities. Over the past 20 years, ISA concepts have been created and utilized in a wide range of industries. To carry out its duties, an ISA can interact and bargain with other ISAs as well as access resources. The requirements and design of the building systems are specified using the structural concept of IAS.

## 9.2 System Approach for Complex System Study

Most real-world issues pertaining to the environment and sustainable development can be categorized as complex composite issues. As such, they have certain unique characteristics that necessitate the study of these issues using interdisciplinary techniques. A system is the integration of linked elements and components that interact with the environment as a whole and give rise to new properties through informational, mechanical, mechanical, energy, and other exchanges. Any component that is removed from the system loses its unique qualities and transforms into a variety of parts or assemblies. To effectively research complex systems, one must adhere to the following system analysis principles: 1. A system description. Characterization of its primary characteristics and aspects. 2. Examination of the relationships between system components, including their functional relationships and informational, physical, dynamic, and temporal interactions. 3. Analysis of how a system interacts with its surroundings, including other systems and the natural world. 4. System division and dismantling. Partitioning indicates the extraction of parallel system parts, whereas decomposition assumes the extraction of a sequence of system parts. These techniques may be based on content analysis, which divides the system into sections based on function analysis or physical partitioning, or cluster analysis, which is an iterative process of integrating system pieces into groups. 5. Analysis of every subsystem or component of the system using the best available tools (multidisciplinary approaches, problem-solving techniques, expert advice, knowledge discovery tools, etc.) 6. Combining the outcomes from the previous phase and creating a unified body of knowledge about the system. Formal design techniques, multi-criteria optimization techniques, decision-based and hierarchical design techniques, artificial intelligence approaches, case-based reasoning, and other techniques like hybrid methods can all be used to synthesize knowledge and compose an entire system model [4].

## 9.3 Earth-to-Air Heat Exchanger for Sustainable Buildings

Scientists from all around the world have been studying energy conservation, sophisticated energy conversion methods, and hybrid energy storage systems as a result of the conflict between the worsening energy crisis and the steadily rising energy demand. Buildings make up over 40% of the world's total primary energy use, according to the research in Reference [5]. The energy consumption of the heating, ventilation, and air conditioning (HVAC) systems accounts for the biggest sharing percentage of the building's overall energy consumption [6], particularly given the rising demands of modern civilization on the indoor thermal environment.

Renewable energy sources, such as solar and geothermal, are promising in terms of supporting renewable and sustainable development and can help lower the primary energy consumption in HVAC systems [7]. Because geothermal energy is abundant, clean, and flexible, its application potential is promising. Furthermore, it may produce renewable energy anywhere in the world and is not limited to any one nation. The earth-to-air heat exchanger (EAHE), a common renewable energy utilization facility for geothermal energy applications in buildings, has gained popularity over the past few decades due to its energy efficiency, energy savings, and environmental friendliness. The EAHE typically comprises of subterranean 3-m network pipes and a blower. The main way that the EAHE functions is that the relatively constant temperature of the subsurface (which is equal to the ground surface's yearly average) allows the subsurface to pre-cool or pre-heat ambient air, which in turn lowers the cooling load on the HVAC system [8]. Earth-to-air heat exchanger systems have been extensively used in a variety of building types over the past few decades, including residential structures, animal housing, and agricultural greenhouses. Li *et al.* [9] methodically combined a solar chimney with an EAHE for a facility building in order to make the most of the renewable energy. Their findings indicate that the mass flow rate of the outdoor air, which is handled by an EAHE within the inside thermal comfort range, can be enhanced to 1000 $m^3/h$ by methodically integrating the solar chimney. While EAHE systems have demonstrated significant cooling capabilities and energy-saving potential in studies, a number of technical issues still need to be resolved before they can be widely used and become more reliable. The comparatively low wintertime outlet air temperature is one of the technological problems. Global research indicates that in the winter, EAHE system output air temperatures are typically below 18°C, meaning that buildings cannot be directly heated by these temperatures. Reheating the EAHE system's outlet air can be accomplished in part by combining it with an electric auxiliary heating system. Furthermore, the over-size heat recovery unit was more effective in damping the annual meteorological oscillation than were the EAHE pipes when they were longer. And, despite the electric auxiliary heating equipment demonstrates reheating potentials, a reasonably considerable amount of electrical energy was still utilized for the direct reheating purpose during the system operating process. Several researchers attempted to install different passive air-based solar thermal devices at the end of EAHE pipes in an effort to lower the amount of direct electric energy consumed. Haghighi and Maerefat [10] replaced the conventional fan and auxiliary electric heater with a solar air chimney at the end of the EAHE pipes in order to lower the electricity usage of the fans and the auxiliary electric heater. The EAHE system may accomplish indoor thermal comfort even in extreme weather conditions (*i.e.*, 0°C ambient temperature and 250 $W/m^2$ solar radiation) with the aid of a solar air chimney. There was an EAHE-building integrated photovoltaic (BIPV) system which had a negligible impact on the EAHE system's outlet air temperature due to the low ambient air temperature and low solar radiation intensity. Peng *et al.* [11] created a hybrid system to enhance the indoor air quality and thermal climate that consists of an EAHE and a sunspace with plat. The system's ability to maintain an indoor air temperature between 15°C and 27°C while using 12% less electricity than the existing system was demonstrated by the results. Phase change materials (PCMs) are highly utilized in passive solar chimney systems to reduce building energy consumption because they are sophisticated energy storage materials with significant latent heat density. Since the EAHE system's integrated PCMs maintain a relatively constant temperature and significant latent heat density throughout the charging and discharging cycles, they have also drawn increasing attention.

Due to its passive character of operation, forecasting the applicability of Earth air heat exchanger (EAHX) is widely dependent on environmental variables. Its long-term functioning and cheap maintenance costs are two good reasons to use it. Important factors influencing the EAHX performance are the building design, soil type, water table, relative humidity, latitude and height of the location, and ambient air temperature. Robust, moist soil that receives a lot of shade generally boosts EAHX efficiency. Earth air heat exchanger (EAHX) is ineffective in achieving thermal comfort in hot, humid regions. Moreover, among its operating constraints are a high-water table, a lack of space, and shallow bedrock. Even with the aforementioned challenges to EAHX use, it remains a compelling substitute to lessen or do away with the requirement for traditional conditioning systems. Combining a solar chimney with an earth-air heat exchanger is a creative idea. As a result, the connected system minimizes summertime peak power demand as well as wintertime energy use. A solar collector system in conjunction with an earth air heat exchanger was proposed by Maerefat and Haghighi [12]. The purpose of the system ventilation was to use the earth as a sink and solar radiation as the heat source. Figure 9.1 depicts the system's schematic. An absorber wall that served as a collecting surface and a glass surface facing south were features of the roof solar chimney layout. Inside the rectangular enclosure, the air was getting hot. The force behind the removal of air from the room was the buoyancy. The pre-cooled air was delivered to the room by the EAHX, which was made up of horizontal pipes dug several meters below the surface of the earth. Based on the Adapted Comfort Standard, the research attempted

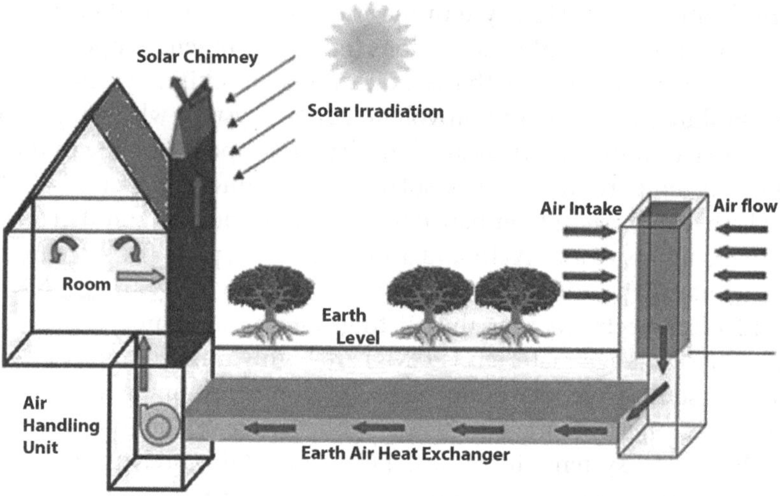


**Figure 9.1** Working of earth coupled solar chimney.

to determine the system's capacity to provide favorable indoor conditions (ACS). Separate mathematical models for EAHX and solar collector (SC) were created. The parametric study demonstrated that using the ideal cooling pipe diameter of 0.5 m resulted in the smallest number of SCs and EAHXs that were needed. More cooling was achieved by extending and widening the conduit. It was found that intense solar radiation caused the ACH number and interior air temperature to rise, while the EAHX values used had to go up. Moreover, the chimney effect's ability to circulate air was diminished in high outside temperatures. In these conditions, increasing the SC count would be beneficial. With the right system adjustments, it was able to achieve thermal comfort even in adverse environmental conditions, such as high air temperature or low sun intensity. However, the solar chimney was no longer appropriate when the outside air temperature reached 45°C or above. It should be mentioned that mold infestation and humidity present additional design challenges for EAHXs. There are claims that it could endanger the residents' health. Twelve EAHX with varying designs, pipe materials, sizes, and ages were examined by Wagner *et al.* [13]. The microbial proliferation in the underground pipes was a worry for the investigation. Remarkably, after passing through the pipe system, the airborne quantities of viable spores and germs, for the most part, even decreased. According to the investigations, EAHX operation was deemed appropriate provided that routine controls were implemented or cleaning facilities were accessible. To eliminate the risk, it is strongly advised that the EAHX pipe have an antibacterial material layer. You will see that having an ideal condensation drain for the subterranean cooling pipes is crucial. Therefore, whether or not an antimicrobial layer is employed, the pipes should be positioned at a 4° slope to guarantee the continuous removal of condensed water from the pipes. In order to completely understand the performance of an EAHX, extra aspects such as the influence of ice-formation in winter could be a subject of the future research. Table 9.1 summarizes a subset of the EAHX-SC topic research. It outlines each case's key conclusions as well as its shortcomings.

## 9.4 EAHE Performance Evaluation: Numerical Method

Calculating the temperature and moisture mass inside the buried tube and circumferential soil, Pfafferott [15] and Santamouris *et al.* [16] extended the work of Puri [17] to figure out an axi-symmetric model of the transient heat and mass transfer phenomena for the prediction of EAHE performance beneath the surface. Simulation of ground heat exchanger along with the air can be carried out simultaneously for the heat and moisture mass transfer using mass and energy conservation equations. The applied boundary conditions are energy and mass balance at the interfaces of soil and surrounding air, and control volume between the ground heat exchanger and air flow.

The energy and mass conservation Equations 9.1 and 9.2 represent moisture transfer with phase change in:

$$\frac{\partial(\rho CT)}{\partial t} = \nabla\left(\left(k + L\rho_l\alpha_v\right)\nabla T\right) + \nabla\left(\left(L\rho_l\alpha_{\Theta,v}\right)\nabla\Theta\right) + q_V \quad (9.1)$$

$$\frac{\partial\Theta}{\partial t} = \nabla\left(\left(\alpha_l + \alpha_v\right)\nabla T\right) + \nabla\left(\left(\alpha_{\Theta,l} + \alpha_{\Theta,v}\right)\nabla\Theta\right) + \frac{\partial K_h}{\partial z} + \Theta_V \quad (9.2)$$

The moisture diffusivities are defined as follows:

$$\alpha_l = K_h\frac{\partial\varphi}{\partial T_z},\ \alpha_{\Theta,l} = K_h\frac{\partial\varphi}{\partial\Theta} \quad (9.3)$$

**Table 9.1** A summary of selected research on EAHX-SC.

| **Method** | **Environment** | **Use** | **Conclusions** | **Restrictions** | **Reference** |
|---|---|---|---|---|---|
| Analytical method | Semi-arid, continental climate | Space cooling | Ideal dimensional specifications for an EAHX-SC system. | 1. Model operates in steady state condition.<br>2. Environmental factors such as soil type and thermal saturation were not taken into consideration.<br>3. Comparing the outcomes of a dynamic operation to the analytically obtained answer may yield contradicting findings. | [12] |
| Experimental method | Hot summer, continental climate | Space cooling | System's overall cooling capacity nearly met the cooling load specified for the building. | Not suited for residential buildings due to insufficient space. | [9] |
| Analytical method | Semi-arid, continental climate | Space heating | Effective dimensions of the system. | 1. Model operates in steady state condition.<br>2. No explanation given on the system's economic viability. | [14] |

$$\alpha_v = \alpha\tau(\Theta_s - \Theta)\frac{1}{\rho_l}\frac{\partial \rho_v}{\partial T_z},\ \alpha_{\Theta,v} = \alpha\tau(\Theta_s - \Theta)\frac{1}{\rho_l}\frac{\partial \rho_v}{\partial \Theta} \tag{9.4}$$

$$\varphi = \varphi_s\left(\frac{\Theta}{\Theta_s}\right)^m,\ K_h = K_{h_s}\left(\frac{\Theta}{\Theta_s}\right)^{2m+3} \tag{9.5}$$

The matrix potential, $\varphi$, and the hydraulic conductivity of soil, $K_h$, are given by the pedo-transfer functions of moisture content by Equation 9.5, m is a constant depending on soil type [18, 19].

The initial temperature of the soil at the depth z (m) and any time t (day) with the value of the parameters taken as mentioned earlier are given by Equation 9.6 [20]. The heat conduction theory based on annual earth below-surface temperature taking a semi-infinite homogenous solid [20] has been mathematically modeled by Kusuda and Achenbach [21], as shown in Equation 9.6.

$$T_g(z,t) = T_m - A_{surf}exp^{\left[-z\left(\frac{\pi}{8760\alpha}\right)^{0.5}\right]}\cos\left[\frac{2\pi}{8760}\left(t - t_0 - \frac{z}{2}\left(\frac{8760}{\pi\alpha}\right)^{0.5}\right)\right] \tag{9.6}$$

## 9.5 Discussion

This chapter presents recent research on developing EAHE and evaluating its performance on a worldwide scale. Earth-to-air heat exchangers (EAHE) can include atmospheres like the Mediterranean, humid subtropical, hot, arid regions, extremely cool, or oceanic. By preheating a building's air during the winter and the opposite during the summer, this helps reduce energy use. They provide a significant reduction in greenhouse gas emissions and the cooling/heating load of buildings [22–25]. Every EAHE system performs better when the depth and length of the buried pipe are increased, but less well when the pipe diameter and air velocity are increased. If the cooling or heating is insufficient, a conventional air conditioner can be connected to the EAHEs' outlet. The thermo-physical behavior of earth-air heat exchangers is simulated by a number of calculation models that may be found in the literature. A well-designed EAHE can lower a typical house's electricity use by 30% [26–28]. The EAHE systems, which have been widely utilized for years, offer decreases in the heating/cooling load of buildings, electricity consumption, chlorofluorocarbon (CFC) and hydrochlorofluorocarbon (HCFC) usage, and greenhouse gas emissions. The thermal performance of an EAHE system typically rises with increasing pipe length and burial depth while falling with increasing pipe diameter and air velocity.

It is possible to draw the conclusion that effective usage of EAHE systems in conjunction with environmentally friendly energy sources and cutting-edge technology will be crucial in reducing energy consumption and protecting the environment not just in India but globally. In light of this, the authors hope that this review study will be of great use to scientists and researchers working on projects involving the passive heating and cooling of buildings, particularly where EAHE systems are included [29–32].

## References

1. Loftness, V., Hartkopf, V., Gurtekin, B., Hua, Y., Qu, M., Snyder, M., Gu, Y., Yang, X., Building Investment Decision Support (BIDSTM) Cost-Benefit Tool to Promote, High Performance Components, in: *Flexible Infrastructures & Systems Integration for Sustainable Commercial Buildings and Productive Organizations*, pp. 12–31, AIA Pilot Report on University Research, 2005.
2. Palme, M., Isalgue, A., Coch, H., Serra, R., Robust design: a way to control energy use from human behaviour in architectural spaces, in: *Proceedings PLEA 2006*, Geneva, September 6-8, pp. 1–311, 2006.
3. Akkermans, H., Being Smart In Information Processing, in: *Technological And Social Challenges And Oppertunities, Proceedings IFIP IIP 2002*, 2002.
4. Santamouris, M., Gaitani, N., Spanou, A., Saliari, M., Giannopoulou, K., Vasilakopoulou, K., Kardomateas, T., Using cool paving materials to improve microclimate of urban areas - Design realization and results of the flisvos project. *Build. Environ.*, 53, 128–136, 2012.
5. Liu, Z., Yu, Z., Yang, T., Qin, D., Li, S., Zhang, G., Haghighat, F., Joybari, M.M., A review on macro-encapsulated phase change material for building envelope applications. *Build. Environ.*, 144, 281–294, 2018.
6. Chua, K.J., Chou, S.K., Yang, W.M., Yan, J., Achieving better energy-efficient air conditioning – A review of technologies and strategies. *Appl. Energy*, 104, 87–104, 2013.
7. Zhou, Y., Zheng, S., Liu, Z., Wen, T., Ding, Z., Yan, J., Zhang, G., Passive and active phase change materials integrated building energy systems with advanced machine-learning based climate-adaptive designs, intelligent operations, uncertainty-based analysis and optimisations: A state-of-the-art review. *Renew. Sustain. Energy Rev.*, 130, 109889, 2020.

8. Soni, S.K., Pandey, M., Bartaria, V.N., Energy metrics of a hybrid earth air heat exchanger system for summer cooling requirements. *Energy Build.*, 129, 1–8, 2016.
9. Li, H., Yu, Y., Niu, F., Shafik, M., Chen, B., Performance of a coupled cooling system with earth-to-air heat exchanger and solar chimney. *Renew. Energy*, 62, 468–477, 2014.
10. Bisoniya, T.S., Kumar, A., Baredar, P., Experimental and analytical studies of earth–air heat exchanger (EAHE) systems in India: A review. *Renew. Sustain. Energy Rev.*, 19, 238–246, 2013.
11. Zhen, P., Wu, D., Tenorio, R., An integrated low-energy ventilation system to improve indoor environment performance of school buildings in the cold climate zone of China. *Build. Environ.*, 182, 107–108, 2020.
12. Maerefat, M. and Haghighi, A., Passive cooling of buildings by using integrated earth to air heat exchanger and solar chimney. *Renew. Energy*, 35, 2316–24, 2010.
13. Wagner, R., Beisel, S., Spieler, A., Vajen, K., Gerber, A., Measurement, modelling and simulation of an earth-to-air heat exchanger in Manburg (Germany), in: *Proceedings of the ISES Europe solar congress*, Kopenhagen, Dänemark, 2000.
14. Haghighi, A. and Maerefat, M., Design guideline for application of earth-to-air heat exchanger coupled with solar chimney as a natural heating system. *Int. J. Low-Carbon Technol.*, 10, 294–304, 2015.
15. Pfafferott, J., Evaluation of earth-to-air heat exchangers with a standardised method to calculate energy efficiency. *Energy Build.*, 35, 971–83, 2003.
16. Santamouris, M., Mihalakakou, G., Balaras, C., Argiriou, A., Asimakopoulos, D., Vallindras, M., Use of buried pipes for energy conservation in cooling of agricultural greenhouses. *Sol. Energy*, 55, 2, 111–24, 1995.
17. Puri, V.M., Heat and mass transfer analysis and modeling in unsaturated ground soils for buried tube systems. *Energy Agric.*, 6, 179–93, 1987.
18. Campbell, J.D., Pore pressures and volume changes in unsaturated soils, PhD Thesis, University of Illinois at Urbana-Champaign, 1973.
19. Clapp, R.B. and Hornberger, G.M., Empirical equations for some soil hydraulic properties. *Water Resour. Res.*, 14, 601–4, 1977.
20. Carslaw, H.S. and Jaeger, J.C., *Conduction of heat in solids*, 2nd ed., Clarendon Press, Oxford, 1959.
21. Kusuda, T. and Achenbach, P.R., Earth temperature and thermal diffusivity at selected stations in United States. *ASHRAE Trans.*, 71, 1965.
22. Kumar, A., Singh, V.P., Meena, C.S., Dutt, N., *Thermal Energy Systems: Design, Computational Techniques and Application*, Ist edition, CRC Press, 2023, https://doi.org/10.1201/9781003395768.
23. Kumar, A., Gori, Y., Dutt, N., Singla, Y.K., Maurya, A. (Eds.), *Advanced Computational Methods in Mechanical and Materials Engineering*, 1st ed., CRC Press, 2021, https://doi.org/10.1201/9781003202233.

24. Meena, C.S., Prajapati, A.N., Kumar, A., Kumar, M., Utilization of Solar Energy for Water Heating Application to Improve Building Energy Efficiency: An Experimental Study. *Buildings*, 12, 2166, 2022, https://doi.org/10.3390/buildings12122166.
25. Singh, V.P., Jain, S., Karn, A., Kumar, A., Experimental Assessment of Variation in Open Area Ratio on Thermohydraulic Performance of Parallel Flow Solar Air Heater. *Arabian J. Sci. Eng.*, 48, 11695–11711, 2023, https://doi.org/10.1007/s13369-022-07525-7.
26. Verma, V., Meena, C.S., Thangavel, S., Kumar, A., Choudhary, T., Dwivedi, G., Ground and solar assisted heat pump systems for space heating and cooling applications in the northern region of India – A study on energy and CO2 saving potential. *Sustain. Energy Technol. Assess.*, 59, 103405, 2023, ISSN 2213-1388, https://doi.org/10.1016/j.seta.2023.103405.
27. Meena, C.S., Kumar, A., Jain, S., Rehman, A.U., Mishra, S., Innovation in Green Building Sector for Sustainable Future. *Energies*, 15, 6631, 2022, https://doi.org/10.3390/en15186631.
28. Pant, G., Meena, C.S., Saxena, A., Kumar, A., Singh, V.P., Dutt, N., Study the Temperature Variation in Alternate Coils of Insulated Condenser Cum Storage Tank: Experimental Study, in: *Advances in Fluid and Thermal Engineering. FLAME 2022.* Lecture Notes in Mechanical Engineering, B.S. Sikarwar, S.K. Sharma, A. Jain, K.M. Singh, (Eds.), Springer, Singapore, 2023, https://doi.org/10.1007/978-981-99-2382-3_52.
29. Saxena, A., Prajapati, A.N., Pant, G., Meena, C.S., Kumar, A., Singh, V.P., Water Consumption Optimization of Hybrid Heat Pump Water Heating System, in: *Recent Advances in Mechanical Engineering. FLAME 2022.* Lecture Notes in Mechanical Engineering, A.K. Shukla, B.P. Sharma, A. Arabkoohsar, P. Kumar, (Eds.), Springer, Singapore, 2023, https://doi.org/10.1007/978-981-99-1894-2_61.
30. Kundu, A., Kumar, A., Dutt, N., Meena, C.S., Singh, V.P., Introduction to Thermal Energy Resources and their Smart Applications, in: *Thermal Energy Systems: Design, Computational Techniques and Applications*, A. Kumar, V.P. Singh, C.S. Meena, N. Dutt, (Eds.), pp. 1–15, CRC Press, Boca Raton, FL, USA, 2023, Chapter 1; http://dx.doi.org/10.1201/9781003395768-1.
31. Dewangan, A.K., Moinuddin, S.Q., Cheepu, M., Sajjan, S.K., Kumar, A., Thermal Energy Storage: Opportunities, Challenges and Future Scope, in: *Thermal Energy Systems: Design, Computational Techniques and Applications*, A. Kumar, V.P. Singh, C.S. Meena, N. Dutt, (Eds.), pp. 17–28, CRC Press, Boca Raton, FL, USA, 2023, Chapter 2; http://dx.doi.org/10.1201/9781003395768-2.
32. Kundu, A., Kumar, A., Dutt, N., Singh, V.P., Meena, C.S., Modelling and Simulation of Thermal Energy System for Design Optimization, in: *Thermal Energy Systems: Design, Computational Techniques and Applications*, A. Kumar, V.P. Singh, C.S. Meena, N. Dutt, (Eds.), pp. 103–140, CRC Press, Boca Raton, FL, USA, 2023, Chapter 7; http://dx.doi.org/10.1201/9781003395768-7.

# 10

# Numerical Study of Solar Air Heater with Semi-Cylindrical Tube Roughness

**Ankush Hedau*** **and S. K. Singal**

*Department of Hydro and Renewable Energy, Indian Institute of Technology Roorkee, Uttarkhand, India*

***Abstract***

In this study, a three-dimensional computational fluid dynamics (CFD)-based numerical investigation is conducted to investigate the thermohydraulic performance (THP) of solar air heater (SAH) using semi-cylindrical tubes as the roughness elements. The range of design parameters is considered as Reynolds numbers (Re) of 3000 to 19,000, tube pitch ratios (P/H) of 1 to 4, and roughness height ratios ($e_r/H$) of 0.2 to 0.5. Numerical results are validated against previous experiment results, which show close agreement. From the simulation results, the maximum value of the Nusselt number is found to be 112.64 at Re of 19,000, which corresponds to $e_r/H$ as 0.5 and P/H as 4, whereas the maximum value of friction factor is calculated to be 0.21 at Re of 3000, for these same values of $e_r/H$ and P/H. However, the maximum THP value is calculated to be 1.45 at Re of 3000, $e_r/H$ as 0.2, and P/H as 4. The outcome of the research may be useful to researchers for further studies and industries to develop efficient SAH for drying applications.

***Keywords*:** Solar air heater, computational fluid dynamics, Nusselt number, friction factor, Reynolds number

## Nomenclature

| | |
|---|---|
| $<\tau\alpha>$ | Transmittance-absorptance product |
| $D_h$ | Hydraulic diameter (m) |

**Corresponding author*: ankush.hedau@gmail.com; ORCID: 0000-0003-0620-9895

Ashwani Kumar, Nitesh Dutt and Mukesh Kumar Awasthi (eds.) Heat Transfer Enhancement Techniques: Thermal Performance, Optimization and Applications, (233–250) © 2025 Scrivener Publishing LLC

| | |
|---|---|
| $e_b/H$ | Relative blockage height ratio |
| $e_r/H$ | Relative roughness height ratio |
| f | Friction factor |
| H | Duct height (m) |
| $f/f_s$ | Friction factor increment |
| L | Test section length (m) |
| Nu | Nusselt number |
| $P/e_r$ | Pitch-to-rib height ratio |
| $Nu/Nu_s$ | Nusselt number enhancement |
| V | Air velocity (m/s) |
| P/H | Tube pitch ratio |
| Re | Reynolds number |
| t | Tube thickness (m) |
| W | Duct width (m) |
| W/H | Aspect ratio |
| h | Heat transfer coefficient (W/m$^2$ °C) |
| k | Thermal conductivity (W/m °C) |
| $\Delta P$ | Pressure drop within the duct (Pa) |
| $T_{am}$ | Air mean temperature (°C) |
| $\dot{m}$ | Air flow rate (kg/s) |
| $C_p$ | Air-specific heat (J/kg °C) |

## Abbreviations

| | |
|---|---|
| ANSYS | Analysis of systems |
| CFD | Computational fluid dynamics |
| GIT | Grid independency test |
| RNG | Renormalization group |
| SAH | Solar air heater |
| THP | Thermohydraulic performance |

## 10.1 Introduction

Several renewable and non-renewable energy sources are used to reduce the gap between the energy supply and energy demand. Many non-renewable sources: coal, natural gas, petroleum oil, etc., are used to meet the energy demand and produce various harmful impacts, such as greenhouse gas emissions and global warming. To reduce greenhouse gas emissions and global warming, the world is moving from non-renewable to renewable energy sources such as solar, biomass, and wind. Solar energy is a high potential to produce thermal and electrical energy free of cost by using the solar collector. Solar air heaters (SAHs) are one of the ways to capture sun energy for different low-temperature purposes such as crop drying, greenhouse heating, and space cooling (Hedau & Saini, 2023b). As a result of creating a laminar boundary layer over the length of the absorbing plate, the rate of heat transfer is low. The various roughness geometries, such as ribs (Dutt *et al.*, 2023; Julker *et al.*, 2014; A. Kumar *et al.*, 2020; R. Kumar & Goel, 2021), baffles (Karwa *et al.*, 2005; Tandel & Modi, 2022), jet impingement (Chauhan & Thakur, 2014; S. Yadav & Saini, 2020), and dimple protrusion (Prasad & Saini, 1988; Saini & Verma, 2008) are used to increase the heat transfer rate inside the rectangular duct of the SAH.

Singh *et al.* (2015) investigated the fluid flow and heat transfer behavior of SAH with four different transverses ribs: trapezoidal, saw tooth, circular, and square. It was observed that the circular cross-section ribs provide a lower increment of friction factor compared to square and trapezoidal cross-sections. A similar experimental investigation was performed by Sah and Mandapati (2020) to determine the thermal performance of the double pass SAH using four different ribs such as rectangular, triangular, semi-circular, and modified arc. It was concluded that the semi-circular ribs provide a higher average temperature difference and thermal efficiency than rectangular and triangular ribs. Ahn (2001) experimentally concluded that the circular ribs in the rectangular duct provide higher thermohydraulic performance (THP) in the range of approximately 1.8 to 2.8 for the Reynolds number varying between 10,000 to 60,000. Kumar and Goel (2021) compared the THP of the SAH with circular and semi-circular ribs as the roughness elements. In comparison to smooth duct, the maximum THP of the circular and semi-circular ribs was reported to be 1.6 and 1.79, respectively for the same range of parameters studied.

The friction and turbulent flow behavior of the rectangular duct with transverse semi-circular ribs as roughness elements were experimentally investigated by the Nine (2014) at roughness height ratio ($e_r/H$) as 0.116, pitch to rib height ratio ($P/e_r$) of 8 to 14 and Re of 15000 to 30000. The maximum thermal enhancement ($Nu/Nu_s$) was observed for the $P/e_r$ value of 10 due to higher turbulent kinetic energy produced by the recirculation and reattachment of the secondary flows between the ribs. Yadav and Bhagoria (2014) investigated the turbulence flow behavior of the SAH using transverse semi-circular ribs and validated simulation result with the experimental result of Nine (2014). The numerical investigation was conducted at $e_r/H$ of 0.021 to 0.06, $P/e_r$ of 14.29, and Re from 3800 to 18000. For the considered range of the parameter investigated, transverse semi-circular ribs attain good heat transfer enhancement ($Nu/Nu_s$) of approximately

2.34 times and friction factor enhancement ($f/f_s$) of approximately 1.92 times than the smooth duct.

Kumar (2022) analyzed the fluid flow and heat transfer behavior of curved double pass SAH using semi-circular ribs and arched baffles. The arched baffles accelerate the flow and increase the intensity of turbulence inside the rectangular duct. The arched baffles increase the heat transfer enhancement in the range of 20% to 28% higher than the without arched baffles. Hedau and Saini (2023a) numerically investigated the THP of the DPSAH using semi-circular tubes and perforated blocks as the roughness elements. The maximum THP value was found to be 2.51 at open area ratio of 30%, Re of 3000, $e_b/H$ as 0.4, $e_r/H$ as 0.3. Manjunath (2018) used sinusoidal absorbing plate to study the heat transfer behavior of SAH. The study was performed by varying the roughness height of the sinusoidal corrugations ($e_r$) of 7.5 mm to 20 mm ($H/e_r$: 4 to 1.5). The maximum thermal efficiency was reported to be 68.7 % for the $H/e_r$ of 3 and Re of 13500.

From the literature review of SAH having circular and semicircular ribs, it was noted that the THP was affected by the roughness-height ratios ($e_r/H$) and tube pitch ratios (P/H). Furthermore, semi-circular ribs perform better in terms of THP than circular and square ribs. It has been found that semi-circular tubes with higher roughness height ($e_r$) have better thermal efficiency and THP in the research conducted by Manjunath *et al.* (2018) and Hedau and Saini (2023a). To predict the frictional loss, the pitch between the two tubes needs to be varied in the ranges investigated by the previous researchers. Further research is required to be examined on the semi-circular geometry with a higher value of roughness height ($e_t$). The given research is focused on the THP improvement of the SAH using different values of the roughness heights of the semi-cylindrical tubes ($e_r$) and pitch between the two tubes (P).

The main objectives of the present studies are:

i. To analyze the effect of design parameters such as tube pitch ratios (P/H) and roughness height ratios ($e_r/H$) on Nusselt number and friction factor corresponding to different operating conditions of Re.
ii. To optimize the performance in terms of THP for the considered range of the design parameters and operating conditions.

## 10.2 Numerical Simulation

The computational fluid dynamics (CFD)-based numerical investigation is conducted in this study to analyze the performance of SAH with transverse semi-cylindrical tubes as the roughness elements. The CFD simulation is carried out in three steps includes the pre-processing, processing and post processing.

### 10.2.1 Preprocessing

The design modeler is used to create the rectangular duct with an inlet length of 600 mm, test section length of 1000 mm, and exit section length of 550 mm. The inlet length ($\geq 5\sqrt{WH}$) and exit length ($\geq 2.5\sqrt{WH}$) of the rectangular duct are selected as per the ASHRAE standard (Hale *et al.*, 1986; Standard, 1977). The length of the test section is chosen to satisfy the Dittus-Boelter equation criteria, *i.e.*, $L/D_h>10$. The width and duct height of the airflow passage is taken to be 350 mm and 35 mm, respectively. The design parameter of the SAH is listed in Table 10.1.

The properties of the absorbing plate and the air is listed in Table 10.2.

The THP of the proposed geometry is evaluated at two design parameters: roughness height ratios ($e_r/H$) and tube pitch ratios (P/H). The Reynolds number is used as the operating parameters and the range to

**Table 10.1** Design parameter of the SAH with semi-cylindrical tubes as the roughness elements.

| **Parameters** | **Values** |
|---|---|
| Test section length (L) | 1000 mm |
| Outlet section | 550 mm |
| Inlet section | 600 mm |
| Duct width (W) | 350 mm |
| Duct height (H) | 35 mm |
| Hydraulic diameter ($D_h$) | 63.63 mm |
| **Transverse cylindrical tubes** | |
| Number of semi-circular tubes | 7 |
| Tube width (W) | 350 mm |
| Roughness height ratios ($e_t/H$) | 0.2–0.5 |
| Tube thickness (t) | 1 mm |
| Tube pitch ratio (P/H) | 1–4 |

**Table 10.2** Physical properties of materials.

| Property | Aluminum | Air |
|---|---|---|
| Dynamic viscosity, μ (Pa.s) | - | 1.86e–5 |
| Thermal conductivity, k (W/m °C) | 202.4 | 0.026 |
| Specific heat $C_p$ (J/kg °C) | 871 | 1006 |
| Density, ρ (kg/m³) | 2716 | 1.192 |

be considered for the numerical analysis is 3000 to 19,000. The parameters of the investigated geometry are presented in Figure 10.1. Based on the previous research investigation, the range of parameters for numerical investigation was selected. The range of design parameters are described in Table 10.3.

Further, the fine and uniform meshing is created using the fluent meshing in order to simulate the fluid flow and heat transfer. The edge sizing with number and divisions and hard behaviors is used to create the fine uniform meshing of the proposed geometry. The inflation layer is provided on the absorbing plate to accurately predict the heat transfer within the rectangular duct. Furthermore, the grid independency test is conducted at Re as 11,000, P/H as 4, and $e_r/H$ of 0.2 in order to obtain accurate results for the values of friction factor and Nusselt number. The result of the grid independency test (GIT) is shown in Table 10.4.

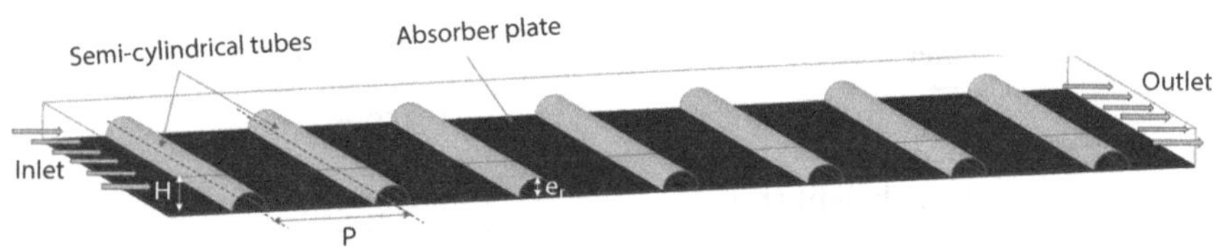


**Figure 10.1** Investigation parameters of the absorbing plate.

**Table 10.3** Range of the design parameters.

| Sr. no. | Design parameter | Range of parameter |
|---|---|---|
| 1 | Reynolds number (Re) | 3000–19000 |
| 2 | Tube pitch ratios (P/H) | 1–4 |
| 3 | Roughness height ratios ($e_r$/H) | 0.2–0.5 |

**Table 10.4** Grid independency test.

| Number of elements | Nu | f | Percentage variation in Nu | Percentage variation in f |
|---|---|---|---|---|
| 851,741 | 76.25 | 0.0328 | | |
| 3,154,782 | 70.54 | 0.0318 | 7.49% | 3.05% |
| 6,548,972 | 67.89 | 0.0310 | 3.76% | 2.52% |
| 9,841,573 | 66.52 | 0.0306 | 2.02% | 1.29% |
| 13,548,720 | 65.93 | 0.0303 | 0.89% | 0.98% |
| 15,540,268 | 65.42 | 0.0301 | 0.77% | 0.66% |

### 10.2.2 Processing

The three governing equations (mass, energy, and momentum equation) are used to predict the performance of the proposed geometry.

Continuity equation

$$\frac{\partial(\rho u_j)}{\partial x_j} = 0 \tag{10.1}$$

Momentum equation

$$\frac{\partial}{\partial x_i}(\rho u_i u_j) + \frac{\partial P}{\partial x_i} = \frac{\partial}{\partial x_i}\left[\mu\left(\frac{\partial u_i}{\partial x_j} + \frac{\partial u_j}{\partial x_i}\right)\right] + \frac{\partial}{\partial x_j}\left[\mu_t\left(\frac{\partial u_i}{\partial x_j} + \frac{\partial u_j}{\partial x_i}\right)\right] \tag{10.2}$$

Energy equation

$$\frac{\partial}{\partial x_i}(\rho u_j T) - \frac{\partial}{\partial x_j}\left[(\Gamma + \Gamma_t)\frac{\partial T}{\partial x_j}\right] = 0 \tag{10.3}$$

These governing equations are solved in the steady state regime by the finite volume approach based on the boundary conditions presented in Table 10.5.

**Table 10.5** Boundary condition.

| **Parameter** | **Boundary conditions** | **Values** |
|---|---|---|
| Top and sidewall | Wall | Insulated |
| Absorbing plate | Wall | 880 $Wm^{-2}$ |
| Outlet | Pressure outlet | 101,325 Pa (Atm.) |
| Inlet | Velocity inlet | 0.735–4.66 m/s (3000–19000) |

In real conditions, solar radiation of 1000 $W/m^{-2}$ is provided on the absorbing plate through the glass cover. The actual solar radiation of the absorbing plate is calculated by considering the transmittance-absorptance product (<τα>) of absorbing plate and glass cover. So, in numerical simulation, solar radiation is exposed to the constant solar radiation of 880$W/m^{-2}$ by considering the transmissivity-absorptivity product (<τα>) of equal to 0.88 (Duffie, John A.; Beckman, 2005). The following assumptions were considered during the numerical simulation of the proposed geometry:

i. Flow is turbulent, incompressible, steady and three-dimensional.
ii. Air and absorbing plate properties remain constant at operating temperatures.
iii. Fluid outlet pressure equals atmospheric pressure (101,325 Pa).
iv. The sidewall and top wall of the SAH are adiabatic, and the absorbing plate is exposed to solar radiations of 880 $W/m^{-2}$.
v. The radiation heat transfer is negligible.

The second-order upwind scheme with the SIMPLE algorithm is used for all the equations to achieve the accurate numerical results of the proposed geometry. For the convergence of the equations, the residual selected for the continuity equation is in the order of $10^{-3}$, and for the other equations is in the order of $10^{-6}$. The renormalization group (RNG) k-ε with enhanced wall treatment is employed as the turbulence model for the analysis (Yadav & Saini, 2020).

### 10.2.3 Postprocessing

In postprocessing, the area weighted average of outlet air ($T_o$), inlet air ($T_i$) and absorbing plate temperature ($T_{ab}$), and pressure drop ($\Delta P$) is noted in order to calculate the Nusselt number and friction factor. Additionally, the fluid flow and heat transfer within the rectangular duct are visualized in the postprocessing by plotting the velocity, temperature, and turbulence kinetic energy contours. The velocity profile at any plane and section inside the rectangular duct can also be used to predict the behavior of the turbulence.

### 10.2.4 Data Reduction

Nusselt number, friction factor, and THP value are calculated based on the following equation after data collection.

$$Nu = \frac{hD_h}{k} \tag{10.4}$$

$$f = \frac{(\Delta P) \times D_h}{2\rho L V^2} \tag{10.5}$$

$$THP = \frac{\left( Nu / Nu_s \right)}{\left( f / f_s \right)^{1/3}} \tag{10.6}$$

The value of h is calculated using the equation below as,

$$\dot{m}C_p (T_o - T_i) = hA_{ab} (T_{ab} - T_{am}) \tag{10.7}$$

## 10.3 Validation

The numerical result of the SAH having semi-cylindrical tubes as the roughness elements is validated with the experimental result of Nine *et al.* (2014). The numerical validation is conducted for the friction factor values corresponds to $P/e_r$ of 14, $e_r/H$ as 0.116, and Re from 10,400 to 24,000, as shown in Figure 10.2. The numerical results show a good prediction with friction factor values of Nine *et al.* (2014) with a mean absolute percentage error of 7.8%. The error appears due to the numerical simulation,

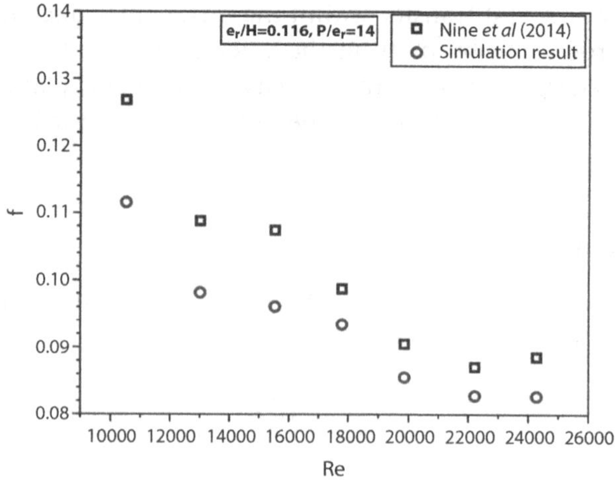


**Figure 10.2** Validation of friction factor values of simulation result with experimental result (Nine *et al.*, 2014).

inaccuracy in the experimental measuring and indicating instruments, and truncation errors.

## 10.4 Results and Discussions

The effects of roughness height ratios ($e_r/H$) and tube pitch ratios (P/H) on the Nusselt number and friction factor is examined at various operating conditions of Re and as described in the next sub-section.

### 10.4.1 Effect of Roughness Height Ratio ($e_r/H$)

The variation of Nusselt number and friction factor with Re for different values of $e_r/H$ is presented in Figures 10.3 and 10.4. The Nusselt number increases in the range of 28 to 112.64 with increase in $e_r/H$ from 0.2 to 0.5 and Reynolds number from 3000 to 19,000. However, the friction factor increases in the range of 0.0253 to 0.21 with increase in $e_r/H$ from 0.2 to 0.5, and it decreases with increase in Re from 3000 to 19,000. The maximum Nusselt number is calculated to be 112.64 at Re as 19,000, $e_r/H$ as 0.5 and P/H as 4, while maximum friction factor is calculated to be 0.21, at Re as 3000, $e_r/H$ as 0.5, and P/H as 4. For the range of design parameters, the maximum increment in Nusselt number ($Nu/Nu_s$) is found to be in the

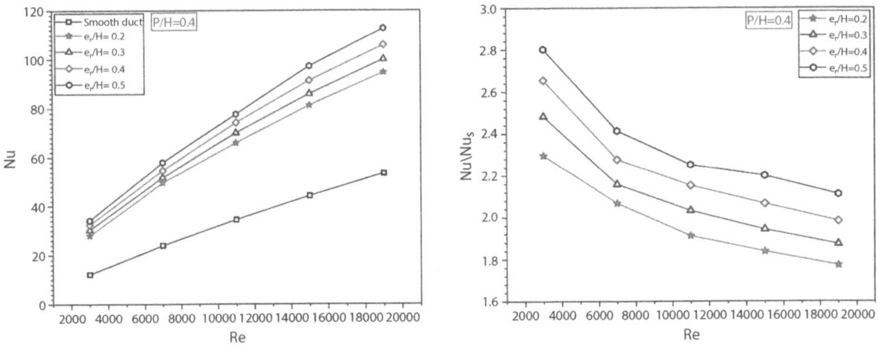


**Figure 10.3** Variation of Nu and Nu/Nu$_s$ with Re.

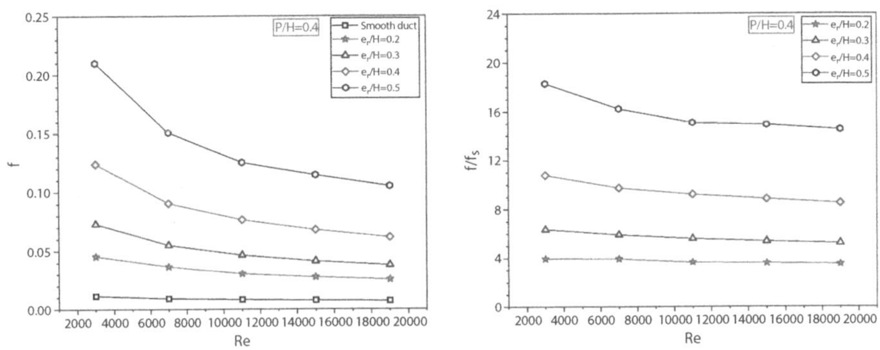


**Figure 10.4** Variation of f and f/f$_s$ with Re.

range of 1.77 to 2.8, and the maximum increment in friction factor (f/f$_s$) is found to be 3.5 to 18.29, as plotted in Figures 10.3 and 10.4, respectively.

The heat transfer increases due to two phenomena: (1) with increasing the value of e$_r$/H, the heat transfer surface area of the absorbing plate increases, thus exchanging large heat transfer between the absorbing plate and air, as shown in Figure 10.5 (b). In the event of an increase in Re at same values of e$_r$/H, the flow separation and reattachment near the edge and tip of semi-cylindrical tubes increase thus increasing the intensity of turbulence, as shown in Figure 10.6. This results in a break the thermal boundary layer and enhance the heat transfer. With an increase in e$_r$/H from 0.2 to 0.5, the frictional loss increases. This is a result of increasing flow separation and flow recirculation within the rectangular ducts.

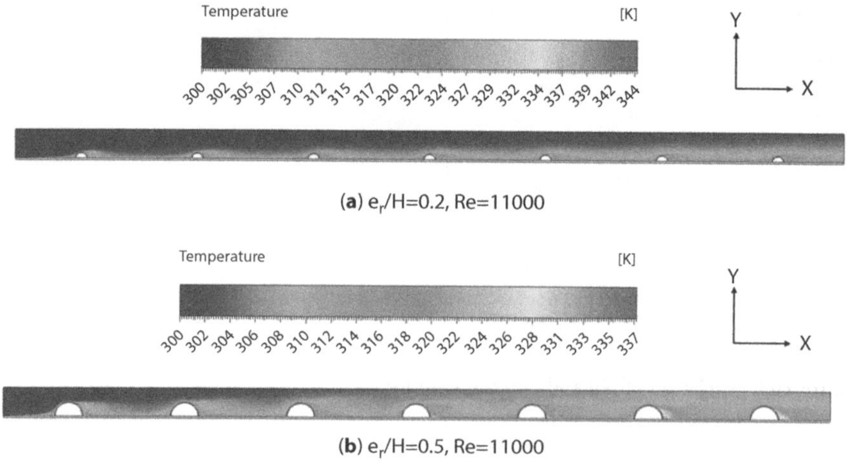


**Figure 10.5** Temperature contours.

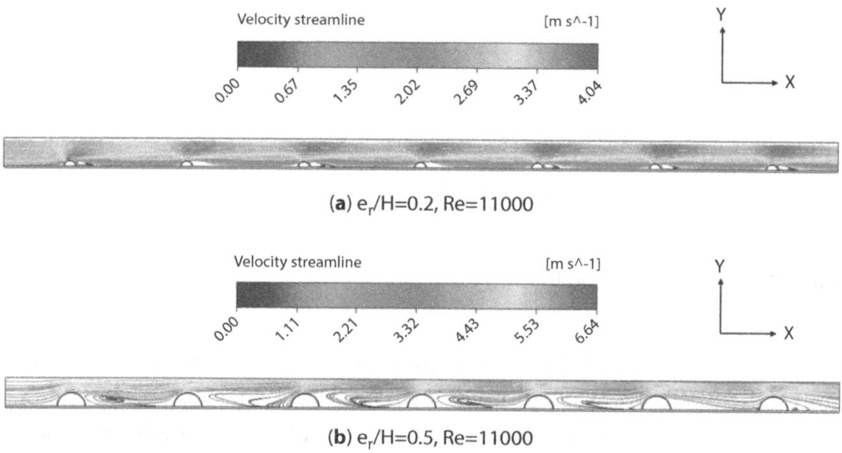


**Figure 10.6** Velocity contours.

## 10.4.2 Effect of Tube Pitch Ratio (P/H)

The variation of Nusselt number and friction factor with Re for different values of P/H is shown in Figures 10.7 and 10.8. The Nusselt number and friction factor decrease with increasing the ratio of P/H from 1 to 4. However, the Nusselt number increases and the friction factor decreases with increasing the Reynolds number from 3000 to 19,000. The maximum Nusselt number is calculated as 98.44 at Re as 19,000, P/H as 1 and $e_r/H$ of 0.2, while the maximum friction factor is calculated to be 0.0724, at Re as

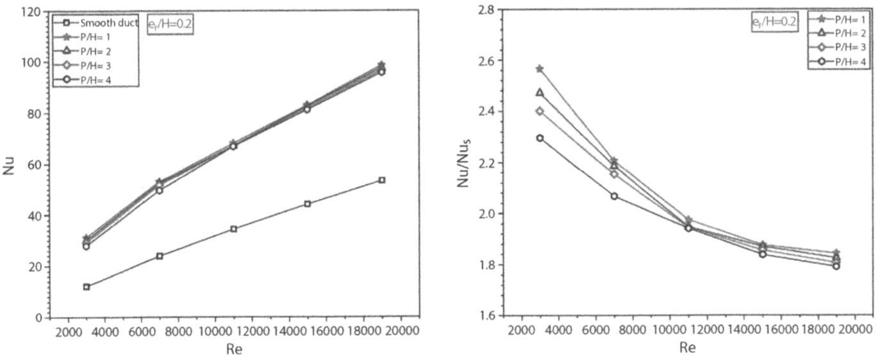


**Figure 10.7** Variation of Nu and $Nu/Nu_s$ with Re.

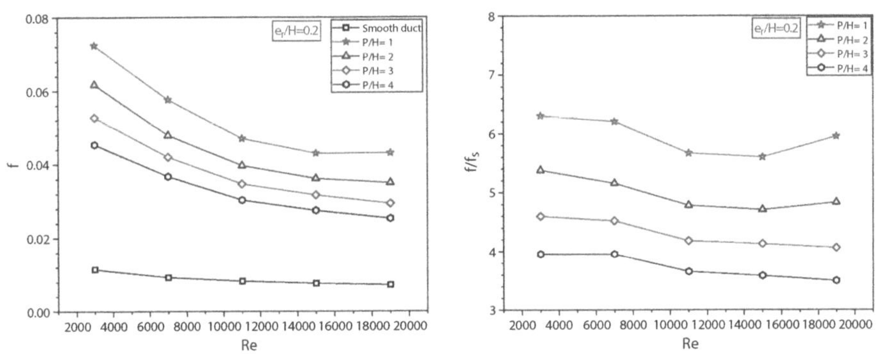


**Figure 10.8** Variation of f and $f/f_s$ with Re.

3000, P/H as 1, and $e_r/H$ of 0.2. For the considered range of design parameters, the maximum increment in Nusselt number ($Nu/Nu_s$) is found to be in the range of 1.77 to 2.57, and the maximum increment in friction factor ($f/f_s$) is found to be 3.5 to 6.3, as plotted in Figures 10.7 and 10.8, respectively.

The heat transfer increase with decrease in P/H from 4 to 1. The reason is due to increasing the separation and recirculation of secondary flow between the semi-cylindrical tubes along the rectangular duct, as presented in Figure 10.9. Further, the effect of reattachment between primary flow and secondary flow increase with increase in Re which increases the turbulence intensity. This results in the secondary flow coming in more contact with the absorbing plate which covers the hot spot region, as shown in Figure 10.10. This results in a reduction of the absorbing plate temperature which simultaneously increases the heat transfer ($Nu/Nu_s$). Similarly, the friction factor increment increases with a decrease in P/H from 4 to 1.

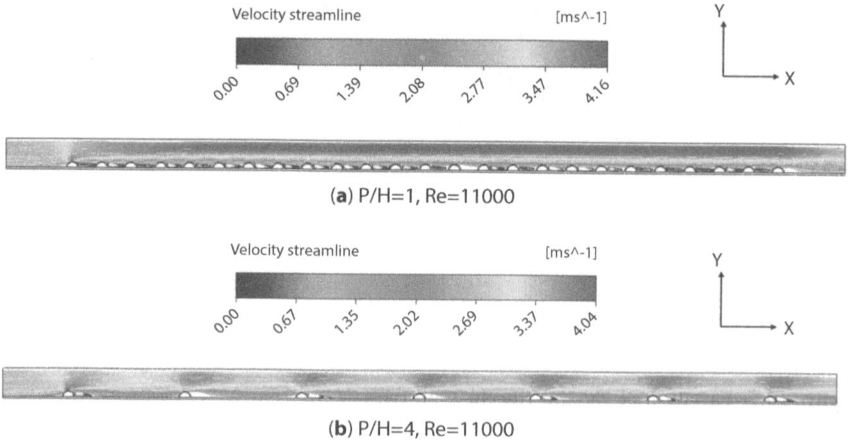


**Figure 10.9** Velocity contours.

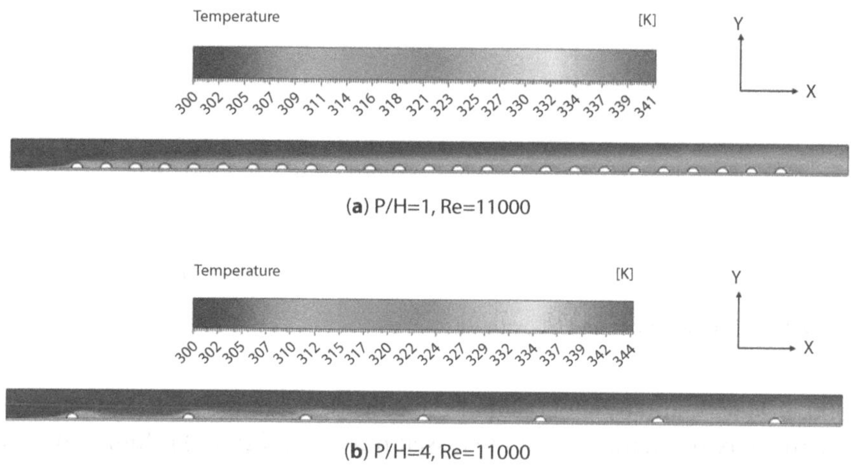


**Figure 10.10** Temperature contours.

This is because of higher flow recirculation (flow reversal) which causes higher frictional drag resistance and friction factor, as presented in Figure 10.9. Further, frictional resistance decreases with P/H from to 1 to 4 which results in lower friction factor increment ($f/f_s$).

### 10.4.3 Thermohydraulic Performance (THP)

For different values of P/H, the plot of THP versus Re is presented in Figure 10.11. The THP increases with increase in P/H from 1 to 4. This is because of

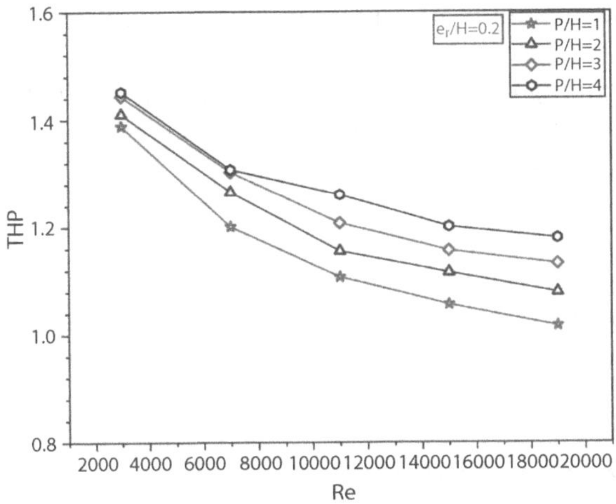


**Figure 10.11** THP vs. Re.

the reduction in the separation and recirculation of secondary flows which results in a small amount of friction factor increment ($f/f_s$). In comparison to $f/f_s$, heat transfer enhancement ($Nu/Nu_s$) is higher, which is due to recirculation flow covering the hot spot region and reducing the absorbing plate temperature. Further, the THP decreases with increasing the Reynolds number from 3000 to 19,000. This is due to the reduced value of $Nu/Nu_s$ in penalty with large increment $f/f_s$. The maximum value of THP is observed to be 1.45 at design parameters of P/H as 4, $e_r/H$ as 0.2, and Re of 3000.

## 10.5 Conclusions

The THP characteristics of the SAH are studied using numerical investigation at various design parameters of roughness elements such as tube pitch ratios (P/H) of 1 to 4, roughness height ratios ($e_r/H$) of 0.2 to 0.5, and under various values of Re from 3000 to 19000. From the numerical investigation, the following conclusion is summarized.

i. The roughness height of the semi-cylindrical tubes ($e_r$) was found to be a great influence on the THP of the SAH.
ii. The maximum value of the Nusselt number is found to be 112.64 at Re of 19,000, and the maximum value of the friction factor is calculated to be 0.21 at Re as 3000, corresponding to the $e_r/H$ as 0.5 and P/H as 4.

iii. The maximum increment in Nusselt number and friction factor is found to be 2.8 and 18.29, respectively, corresponding to $e_r/H$ as 0.5, P/H as 4, and Re of 3000.
iv. The maximum value of THP is observed as 1.45, at Re as 3000, P/H of 4, and $e_r/H$ of 0.5.

For the researchers, this research could prove useful for further studies and industries to develop efficient SAH for drying applications. Furthermore, researchers should incorporate storage material inside semi-cylindrical tubes to improve the thermal efficiency during the non-sunshine hours.

## Declaration of Competing Interest

The authors declare that they have no known competing financial interests or personal relationships that could have appeared to influence the work reported in this chapter.

## Data Availability

Data will be made available on request.

## Acknowledgement

The authors acknowledge the Ministry of Education (MoE), Government of India for providing the financial support in the form of fellowship and research grant (Grant No: PM-31-22-675-414). We also acknowledge the National Supercomputing Mission (NSM) for providing computing resources of 'PARAM Ganga' at IIT Roorkee, which is implemented by C-DAC and supported by the Ministry of Electronics and Information Technology (MeitY) and Department of Science and Technology (DST), Government of India.

## References

Ahn, S.W., The effects of roughness types on friction factors and heat transfer in roughened rectangular duct. *Int. Commun. Heat Mass Transfer*, *28*, 7, 933–942, 2001, https://doi.org/10.1016/S0735-1933(01)00297-4.

Chauhan, R. and Thakur, N.S., Investigation of the thermohydraulic performance of impinging jet solar air heater. *Energy*, *68*, 255–261, 2014, https://doi.org/10.1016/J.ENERGY.2014.02.059.

Duffie, J.A. and Beckman, W.A., *Solar Engineering of Thermal Process*, p. 468, 2005.

Dutt, N., Hedau, A.J., Kumar, A., Awasthi, M.K., Varun, Singh, P., Dwivedi, G., Thermo-hydraulic performance of solar air heater having discrete D-shaped ribs as artificial roughness. *Environ. Sci. Pollut. Res.*, *1*, 1–22, 2023, https://doi.org/10.1007/s11356-023-28247-9.

Dutt, N., Hedau, A.J., Kumar, A., Awasthi, M.K., Thermo-hydraulic performance investigation of solar air heater duct having staggered D-shaped ribs: Numerical approach. *Heat Transfer*, 53, 3, 1501–1531, 2024, https://doi.org/10.1002/htj.22998.

Hale, Hale, S.J.B., Hale, *Ashrae: Methods of testing to determine the thermal performance of solar collectors*, 1986.

Hedau, A. and Saini, R.P., Thermo-hydraulic performance of double pass solar air heater duct having semi-circular tubes and perforated blocks as artificial roughness. *Renewable Energy*, *205*, October 2022, 543–562, 2023a, https://doi.org/10.1016/j.renene.2023.01.087.

Hedau, A. and Saini, R.P., A review on artificial roughened solar air heaters with and without thermal energy storage. *Int. J. Green Energy*, 1–35, 2023b, https://doi.org/10.1080/15435075.2023.2194387.

Nine, M., Lee, H., Chung, H., Ji, M., Jeong, H., Turbulence and pressure drop behaviors around semi-circular ribs in a rectangular channel. *Therm. Sci.*, *18*, 2, 419–430, 2014, https://doi.org/10.2298/TSCI111022142N.

Karwa, R., Maheshwari, B.K., Karwa, N., Experimental study of heat transfer enhancement in an asymmetrically heated rectangular duct with perforated baffles. *Int. Commun. Heat Mass Transfer*, *32*, 1–2, 275–284, 2005, https://doi.org/10.1016/J.ICHEATMASSTRANSFER.2004.10.002.

Kumar, A., Akshayveer, Singh, A.P., Singh, O.P., Efficient designs of double-pass curved solar air heaters. *Renew. Energy*, *160*, 1105–1118, 2020, https://doi.org/10.1016/J.RENENE.2020.06.115.

Kumar, A., Akshayveer, Singh, A.P., Singh, O.P., Investigations for efficient design of a new counter flow double-pass curved solar air heater. *Renew. Energy*, *185*, 759–770, 2022, https://doi.org/10.1016/J.RENENE.2021.12.101.

Kumar, R. and Goel, V., Unconventional solar air heater with triangular flow-passage: A CFD based comparative performance assessment of different cross-sectional rib-roughnesses. *Renew. Energy*, *172*, 1267–1278, 2021, https://doi.org/10.1016/J.RENENE.2021.03.068.

Manjunath, M.S., Karanth, K.V., Sharma, N.Y., Numerical investigation on heat transfer enhancement of solar air heater using sinusoidal corrugations on absorber plate. *Int. J. Mech. Sci.*, *138–139*, 219–228, 2018, https://doi.org/10.1016/j.ijmecsci.2018.01.037.

Prasad, B.N. and Saini, J.S., Effect of artificial roughness on heat transfer and friction factor in a solar air heater. *Sol. Energy*, *41*, 6, 555–560, 1988, https://doi.org/10.1016/0038-092X(88)90058-8.

Sah, N. and Mandapati, M.J.K., Thermal performance of a double-pass solar air heater (SAH) with ribbed absorber surface – an experimental study. *World J. Eng.*, *17*, 3, 373–380, 2020, https://doi.org/10.1108/WJE-08-2019-0217.

Saini, R.P.A. and Verma, J., Heat transfer and friction factor correlations for a duct having dimple-shape artificial roughness for solar air heaters. *Energy*, *33*, 8, 1277–1287, 2008.

Singh, S., Singh, B., Hans, V.S., Gill, R.S., CFD (computational fluid dynamics) investigation on Nusselt number and friction factor of solar air heater duct roughened with non-uniform cross-section transverse rib. *Energy*, *84*, 509–517, 2015, https://doi.org/10.1016/J.ENERGY.2015.03.015.

Standard, A. and others, Methods of testing to determine the thermal performance of solar collectors. ANSI/ASHRAE, 93–1986, 1977.

Tandel, H.U. and Modi, K.V., Experimental assessment of double-pass solar air heater by incorporating perforated baffles and solar water heating system. *Renew. Energy*, *183*, 385–405, 2022, https://doi.org/10.1016/J.RENENE.2021.10.087.

Yadav, A.S. and Bhagoria, J.L., A Numerical Investigation of Turbulent Flows through an Artificially Roughened Solar Air Heater. *Int. J. Comput. Methodol.*, *65*, 7, 679–698, 2014, https://doi.org/10.1080/10407782.2013.846187.

Yadav, S. and Saini, R.P., Numerical investigation on the performance of a solar air heater using jet impingement with absorber plate. *Sol. Energy*, *208*, August, 236–248, 2020, https://doi.org/10.1016/j.solener.2020.07.088.

# 11

# Design and Analysis of Solar Tracking System for PV Thermal Performance Enhancement

**Bhupender Singh[1], Preet Kaur[2], Ashok Kumar Yadav[3]*, Mukesh Kumar Awasthi[4] and Ashwani Kumar[5]**

[1]*Department of Mechanical Engineering, YMCA University of Science and Technology, Faridabad, Haryana, India*
[2]*Department of Electronics and Communication Engineering, YMCA University of Science and Technology, Faridabad, Haryana, India*
[3]*Department of Mechanical Engineering, Raj Kumar Goel Institute of Technology, Ghaziabad, India*
[4]*Department of Mathematics, Babasaheb Bhimarao Ambedkar University, Lucknow, India*
[5]*Department of Mechanical Engineering, Technical Education Department, Uttar Pradesh, Kanpur, India*

## *Abstract*

Growing energy demand and environmental concerns have led to a considerable increase in the relevance of natural energy sources as a substitute for conventional energy sources. The renewable energy sector is rapidly emerging as a new growth area for many countries due to the immense potential it offers both environmentally and economically. Solar energy is an essential major source of energy, especially for rural areas.

The development and assessment of a solar tracking system capable of optimizing the solar panel's power output is the primary goal of this research. A solar panel, two servo motors, four light-dependent resistors (LDRs), an Arduino Uno R3 microprocessor, and other electronic parts make up the solar tracking system's

**Corresponding author*: ashokme015@gmail.com
Ashok Kumar Yadav: ORCID: 0000-0002-7350-7641
Mukesh Kumar Awasthi: ORCID: 0000-0002-6706-5226
Ashwani Kumar: ORCID: 0000-0003-4099-935X

Ashwani Kumar, Nitesh Dutt and Mukesh Kumar Awasthi (eds.) Heat Transfer Enhancement Techniques: Thermal Performance, Optimization and Applications, (251–268) © 2025 Scrivener Publishing LLC

architecture. The system's principal controller is an Arduino microcontroller, which senses sunlight using a light-dependent resistor. A servo motor is used to move the solar panel so that it faces the sun. The solar tracking system prototype was constructed and put to the test. The data results are collated, and a graphical method is used to compare the performance of a static solar panel with a solar tracking system. As a result, tracking solar panels are able to produce greater current, voltage, and power than stationary solar panels. After the tracking system was put in place, the solar panel's total efficiency rose.

***Keywords:*** Solar tracking, Arduino, microcontroller, solar panel, efficiency, LDR

## 11.1 Introduction

Numerous scientific studies have been conducted in the area of solar energy conversion in general as well as in one particularly specific area within it, which is photovoltaic (PV) technology. In this instance, designing, optimizing, and implementing systems based on this technology are pressing concerns since, when done well, they can improve solar energy extraction [1–4]. Since maximum efficiency is attained when the incident light is perpendicular to the PV panels' cells, one approach attempted is to increase the amount of solar radiation captured—this is known as "solar tracking." Therefore, the climate variables of solar radiation, ambient temperature, and wind speed, as well as the system's matching of the load and the solar panels' proper location, all affect how energy-efficient these systems are. Because most solar panels are employed in a fixed position, their full power production is not regularly produced [5–9].

Studying the issue of PV orientation which necessitates the use of a sun tracker attached to the PV system was important in order to maximize the efficiency of PV panels. With the use of a solar tracker, one may monitor the sun's position throughout the day and modify the solar panel's angle to ensure that it is always at its typical, regular angle [10–14]. The conversion and degree of absorption may be improved, and mechanical steering devices can be used to increase the incident radiation rate in order to maximize the useable energy rate [15–19].

A PV installation mounted atop a motor-based supporting structure is called a solar tracker. It enables the solar panels to be pointed towards the sun throughout the day in order to get as much sunlight as possible. As a result, based on the kind of control, the sensitivity of the sensors, and the positioning method they use, solar trackers may be divided into two major groups [20–23]. Additionally, two tracking methods can be used to differentiate them: the sun tracking system, which is based on the orientation of solar panels throughout the day to better exploit the PV cells, and the MPPT (maximum power point tracking) method, which is based on an algorithm to find the maximum power curve of the PV panel [24, 25].

A great deal of study has been done to enhance and boost solar energy conversion to electrical energy output. We might mention the studies conducted by F. R. Rubio *et al.* and Robert H. Dold [6]. They described a two-axis solar tracker that measures azimuth and elevation movement in 2007 and is resistant to harsh weather. The potential benefits of a straightforward solar tracking system using a stepper motor and light sensor were mentioned by J. Rizk [6] in 2008. A solar tracking system was developed, put into practice, and the results of the experimental testing were quite conclusive. The benefits and drawbacks of various sun monitoring methods were covered by Hossein Mousazadeh in 2009 [6]. It was discovered that polar and azimuth/elevation axes were the most efficient means of tracking the sun. According to study done in 2010, Nader Barsoum [6] stated that a single-axis sun tracking system may boost energy output by around 20%, while a dual-axis tracking system can boost output by 40%. In comparison to a stationary collector of the same size, the sun tracking system allowed solar radiation to continually strike the flat plate collector perpendicularly, as reported by L. Kancevica [6] in 2012. This resulted in an average of 1.4 times more thermal energy being generated. In 2013, Mostefa Ghassoul [6] provided details on building and evaluating an inventive, reasonably priced sun monitoring system.

Thus, the study findings reported in the article [7–11] verify that solar panel orientation may raise conversion system efficiency by 20% to 50%, resulting in a notable increase in energy generation. An important component of solar systems, the three-phase inverter, was the subject of a research in Reference [11], which gave us the concept for the conditioning and modelling of the motor that we employed in our actual prototype. Since then, a lot of study has been done on the design and optimization of the sun tracking technique to maintain the active surface of the solar module perpendicular to the sun's rays in order to maximize the amount of sunlight that the cells get. 2018 saw the writers of the article in Reference [12] provide a thorough analysis of the possibilities of the solar tracking system. Their description aids in providing a general understanding of the kinds, construction, design parameters, and procedures of drive systems encompassing many applications. They discussed the single-axis and dual-axis solar tracking systems, which are the two primary varieties of solar tracking systems. According to their publication, 41.58% of current research papers reported on dual-axis tracking systems, whereas 42.57% of studies explored and presented single-axis tracking systems. Azimuth and elevation tracking, horizontal tracking, azimuth tracking 10%, polar tracking 16.67%, and utilization 4.44% were all achieved in the sun tracking approaches as well. This inspired us to keep refining the modelling outcomes of the various blocks in our technique as well as the modelling outcomes of the other blocks [13]. A number of electronic systems were created and tested by References [14–25] for the purpose of controlling and piloting the bi-axial solar trackers of PV power plants. The tracking is controlled, as in our instance, by a PC program featuring an intuitive graphical user interface. By calculating the sun's circadian orbit, the program is able to move the solar panels in a way that maintains their surface perpendicular to the sun's beams, so increasing the efficiency of energy generation. For instance, Reference [14] described and provided support for a novel PV sun tracker design approach. This methodology makes use of the benefits that the latitude of the installation zone affords in terms of the PV panel's efficiency and orientation. They validated the suggested design process through experiments.

We were able to identify the solar tracking systems that offer advantages including increased efficiency, tracking precision, ease of installation, and cost-effectiveness by examining the scientific content of Reference [16]. According to their research, double axis solar tracking (ST)—polar-axis and azimuth/elevation—with dynamic closed-loop feedback control, solar movement models, and feedback control is the most efficient. It often yields an increase in energy return of more than 40% when compared to fixed PV panels. They also demonstrated how big systems save a lot of money and materials. The moving fixtures typically use very little energy (2–5% of the total energy gathered), however if optimization is not done, this might increase. For this reason, from the beginning of system development, we have optimized all power-saving hardware and software settings to prevent material waste and excessive power consumption by tracking units.

An overview of the developments made in the field of solar tracking systems is provided by a different research by Reference [17]. The performance study of dual-axis solar tracking systems with various, recently developed designs and approaches was highlighted. The following are some conclusions from their paper: in general, dual-axis solar tracking systems outperform their fixed and single-axis equivalents in terms of efficiency. Because these systems include spinning parts, they also require maintenance. These systems also have more intricate control and design processes. The functioning of the solar system is influenced by several physical variables, including the choice of tracking method [26–31].

A sun tracking solar panel system using Arduino is a fascinating application of technology that maximizes the efficiency of solar energy collection. This innovative approach ensures that solar panels are always optimally positioned to receive the maximum amount of sunlight, resulting in increased energy generation [32–37].

In a conventional fixed solar panel setup, the panels are mounted at a fixed angle to the ground. This angle is typically set based on the latitude of the installation location. While this arrangement works well to some extent, it doesn't account for the changing position of the sun throughout the day and across seasons. As a result, there are times when the panels are not aligned optimally with the sun's rays, leading to reduced energy output [38–40].

A sun tracking solar panel system addresses this limitation by employing sensors and motors to continuously adjust the orientation of the solar panels [41, 42]. By tracking the sun's movement, the system ensures that the panels are always facing directly towards the sun, maximizing the amount of solar energy captured. This dynamic approach significantly boosts the overall energy production of the solar installation [43–48].

The core components of a sun tracking solar panel system include light sensors, motors, and an Arduino microcontroller. Light sensors, also known as photodiodes or LDRs (light dependent resistors), are used to detect the intensity of sunlight. These sensors feed data to the Arduino, which processes the information and determines the current position of the sun relative to the panels. The Arduino then commands the motors responsible for adjusting the position of the solar panels. These motors can be either linear actuators or servo motors, depending on the complexity of the system. Linear actuators allow for smooth movement in two dimensions, while servo motors offer precise rotational adjustments [49–55].

The heart of the system lies in the programming logic implemented on the Arduino. The code reads the data from the light sensors, calculates the sun's position based on that data, and compares it with the optimal angle for solar energy capture. If the detected angle is not aligned with the optimal angle, the code sends commands to the motors, adjusting the panels to the correct orientation [56–62]. Creating a functional sun tracking solar panel system requires a careful consideration of several factors. These include the geographical location of the installation, the design and mechanics of the motorized system, and the power consumption of the motors. Additionally, weather conditions and potential obstacles must be taken into account to ensure the system's durability and longevity [63–67].

While designing and building a sun tracking solar panel system using Arduino can be an exciting project, it also involves challenges. Calibration of the system, synchronization of the sensors and motors, and optimizing the code for accuracy are among the key hurdles to overcome. However, numerous online resources, tutorials, and Arduino libraries are available to guide enthusiasts through each step of the process [68–71].

## 11.2 Background and Motivation

Traditional fixed solar panel installations are mounted at a fixed angle relative to the ground, typically based on the latitude of the installation site. While this configuration provides reasonable energy production, it falls short of capturing the maximum potential energy due to the sun's dynamic movement. As the sun changes its position in the sky throughout the day and across seasons, fixed panels may not always be optimally aligned with the sun's rays [72–75].

A sun tracking solar panel system addresses this limitation by continually adjusting the orientation of the solar panels to directly face the sun. This dynamic tracking ensures that the panels receive the highest amount of sunlight possible, leading to increased energy output and improved efficiency [76–79].

### 11.2.1 Types of Solar Tracking System

As per the mode of motions, the solar tracking system is classified into two types [7, 70].

Single-axis solar tracking system: this type of system tracks the movement of the sun along one axis only. The axis of rotation can be either horizontal (azimuth) or vertical (elevation). Single-axis tracking systems follow the sun's path from sunrise to sunset by adjusting the angle of solar panels along a single axis. There are two common configurations:

- Horizontal single-axis tracking: the solar panels rotate around a horizontal axis, following the sun's path as it rises and sets. This is more suitable for locations where the sun's path is primarily in the east-west direction.
- Vertical single-axis tracking: in this configuration, the solar panels rotate around a vertical axis, changing their tilt angle to follow the sun's path as it moves across the sky. This type of tracking is more suitable for regions with significant changes in solar elevation angle throughout the day.

Dual-axis solar tracking system: dual-axis tracking systems are more complex and capable of capturing sunlight by adjusting solar panels along both horizontal and vertical axes. This allows them to track the sun's movement in both azimuth and elevation. Dual-axis trackers provide more precise alignment with the sun's position, resulting in higher energy yields. However, they are typically more expensive and require more sophisticated control mechanisms.

## 11.3 Fundamentals of Arduino-Based Solar Tracking System

The system works by continuously monitoring the intensity of sunlight using LDR sensors, processes data via an Arduino, and employs servo motors for solar panel alignment, optimizing energy capture from sunrise to sunset.

By employing PV cells, the solar panel measures the intensity of light to precisely align itself with the sun's position in the sky. The tracking system consistently fine-tunes the panel's orientation to ensure it remains perpendicular to the sun, thus maximizing the absorption of light and conversion of energy. This proof-of-concept utilizes LDR sensors to detect the intensity of sunlight, and servo motors controlled by an Arduino unit to automate the rotation of the panel [7, 8]. The Arduino Uno board interfaces with the LDR sensor's output to precisely regulate the movement of the motors. Moreover, the system can also be manually controlled using a potentiometer, providing an additional level of flexibility (Figure 11.1).

### 11.3.1 Components of a Sun Tracking Solar Panel System

A sun tracking solar panel system consists of several key components working in harmony [9–13]:

1. Light sensors: also known as photodiodes or LDRs, these sensors measure the intensity of light falling on them. They serve as the system's "eyes," detecting the sun's position relative to the panels.
2. Motors: motors are responsible for physically adjusting the position of the solar panels. Linear actuators or servo motors can be used, depending on the desired level of complexity and accuracy.

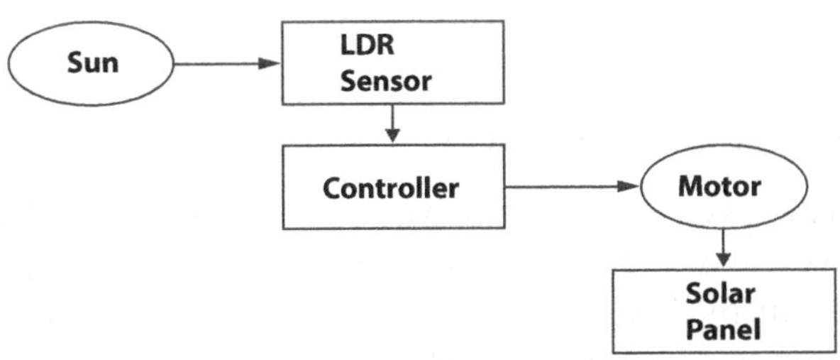


**Figure 11.1** Block diagram of Arduino-based solar tracking system.

3. Arduino microcontroller: the brain of the system, the Arduino processes the data from the light sensors and controls the motors to ensure proper panel alignment. It contains the programming logic that drives the tracking mechanism.

### 11.3.2 Working Procedure of Solar Tracking System

The heart of the sun tracking system lies in its ability to accurately detect the sun's position and adjust the panels accordingly. The process can be broken down into several stages:

1. Sensor data acquisition: the light sensors constantly measure the intensity of sunlight falling on them. This data is read by the Arduino, which interprets the light intensity as an indication of the sun's position.
2. Calculating sun's position: using the data from the light sensors, the Arduino calculates the current position of the sun. This involves comparing the intensity of light detected by the sensors and determining the direction from which the light is coming.
3. Comparison with optimal angle: the Arduino then compares the calculated sun's position with the optimal angle for solar energy capture. This angle is typically based on the time of day, season, and the specific geographical location of the solar panels.
4. Motor control: if the calculated angle deviates from the optimal angle, the Arduino sends commands to the motors. These commands adjust the position of the solar panels to align them precisely with the sun's rays.
5. Continuous tracking: the system operates in a continuous loop, constantly monitoring the sun's position and adjusting the panels accordingly. This ensures that the panels remain optimally aligned throughout the day.

### 11.3.3 Development of Model

The solar tracking system is designed into two subsystems. The initial subsystem involves movement, encompassing the sun tracking solar panel's rotation mechanism. The second subsystem manages tracking actions executed by the sun tracking solar panel (Figure 11.2).

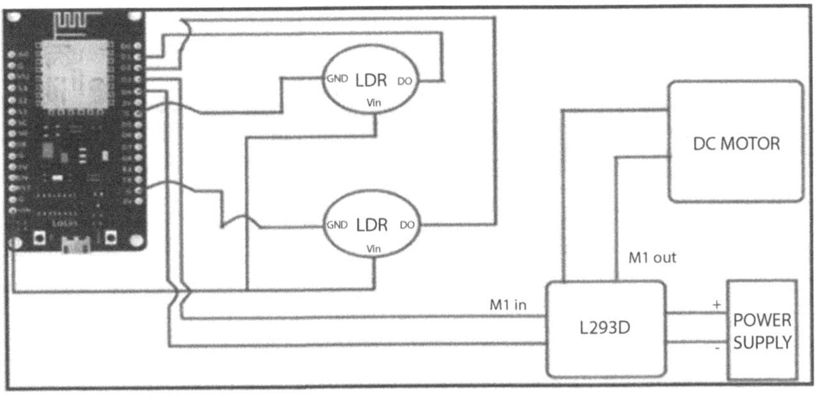


**Figure 11.2** Schematic diagram of model.

### Movement Subsystem

The movement subsystem is in charge of performing all the physical actions of the system. This subsystem consists of a rotating fixture and servo motor.

### Building a Rotating Fixture

Within the framework of this Arduino-based sun tracking solar panel, we incorporate 3D-printed components to create a rotating fixture. These customized elements facilitate the necessary movements for optimal solar panel orientation. While our approach employs 3D printing, your project offers flexibility, allowing for the utilization of materials such as cardboard, wood, or plastic to craft the fixture.

### Tracking Subsystem

The tracking subsystem acts as a controller of the final product such that it decides the actions of the solar panel. This subsystem consists of Arduino Uno controller board, photo resistors and circuit components (like resistors and wires).

Light dependent resistors (LDRs), also known as photo resistors, are used to measure the intensity of light falling on the solar panel. Light dependent resistors change their resistance based on the amount of light they receive. When exposed to brighter light, their resistance decreases, and when exposed to lower light levels, their resistance increases. In this system, the LDRs serve as light intensity sensors, detecting changes in the ambient light conditions caused by the sun's movement. As the intensity of sunlight changes throughout the day due to the sun's movement, the LDR sensors detect these changes. The Arduino Uno board processes the sensor

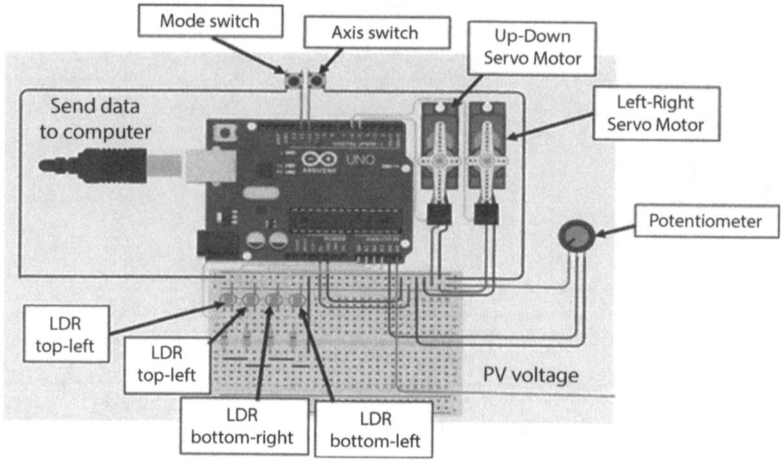


**Figure 11.3** Arduino-based solar tracking system.

data and calculates the optimal position for the solar panel to be perpendicular to the incoming sunlight. It then sends control signals to the servo motors to adjust the panel's orientation accordingly (Figure 11.3).

## 11.4 Benefits and Challenges

The benefits of a sun tracking solar panel system are substantial. By consistently facing the sun, panels capture more sunlight, resulting in higher thermal performance. This increased efficiency translates into a quicker return on investment for solar installations, making them more economically viable.

However, challenges do exist. The system's accuracy relies on precise calibration of sensors and motors. External factors like cloud cover, shadows, and obstacles can affect the accuracy of the tracking system. Moreover, the power consumption of the motors must be balanced with the energy gained from improved sunlight exposure.

## 11.5 Conclusions

This dual axis Arduino-based solar tracking system underscores the critical importance of advancing solar energy harnessing in the face of escalating energy demands and environmental concerns. Solar power's potential to revolutionize our energy landscape cannot be overstated, but the challenge of maximizing solar panel efficiency persists. This project directly addresses this challenge by leveraging technology like the Arduino controller board to optimize solar energy capture. Through dynamic panel orientation adjustments and periodic evaluation, the system ensures optimal sunlight interception, leading to enhanced energy absorption and overall efficiency. By pushing the boundaries of solar energy utilization, this contributes significantly to the ongoing global pursuit of sustainable energy solutions.

## References


1. Hua, Z., Ma, C., Lian, J., Pang, X., Yang, W., Optimal capacity allocation of multiple solar trackers and storage capacity for utility-scale photovoltaic plants considering output characteristics and complementary demand. *Appl. Energy*, 238, 721–733, 2019.
2. Gómez-Uceda, F.J., Moreno-Garcia, I.M., Jiménez-Martínez, J.M., López-Luque, R., Fernández-Ahumada, L.M., Analysis of the Influence of Terrain Orientation on the Design of PV Facilities with Single-Axis Trackers. *Appl. Sci.*, 10, 8531, 2020.
3. Chong, K. and Wong, C., General Formula for on-axis sun-tracking system and its application in improving tracking accuracy of solar collector. *Sol. Energy*, 83, 298–305, 2009.
4. Ferroudji, F., Conception et Optimisation en Dynamique Forcée D'un Nouveau Prototype de Système de Suiveur Solaire à Deux axes, Ph.D. Thesis, University of Batna, Batna, Algeria, 2015.
5. Abu-Khadera, M., Badranb, O., Abdallah, S., Evaluating multi-axes sun-tracking system at different modes of operation in Jordan. *Renew. Sustain. Energy Rev.*, 12, 864–873, 2008.
6. Awasthi, A., Shukla, A.K., M Manohar, S.R., Dondariya, C., Shukla, K.N., Porwal, D., Richhariya, G., Review on sun tracking technology in solar PV system. *Energy Rep.*, 6, 392–405, 2020.
7. Khujamatov, K.E., Khasanov, D.T., Reypnazarov, E.N., Modeling and Research of Automatic Sun Tracking System on the bases of IoT and Arduino UNO, IEEE, 2019.
8. Zolkapli, M., AI-Junid, S.A.M., Othman, Z., Manut, A., Mohd Zulkifli, M.A., High-Efficiency Dual-Axis Solar Tracking Development using Arduino. *2013 International Conference on Technology, Informatics, Management.*
9. Juang, J.-N. and Radharamanan, R., Design of a Solar Tracking System for Renewable Energy. *2014 Conference of the American Society for Engineering Education.*

10. Lande, A. *et al.*, Design of Arduino Based Solar Tracker for Renewable Energy. *Int. Res. J. Eng. Technol.*, 09, 12, 795–978, 2022.
11. B.K.C., Study and analysis of Arduino based solar tracking panel. *Int. Res. J. Eng. Technol.*, 07, 02, 2816–2820.
12. Pawar, P., Solar Tracking System Using Arduino. *Int. J. Sci. Eng. Res.*, 9, 2, 102–104, 2018.
13. Kaur, T., Arduino based Low Cost Active Dual Axis Solar Tracker. *1st IEEE International Conference on Power Electronics, Intelligent Control and Energy Systems (ICPEICES-2016)*, pp. 1–5, 2016.
14. Yadav, A.K., Ahmed, A., Singh, A., Predictive Modeling and Optimization of Chlorella vulgaris Microalgae Biodiesel Production: Assessing Performance and Emission Characteristics on a Raw Biogas-Powered Diesel Engine. *J. Environ. Inform.*, accepted-2023.
15. Yadav, A.K., Ahmed, A., Singh, A., Agbulut, U., A Hybrid RSM-GA-PSO Approach on Optimization of Process intensification of Linseed Biodiesel Synthesis Using an Ultrasonication Reactor: Enhancing Fuel Properties and Engine Characteristics with Ternary Fuel Blends. *Energy*, 2023, DOI: 10.1016/j.energy.2023.129077.
16. Dewangan, A., Mallick, A., Yadav, A.K., Islam, S., Agbulut, U., Production of oxy-hydrogen gas and the impact of its usability on CI Engine combustion, performance, and emission behaviors. *Energy*, 278, 127937.
17. Yadav, A.K., Ahmed, A., Singh, A., A Comprehensive Machine Learning-Coupled Response Surface Methodology Approach for Predictive Modeling and Optimization of Biogas Potential in Anaerobic Co-Digestion of Organic Waste. *Biomass Bioenergy*, 2023, https://doi.org/10.1016/j.biombioe.2023.106995.
18. Yadav, A.K., Khan, M.E., Pal, A., Dubey, A.M., Performance and emission characteristics of a transportation diesel engine operated with non-edible vegetable oils biodiesel. *Case Stud. Therm. Eng.*, 8, 236–244, 2016.
19. Khan, O., Kumari, P., Yadav, A.K., Idrisi, M.J., Modelling of Compression Ignition Engine by Soft Computing Techniques (ANFIS-NSGA-II and RSM) to Enhance the Performance Characteristics for Leachate Blends with Nano-additives. *Sci. Rep.*, 13, 15429, 2023.
20. Yadav, A.K., Khan, O., Khan, M.E., Utilization of High FFA Landfill Waste (Leachates) as a Feedstock for Sustainable Biodiesel Production: Its Characterization and Engine Performance Evaluation. *Environ. Sci. Pollut. Res.*, 25, 32, 32312–32320.
21. Yadav, A.K., Dewangan, A., Mallick, A., Effect of n-Butanol and Diethyl Ether Additives on Performance and Emission Characteristics of a Diesel Engine Fuelled with Diesel-Pongamia Biodiesel Blends. *J. Energy Eng.*, 144, 6, 04018062, 2018.
22. Dubey, A.M., Ram, R., Yadav, A.K., Ion cell performance using single particle representation of battery electrode. *J. Electroanal. Chem.*, 768, 11–17, 2016.
23. Yadav, A.K., Khan, T.A., Khan, T.A., A hydrodynamic cavitation-assisted system for optimization of biodiesel production from green microalgae oil

using a genetic algorithm and response surface methodology approach. *Environ. Sci. Pollut. Res.*, 29, 49465–49477, 2022.
24. Ahmed, A., Yadav, A.K., Singh, A., Multi-Response Optimization to Improve the Performance and Emissions Characteristics of A VCR Engine Fueled with Microalgae Spirulina (L.): A Response Surface Methodology Approach Coupled with Genetic Algorithm. *Environ. Dev. Sustain.*, Doi:10.1007/s10668-023-04016-z.
25. Singh, D., Singh, S., Yadav, A.K., Khan, O., Dewangan, A., From Theory to Practice: A Sustainable Solution to Water Scarcity by Using a Hybrid Solar Distiller with a Heat Exchanger and Aluminum Oxide Nanoparticles. *ACS Omega*, 37, 8, 33543–33553, 2023.
26. Dewangan, A., Mallick, A., Yadav, A.K., Ahmad, A., Alqahtani, D., Combined Effect of Operating Parameters and Nano Particles on Performance of a Diesel Engine: Response Surface Methodology coupled Genetic algorithm Approach. *ACS Omega*, 8, 27, 24586–24600, 2023.
27. Yadav, A.K. and Pal, A., Experimental Studies on Utilization of Prunus Armeniaca L. (Wild Apricot) Biodiesel as an Alternative Fuel for CI Engine. *Waste Biomass Valorization*, 9, 10, 1961–1969, 2018.
28. Yadav, A.K., Khan, M.E., Pal, A., Experimental Investigations of Performance and Emissions Characteristics of Kusum (Schleichera Oleosa) Biodiesel in a Multi-Cylinder Transportation Diesel Engine. *Waste Biomass Valorization*, 8, 4, 1331–1341, 2017.
29. Ahmed, A., Yadav, A.K., Singh, A., Process optimization of spirulina microalgae biodiesel synthesis using RSM coupled GA technique: a performance study of a biogaspowered dualfuel engine. *Int. J. Environ. Sci. Technol.*, 2023, https://doi.org/10.1007/s13762-023-04948-z.
30. Ahmed, A., Yadav, A.K., Singh, A., Application of machine learning and genetic algorithm for prediction and optimization of Biodiesel Yield from Waste Cooking Oil. *Korean J. Chem. Eng.*, 2023, DOI: 10.1007/s11814-023-1489-9.
31. Ahmed, A., Yadav, A.K., Singh, A., Enhancement of Biogas Yield from Dual Organic Waste Using Hybrid Statistical Approach and Its Effects on Ternary Fuel Blend (Biodiesel/n-butanol /Diesel) Powered Diesel Engine. *Environ. Prog. Sustain. Energy*, doi.org/10.1002/ep.14163.
32. Yadav, A.K., Khan, M.E., Pal, A., Ultrasonic Assisted Production of Biodiesel from Karabi Oil using Heterogeneous Catalyst. *Biofuels*, 9, 101–112, 2018.
33. Yadav, A.K., Khan, T.A., Khan, T.A., Kumar, S., Methyl ester of Gmelina arborea oil as a substitute for petroleum diesel: an experimental study on its performance and emissions in a diesel engine. *Energy Sources Part A*, 43, 11, 1307–1314, 2021.
34. Yadav, A.K., Khan, M.E., Pal, A., Biodiesel Production from Nerium Oleander (Thevetia Peruviana) Oil through Conventional and Ultrasonic Irradiation Methods. *Energy Sources Part A*, 38, 23, 3447–3452, 2016.
35. Khan, O., Yadav, A.K., Khan, M.E., Sharma, D., The Ultrasonic Assisted Optimization of Biodiesel Production from Eucalyptus Oil. *Energy Sources Part A*, 39, 1556–7036, 2017.

36. Dewangan, A. and Yadav, A.K., Wax Deposition during Production of Waxy Crude Oil and Its Remediation. *Pet. Sci. Technol.*, 35, 18, 1831–1838, 2017.
37. Dewangan, A., Yadav, A.K., Mallick, A., Current Scenario of Biodiesel Development in India: Prospects and Challenges. *Energy Sources Part A*, 40, 20, 2494–2501, 2018.
38. Khan, I.A., Singh, S.K., Yadav, A.K., Efficient Production of Biodiesel from Cannabis Sativa oil using Intensified Transesterication (Hydrodynamic Cavitation) Method. *Energy Sources Part A*, 42, 20, 2461–2470, 2020.
39. Khan, O., Yadav, A.K., Khan, M.E., Characterization of Bio-Ethanol Obtained from Eichhornia Crassipes Plant; Its Emission and Performance Analysis on CI Engine. *Energy Sources Part A*, 43, 14, 1793–1803, 2021.
40. Ahmed, A., Yadav, A.K., Singh, A., Optimization of Cavitation-Assisted Biodiesel Production and Fuel Properties from Neochloris Oleoabundans Microalgae oil using Genetic Algorithm and Response Surface Methodology. *J. Process Mech. Eng.: Part E*, doi.org/10.1177/09544089231159832.
41. Yadav, M., Karimi, M.N., Yadav, A.K., Effect of Graphene Oxide Nanoparticles Dispersed Biodiesel on Combustion, Performance, and Emission Characteristics of a Diesel Engine. *J. Process Mech. Eng.: Part E*, Accepted 2023.
42. Das, D., Yadav, A.K., Ahmed, A., Numerical Simulation of the Combustion Chamber and Performance Analysis of Microalgae-Based Biodiesel/Diesel-Powered CI Engine: An Axisymmetric Flow Approach. *J. Process Mech. Eng.: Part E*, accepted, 2023.
43. Singh, S., Singh, B., Kumar, S., Yadav, A.K., Temperature-dependent dynamic hysteresis scaling of ferroelectric hysteresis parameters of lead free [(Ba 0.825 + xCa0.175-x) (Ti1- x Snx)O3]ceramics. *Ferroelectrics*, 551, 133–142, 2019.
44. Ahmed, A., Yadav, A.K., Singh, A., Optimization of Biogas Yield from Anaerobic Co-digestion of Dual waste for environmental sustainability: ANN, RSM, and GA, Approach. *Int. J. Oil Gas Coal Technol.*, 33, 1, 75, 2023.
45. Yadav, A.K., Khan, M.E., Pal, A., A Comparative Study on Ultrasonic Cavitation and Mechanical Stirring Method towards Efficient Production of Biodiesel from Non-Edible Oils and Performance Testing on a C.I. Engine. *Int. J. Environ. Waste Manage.*, 18, 349–367, 2016.
46. Yadav, A.K., Khan, M.E., Pal, A., Performance and Emission Characteristics of a Stationary Diesel Engine fuelled by Schleichera Oleosa oil Methyl Ester (SOME) Produced through Hydrodynamic Cavitation Process. *Egypt. J. Pet.*, 27, 89–93, 2018.
47. Vinay, Singh, B., Yadav, A.K., Optimization of Performance and Emission Characteristics of CI Engine Fuelled with Mahua Oil Methyl Ester-Diesel Blend using Response Surface Methodology. *Int. J. Ambient Energy*, 44, 6, 674–85, 2020.
48. Yadav, A.K., Khan, M.E., Pal, A., Performance, Emission and Combustion Characteristics of an Indica Diesel Engine Operated with Yellow Oleander

(Thevetia Peruviana) Oil Biodiesel Produced through Hydrodynamic Cavitation Method. *Int. J. Ambient Energy*, 39, 4, 365–371, 2018.
49. Yadav, A.K., Pal, A., Ghosh, U., Gupta, S.K., Comparative Study of Biodiesel Production Methods from Yellow Oleander Oil and its Performance Analysis on an Agricultural Diesel Engine. *Int. J. Ambient Energy*, 40, 2, 152–157, 2019.
50. Khan, I.A., Singh, S.K., Yadav, A.K., Sharma, D., Enhancement in the Performance of a Diesel Engine Fueled with Pongamia Methyl Ester and n-Butanol as Oxygenated Additive. *Int. J. Ambient Energy*, 40, 8, 842–846, 2019.
51. Ahmed, A., Yadav, A.K., Singh, A., Modelling and optimisation of VCR diesel engine parameters fuelled with microalgae methyl-ester: a RSM-coupled Taguchi approach. *Int. J. Ambient Energy*, 44, 1, 2498–2506, 2023.
52. Ahmed, A. and Yadav, A.K., Biodiesel Production from Mahua Oil: Characterization, Optimization, and Modeling with a Hybrid Statistical Approach. *Int. J. Ambient Energy*, 2023, Doi: 10.1080/01430750.2023.2262468.
53. Khan, T.A., Yadav, A.K., Khan, M.E., Experimental Studies on Utilization of Neochloris Oleoabundans Microalgae Biodiesel as an Alternative Fuel for Diesel Engine. *Int. J. Ambient Energy*, 44, 1, 115–123, 2023.
54. Singh, D., Yadav, A.K., Kumar, A., Samsher, Energy matrices and life cycle conversion analysis of N-identical hybrid double slope solar distiller unit using Al2O3 nanoparticle. *J. Water Environ. Nanotechnol.*, 8, 3, 267–284, 2023.
55. Khan, M.E., Yadav, A.K., Khan, T.A., A comprehensive review on performance, emission and combustion characteristics of diesel engines fuelled with algae oil and methyl ester. *WEENTECH Proc. Energy*, 6, 2, 160–168, 2020.
56. Yadav, A.K., Singh, S., Gupta, G., Ghosh, U., Generation of Electricity through Staircase. *Adv. Sci. Lett.*, 20, 1303–1306, 2014.
57. Kumar, A., Singh, V.P., Meena, C.S., Dutt, N., *Thermal Energy Systems: Design, Computational Techniques and Applications*, Taylor & Francis (CRC Press), 2023, https://doi.org/10.1201/9781003395768.
58. Kumar, R., Prasad, A., Kumar, A., *Sustainable Smart Manufacturing Processes in Industry 4.0*, Taylor & Francis (CRC Press), 2023, ISBN: 9781003436072, https://doi.org/10.1201/9781003436072.
59. Sharma, V.K., Kumar, A., Gupta, M., Kumar, V., Sharma, D.K., Sharma, S., *Additive Manufacturing in Industry 4.0: Methods, Techniques, Modeling, and Nano Aspects*, Taylor & Francis (CRC Press), 2022, https://doi.org/10.1201/9781003360001.
60. Dutt, N., Hedau, A.J., Kumar, A., Awasthi, M.K., Singh, V.P., Dwivedi, G., Thermo-hydraulic performance of solar air heater having discrete D-shaped ribs as artificial roughness. *Environ. Sci. Pollut. Res.*, 2023, https://doi.org/10.1007/s11356-023-28247-9.

61. Kr. Kushwaha, P., Kr. Sharma, N., Kumar, A., Meena, C.S., Recent Advancements in Augmentation of Solar Water Heaters Using Nanocomposites with PCM: Past, Present, and Future. *Buildings*, 13, 79, 2023, https://doi.org/10.3390/buildings13010079.
62. Singh, V.P., Jain, S., Karn, A., Kumar, A., Experimental Assessment of Variation in Open Area Ratio on Thermohydraulic Performance of Parallel Flow Solar Air Heater. *Arabian J. Sci. Eng.*, 2022, https://doi.org/10.1007/s13369-022-07525-7.
63. Meena, C.S., Prajapati, A.N., Kumar, A., Kumar, M., Utilization of Solar Energy for Water Heating Application to Improve Building Energy Efficiency: An Experimental Study. *Buildings*, 12, 2166, 2022, https://doi.org/10.3390/buildings12122166.
64. Singh, V.P., Jain, S., Kumar, A., Mishra, S., Sharma, N.K., Heat transfer and friction factor correlations development for double pass solar air heater artificially roughened with perforated multi-V ribs. *Case Stud. Therm. Eng.*, 39, 102461, 2022, ISSN 2214-157X, https://doi.org/10.1016/j.csite.2022.102461.
65. Kundu, A., Kumar, A., Dutt, N., Meena, C.S., Singh, V.P., Introduction to Thermal Energy Resources and their Smart Applications, in: *Thermal Energy Systems: Design, Computational Techniques and Applications*, A. Kumar, V.P. Singh, C.S. Meena, N. Dutt, (Eds.), pp. 1–15, CRC Press, Boca Raton, FL, USA, 2023, Chapter 1; http://dx.doi.org/10.1201/9781003395768-1.
66. Dewangan, A.K., Moinuddin, S.Q., Cheepu, M., Sajjan, S.K., Kumar, A., Thermal Energy Storage: Opportunities, Challenges and Future Scope, in: *Thermal Energy Systems: Design, Computational Techniques and Applications*, A. Kumar, V.P. Singh, C.S. Meena, N. Dutt, (Eds.), pp. 17–28, CRC Press, Boca Raton, FL, USA, 2023, Chapter 2; http://dx.doi.org/10.1201/9781003395768-2.
67. Kundu, A., Kumar, A., Dutt, N., Singh, V.P., Meena, C.S., Modelling and Simulation of Thermal Energy System for Design Optimization, in: *Thermal Energy Systems: Design, Computational Techniques and Applications*, A. Kumar, V.P. Singh, C.S. Meena, N. Dutt, (Eds.), pp. 103–140, CRC Press, Boca Raton, FL, USA, 2023, Chapter 7; http://dx.doi.org/10.1201/9781003395768-7.
68. Singh, V.P., Jain, S., Karn, A., Kumar, A., Dwivedi, G., Meena, C.S., Recent Developments and Advancements in Solar Air Heaters: A Detailed Review. *Sustainability*, 14, 12149, 2022, https://doi.org/10.3390/su141912149.
69. Meena, C.S., Kumar, A., Jain, S., Ur Rehman, A., Mishra, S., Innovation in Green Building Sector for Sustainable Future. *Energies*, 15, 6631, 2022, https://doi.org/10.3390/en15186631.
70. Singh, V.P., Jain, S., Karn, A., Kumar, A., Dwivedi, G., Meena, C.S., Cozzolino, R., Mathematical Modeling of Efficiency Evaluation of Double-Pass Parallel Flow Solar Air Heater. *Sustainability*, 14, 10535, 2022, https://doi.org/10.3390/su141710535.
71. Meena, C.S., Kumar, A., Roy, S., Cannavale, A., Ghosh, A., Review on Boiling Heat Transfer Enhancement Techniques. *Energies*, 15, 5759, 2022, https://doi.org/ 10.3390/en15155759.

72. Singh, V.P., Jain, S., Kumar, A., Establishment of Correlations for the Thermo-Hydraulic Parameters Due to Perforation in a Multi-V Rib Roughened Single Pass Solar Air Heater. *Exp. Heat Transfer*, 2022, DOI: 10.1080/08916152.2022.2064940.
73. Prajapati, Y.K., Gupta, P.K., Kumar, A., Free Surface Undulation and Air Entrainment in a Rectangular Tank. *Int. J. Appl. Eng. Res.*, 6, 4, 409–419, 2011.
74. Pant, G., Meena, C.S., Saxena, A., Kumar, A., Singh, V.P., Dutt, N., Study the Temperature Variation in Alternate Coils of Insulated Condenser Cum Storage Tank: Experimental Study, in: *Advances in Fluid and Thermal Engineering. FLAME 2022.* Lecture Notes in Mechanical Engineering, B.S. Sikarwar, S.K. Sharma, A. Jain, K.M. Singh, (Eds.), Springer, Singapore, 2023, https://doi.org/10.1007/978-981-99-2382-3_52.
75. Saxena, A., Prajapati, A.N., Pant, G., Meena, C.S., Kumar, A., Singh, V.P., Water Consumption Optimization of Hybrid Heat Pump Water Heating System, in: *Recent Advances in Mechanical Engineering. FLAME 2022.* Lecture Notes in Mechanical Engineering, A.K. Shukla, B.P. Sharma, A. Arabkoohsar, P. Kumar, (Eds.), Springer, Singapore, 2023, https://doi.org/10.1007/978-981-99-1894-2_61.
76. Verma, V., Meena, C.S., Thangavel, S., Kumar, A., Choudhary, T., Dwivedi, G., Ground and solar assisted heat pump systems for space heating and cooling applications in the northern region of India – A study on energy and CO2 saving potential. *Sustain. Energy Technol. Assess.*, 59, 103405, 2023, ISSN 2213-1388, https://doi.org/10.1016/j.seta.2023.103405.
77. Awasthi, M.K., Dutt, N., Kumar, A., Kumar, S., Electrohydrodynamic capillary instability of Rivlin–Ericksen viscoelastic fluid film with mass and heat transfer. *Heat Transfer*, 1–19, 2023, doi: 10.1002/htj.22944.
78. Pathak, A., Kumar, A., Kumar, A., Kumar, A., Francis Luther King, M., Application of Laser Technology in the Mechanical and Machine Manufacturing Industry, in: *Laser-based Technologies for Sustainable Manufacturing*, A. Kumar, A. Kumar, A. Kumar (Eds.), pp. 107–155, CRC Press, Boca Raton, FL, USA, 2023, Chapter 06; DOI: 10.1201/9781003402398-6.
79. Dutt, N., Hedau, A., Awasthi, M.K., Kumar, A., Meena, C.S., Thermo-hydraulic performance investigation of solar air heater duct having staggered D-shaped ribs: Numerical approach. *Heat Transfer*, 53, 3, 1501–1531, 2024. https://doi.org/10.1002/htj.22998.

# 12

# An Overview on Thermal Characterization of Lithium-Ion Batteries for Enhancing the Durability

**Vikas Chaubey[1]*, Atul Kumar[2], Shailendra Sinha[3] and Rakesh Verma[1]**

[1]*Department of Mechanical Engineering, I.E.T, DSMNRU, Uttar Pradesh, India*
[2]*CoEMTD, I.I.T BHU, Varanasi, U. P., India*
[3]*Department of Mechanical Engineering, I.E.T, Lucknow, India*

***Abstract***

This comprehensive chapter provides a thorough overview of the thermal characterization of lithium-ion batteries (LIBs), emphasizing the enhancement of their long-term durability and sustainability. With LIBs serving as the power source for a diverse array of applications, including portable electronics, electric vehicles, and renewable energy systems, ensuring their reliability and environmental compatibility has become a top priority.

The chapter starts by delving into the fundamental principles underlying thermal characterization techniques, such as differential scanning calorimetry (DSC), thermal gravimetric analysis (TGA), and infrared thermography. It clarifies their essential roles in evaluating crucial thermal parameters, including specific heat capacity, thermal conductivity, and thermal stability within LIB components.

Acknowledging the intricate and multi-layered composition of LIBs, the chapter navigates through the inherent challenges associated with thermal characterization. These encompass complexities like electrode–electrolyte interactions, heat generation during charge–discharge cycles, and the impact of temperature on battery aging and degradation. Underscoring the significance of these

**Corresponding author*: vikaschaubey0111@gmail.com; ORCID: 0009-0005-7942-8134
Atul Kumar: ORCID: 0000-0002-2986-8458
Shailendra Sinha: ORCID: 0000-0002-9284-6833
Rakesh Verma: ORCID: 0000-0002-8891-8419

Ashwani Kumar, Nitesh Dutt and Mukesh Kumar Awasthi (eds.) Heat Transfer Enhancement Techniques: Thermal Performance, Optimization and Applications, (269–290) © 2025 Scrivener Publishing LLC

characterizations, it highlights their contribution to bolstering LIB safety and prolonging their operational lifespan.

In summary, this chapter serves as a valuable resource for researchers, engineers, and practitioners seeking a comprehensive understanding of thermal characterization techniques applied to LIBs, particularly concerning the promotion of long-term supportable durability. By shedding light on the pivotal role of thermal analysis in advancing LIB sustainability, it contributes to the ongoing evolution of energy storage technologies while addressing critical environmental concerns.

Furthermore, the chapter highlights recent advancements in thermal management strategies aimed at fortifying LIB durability and sustainability.

***Keywords***: Lithium-ion, durability, characterization, differential scanning calorimetry, thermal gravimetric analysis

## 12.1 Introduction

### 12.1.1 Li-Ion Batteries

In the domain of energy storage, lithium-ion batteries (LIBs) have emerged as pivotal components, showcasing remarkable efficiency, portability, and reliability. These batteries serve as the backbone powering a wide array of devices, from smartphones and laptops to electric vehicles, fundamentally shaping the trajectory of renewable energy. Their inception has marked a transformative milestone in the energy landscape, ushering in a new era of portable power due to their high energy density and rechargeable capabilities [1, 2].

Functionally, LIBs operate based on the movement of lithium ions between electrodes during charging and discharging cycles. This foundational mechanism has enabled diverse applications, positioning these batteries as indispensable components in contemporary technological advancements. Yet, as we aim to maximize their potential, a critical understanding of their thermal behavior becomes indispensable.

This chapter endeavors to delve into the critical realm of thermal characterization within LIBs, shedding light on its pivotal role in bolstering their durability. Through a comprehensive exploration of the inherent thermal properties, behaviors, and challenges associated with these batteries, the objective is to navigate toward reinforcing their resilience and longevity. By meticulously examining techniques for thermal characterization and offering insights into addressing thermal-related degradation, this chapter seeks to fortify the durability of LIBs, ensuring their continued leadership in the energy landscape for the foreseeable future [3].

## 12.2 Thermal Behavior of Li-Ion Battery

The thermal behavior of LIBs stands as a cornerstone in understanding and optimizing their performance and longevity. These batteries exhibit intricate thermal characteristics that significantly influence their operation, safety, and lifespan [4].

During the charging and discharging processes, LIBs generate heat due to internal resistance and electrochemical reactions. This heat generation, if not efficiently managed, can lead to temperature spikes within the battery. Elevated temperatures can accelerate degradation mechanisms, such as electrolyte decomposition, electrode cracking, and accelerated aging of components, ultimately compromising the battery's performance and safety [5].

Moreover, the thermal behavior of LIBs is sensitive to various factors, including the battery's state of charge, current rate, ambient temperature, and cell design. Changes in these parameters can influence the heat generation and dissipation processes within the battery.

Understanding and characterizing the thermal behavior of LIBs involve employing various techniques such as calorimetry, thermal imaging, and thermal modeling. Calorimetry allows for precise measurement of heat generation and absorption during battery operation. Thermal imaging techniques offer real-time visualization of temperature distribution within the battery, aiding in identifying potential hotspots and thermal irregularities [6].

Furthermore, thermal modeling plays a crucial role in predicting and optimizing the thermal behavior of LIBs under different operating conditions. These models simulate heat generation, transfer, and dissipation within the battery, enabling the assessment of various thermal management strategies to enhance performance and durability. A layout for temperature estimation for Li-ion battery is shown in Figure 12.1.

Effectively characterizing and managing the thermal behavior of LIBs is paramount for ensuring their reliability, safety, and extended lifespan [7]. Strategies aimed at efficient heat dissipation, thermal regulation, and design optimization play a pivotal role in enhancing the durability of these energy storage systems.

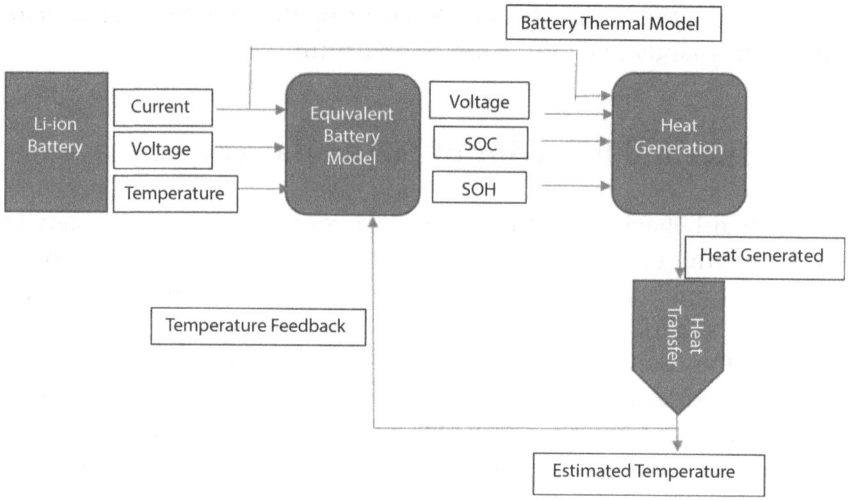


**Figure 12.1** Temperature estimation for Li-ion battery [26].

## 12.3 Heat Generation Mechanism and Thermal Modeling

Heat generation within LIBs arises from various intrinsic processes during charge and discharge cycles. Understanding these mechanisms is critical to effectively model and manage thermal behavior.

### 12.3.1 Ohmic Heating

Resistance within the battery components (electrodes, electrolyte, and current collectors) leads to Ohmic heating. When the battery operates, current flow encounters resistance, generating heat according to Joule's law ($I^2R$), where I is the current and R is the resistance.

### 12.3.2 Faradaic Heating

During charge and discharge, electrochemical reactions at the electrodes cause Faradaic heating. These reactions involve lithium-ion intercalation/deintercalation within the electrode materials, which generates heat due to the energy released or absorbed in these processes.

### 12.3.3 Concentration Polarization

As the battery operates, concentration gradients of lithium ions build up near the electrodes, leading to concentration polarization. Overcoming this polarization requires extra energy, resulting in heat dissipation.

### 12.3.4 Side Reactions

Undesirable reactions at the electrode–electrolyte interface or electrode impurities can occur, causing additional heat generation. These side reactions, like electrolyte decomposition or electrode degradation, contribute to increased heat.

Thermal modeling of LIBs involves complex multiphysics simulations to predict heat generation, distribution, and dissipation within the cell. These models integrate electrochemical, thermal, and transport phenomena [8].

### 12.3.5 1D, 2D, or 3D Thermal Models

These models simulate heat transfer within the battery considering spatial dimensions. They factor in heat conduction through electrodes, current collectors, separator, and electrolyte, accounting for thermal resistances and capacitances.

#### *12.3.5.1 Multiphysics Coupling*

Integrating electrochemical models with thermal ones allows for a comprehensive understanding of how electrochemical reactions generate heat and how this heat affects battery performance. These models capture the interplay between heat generation, chemical reactions, and temperature evolution.

#### *12.3.5.2 Thermal Runaway Prediction Models*

Advanced models predict conditions leading to thermal runaway, where rapid heat generation surpasses heat dissipation, potentially causing catastrophic battery failure. Understanding and preventing thermal runaway is crucial for battery safety.

Design module of lithium-ion battery is shown in Figure 12.2.

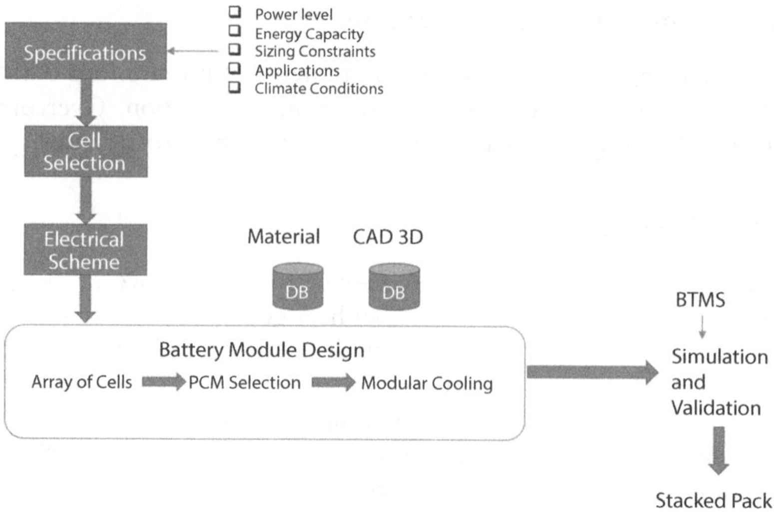


**Figure 12.2** Design module of lithium-ion battery [27].

In summary, thermal modeling of LIBs involves intricate simulations that account for multiple heat generation mechanisms and their interdependencies. Accurate models aid in optimizing battery designs, developing effective thermal management strategies, and ensuring the durability and safety of these energy storage systems [9–11].

## 12.4 The Effect of Temperature on Li-Ion Batteries

Temperature significantly influences the performance, safety, and lifespan of LIBs. The relationship between temperature and battery behavior is multifaceted and crucial for understanding thermal characterization.

### 12.4.1 Optimal Operating Temperature Range

Lithium-ion batteries perform best within a specified temperature range. Operating outside this range can impact their efficiency and lifespan. Typically, an ideal range falls between 15°C and 35°C, but this can vary based on battery chemistry and design.

### 12.4.2 Capacity and Power Output

Low temperatures hinder ion mobility within the battery, reducing its capacity and power output. Conversely, high temperatures enhance ion mobility, improving performance but potentially accelerating degradation mechanisms.

### 12.4.3 Degradation and Aging

Extreme temperatures, both high and low, accelerate the degradation of battery components. High temperatures promote side reactions, leading to electrolyte decomposition and electrode degradation, ultimately reducing the battery's lifespan. Conversely, low temperatures slow down chemical reactions, which can cause increased internal resistance and reduced capacity over time.

### 12.4.4 Safety Concerns

Elevated temperatures can compromise battery safety, increasing the risk of thermal runaway. This phenomenon occurs when the heat generated surpasses the battery's ability to dissipate it, potentially leading to an uncontrollable increase in temperature and catastrophic failure.

### 12.4.5 Thermal Management

Effective thermal management systems are critical to maintaining the battery within the optimal temperature range. These systems utilize cooling or heating mechanisms to regulate the battery's temperature during operation.

### 12.4.6 Charging and Discharging Behavior

Charging an LIB at extreme temperatures can cause irreversible damage. High temperatures during charging can lead to overcharging or thermal instability, while low temperatures can impede the charging process. Discharging at extreme temperatures can reduce available capacity and power output.

Understanding the intricate relationship between temperature and LIB behavior is crucial for devising strategies to enhance their durability. By implementing efficient thermal characterization techniques and developing robust thermal management systems, it is possible to mitigate the detrimental effects of temperature, thereby improving the performance, safety, and longevity of LIBs [12–15].

## 12.5 Thermal Runway Modeling and Safety Tests

Thermal runaway poses a critical risk to LIBs, potentially leading to catastrophic failure. Thermal runway modeling and safety tests are indispensable in understanding and mitigating this hazard, ensuring the safety and durability of these energy storage systems. The life cycle of an LIB is shown in Figure 12.3.

### 12.5.1 Thermal Runaway Modeling

#### *12.5.1.1 Multiphysics Simulations*

Sophisticated computational models integrate electrochemical, thermal, and mechanical aspects to predict the conditions triggering thermal runaway. These models simulate heat generation, transfer, and dissipation, alongside chemical reactions within the battery.

#### *12.5.1.2 Reaction Kinetics*

Modeling reaction kinetics within electrodes and electrolytes aid in identifying potential heat generation mechanisms. Predictive models analyze how different operational conditions, such as current rates, temperatures, and aging, impact the onset of thermal runaway.

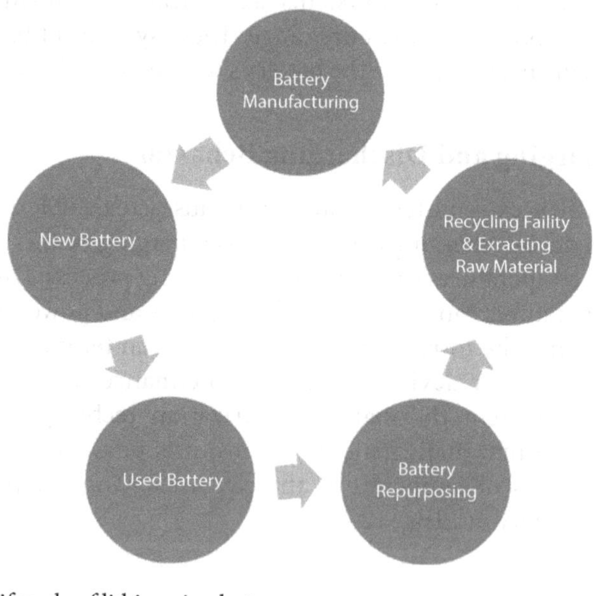


**Figure 12.3** Lifecycle of lithium-ion battery.

#### 12.5.1.3 *Catastrophic Failure Prediction*

Models aim to predict the critical conditions under which battery components fail, causing an uncontrollable rise in temperature. Identifying these conditions helps in designing safety mechanisms to prevent or mitigate thermal runaway.

### 12.5.2 Safety Tests

#### 12.5.2.1 *Accelerated Rate Calorimetry (ARC)*

Accelerated rate calorimetry (ARC) tests subject batteries to controlled thermal abuse, rapidly raising the temperature to simulate extreme conditions. These tests measure heat release rates, aiding in understanding the heat generation behavior and identifying critical temperature thresholds.

#### 12.5.2.2 *Differential Scanning Calorimetry (DSC)*

Differential scanning calorimetry (DSC) tests analyze the thermal behavior of battery materials by measuring heat flow under controlled temperature changes. This helps in assessing the materials' thermal stability and potential risks of thermal decomposition.

#### 12.5.2.3 *Thermal Abuse Testing*

This includes crush, puncture, overcharge, or short-circuit tests under controlled environments to evaluate battery safety under extreme conditions. Observing the response to abuse helps in designing safety features to prevent thermal runaway.

#### 12.5.2.4 *Thermal Imaging and In-Situ Measurements*

Real-time thermal imaging and *in-situ* measurements during abuse tests provide insights into temperature distribution and battery behavior, aiding in validating and refining thermal models.

By combining advanced modeling techniques with rigorous safety tests, researchers and engineers can gain a comprehensive understanding of thermal runaway mechanisms. This knowledge facilitates the development of safer battery designs, effective thermal management strategies, and robust safety protocols to minimize the risk of thermal runaway and ensure the durability and reliability of LIBs.

## 12.6 Interior Electrode Modifications

Interior electrode modifications in LIBs play a pivotal role in enhancing their thermal characteristics, durability, and overall performance. These modifications target the internal structure and composition of the electrodes to address specific challenges related to heat generation, distribution, and dissipation.

### 12.6.1 Nanostructured Materials

Introducing nanostructured materials within electrodes, such as nano-sized particles or thin films, enhances their surface area and promotes faster ion diffusion. This modification improves the battery's efficiency and reduces internal resistance, leading to reduced heat generation during charging and discharging.

### 12.6.2 Coating and Additives

Applying protective coatings or incorporating additives on electrode surfaces helps mitigate side reactions and enhances stability. These modifications minimize electrode degradation, thereby reducing heat generation and improving the overall lifespan of the battery.

### 12.6.3 Conductive Networks

Creating conductive pathways within the electrodes facilitates efficient electron transfer and minimizes resistive losses. Enhanced conductivity leads to reduced heat generation, as it improves the overall efficiency of the battery during operation.

### 12.6.4 Porosity and Structure Control

Adjusting the porosity and structural design of electrodes enables better electrolyte penetration and ion transport. Optimizing these factors enhances the battery's performance by reducing internal heat buildup and improving thermal management.

### 12.6.5 Thermal Conductive Materials

Incorporating materials with high thermal conductivity, such as graphene or carbon nanotubes, within the electrodes aids in efficient heat dissipation.

This modification helps in spreading and dissipating heat more effectively, preventing localized hotspots that can lead to thermal issues.

### 12.6.6 Advanced Composite Structures

Designing electrodes with composite structures, combining different materials with complementary properties, enhances thermal stability and overall performance. These composite structures aim to balance factors like conductivity, stability, and heat dissipation.

Interior electrode modifications hold immense promise in addressing thermal challenges in LIBs. By tailoring the internal composition and structure of electrodes, these modifications aim to optimize performance while mitigating degradation mechanisms that contribute to heat generation, thereby significantly enhancing the durability and reliability of these energy storage systems [16–18].

## 12.7 Exterior Thermal Management System

The exterior thermal management system plays a critical role in regulating the temperature of LIBs, ensuring optimal operating conditions and enhancing their durability. This system focuses on managing heat dissipation and maintaining the battery within the preferred temperature range.

### 12.7.1 Liquid Cooling Systems

These systems use coolants, such as water or specialized fluids, circulated through channels or tubes adjacent to the battery cells. Heat generated during operation is transferred to the coolant, which then dissipates it through a heat exchanger, keeping the battery at an ideal temperature.

### 12.7.2 Air Cooling Systems

Utilizing fans or blowers, air cooling systems draw heat away from the battery cells. Airflow across the battery surface helps in dissipating heat, maintaining the battery within the desired temperature range.

### 12.7.3 Phase Change Materials (PCMs)

Phase change materials (PCMs) absorb and release latent heat during phase transitions. Incorporating these materials in the battery pack helps in regulating temperature by absorbing excess heat during high-demand situations and releasing it when the temperature decreases.

### 12.7.4 Thermal Insulation

Installing insulating materials around the battery pack minimizes heat transfer to the surrounding environment. This helps in maintaining a consistent internal temperature and prevents external temperature fluctuations from affecting the battery.

### 12.7.5 Active Thermal Control Systems

These systems use sensors to monitor battery temperature and regulate the cooling or heating mechanisms accordingly. They provide precise temperature control, optimizing the battery's performance and longevity.

### 12.7.6 Thermal Interface Materials

Enhancing the interface between the battery cells and cooling elements with materials that facilitate efficient heat transfer improves thermal conductivity, ensuring effective heat dissipation.

### 12.7.7 Design Optimization

Efficient placement of the thermal management components within the battery pack, considering airflow, coolant circulation, and heat exchange, is crucial for maximizing their effectiveness.

An effective exterior thermal management system is essential for mitigating thermal-related issues, preventing overheating, and maintaining the battery within the optimal temperature range. By employing these systems, engineers can enhance the durability, safety, and overall performance of LIBs in various applications, from electric vehicles to grid-scale energy storage.

## 12.8 Safety Management Strategy

Developing a robust safety management strategy is crucial for ensuring the reliability and durability of LIBs, especially concerning thermal-related concerns. This strategy encompasses various measures to prevent, monitor, and address potential safety issues associated with battery operation.

**Table 12.1** Lithium-ion battery safety strategy [28].

| | |
|---|---|
| 1. | Follow the manufacturer's instruction for charging, storing, and handling Li-ion battery. |
| 2. | Only use the original equipment manufacturer's charger to charge your e-bike battery. |
| 3. | Always plug the Li-ion battery directly into a wall outlet. Never use power strips or extension cords to charge Li-ion battery. |
| 4. | Never charge the Li-ion battery unattended or overnight. Never leave it plugged in when it is not charging. |
| 5. | Avoid charging or storing Li-ion battery in extreme temperature. |
| 6. | Handle Li-ion battery with care and do not drop or submerge in water. |
| 7. | Do not modify or attempt to repair Li-ion battery. |

Table 12.1 shows LIB safety strategies. These are some of the processes to note for better LIB life.

### 12.8.1 Thermal Monitoring Systems

Implementing comprehensive thermal monitoring systems enables real-time tracking of battery temperature. Sensors strategically placed within the battery pack detect temperature variations, allowing for early detection of abnormal heating.

### 12.8.2 Temperature Thresholds and Alarms

Establishing temperature thresholds and integrating alarm systems that trigger when temperatures surpass safe limits is essential. These alarms alert operators or automated systems, enabling immediate intervention to prevent overheating or thermal runaway.

### 12.8.3 Thermal Shutdown Mechanisms

Incorporating fail-safe mechanisms that trigger a shutdown in case of overheating or abnormal temperature rise is critical. These systems disconnect the battery from the circuit to prevent further thermal escalation and mitigate safety risks.

### 12.8.4 Ventilation and Containment

Designing battery enclosures with proper ventilation and containment systems helps in dissipating heat effectively while containing any potential hazards in the event of thermal runaway.

### 12.8.5 Safety Protocols and Training

Establishing stringent safety protocols and providing training for personnel handling LIBs is essential. Training should cover proper handling, storage, and emergency response procedures in case of thermal incidents.

### 12.8.6 Battery Management Systems (BMS)

Battery management systems (BMS) equipped with thermal protection algorithms continuously monitor and manage battery temperature. They regulate charging rates, discharging limits, and operating conditions to prevent thermal stress on the battery.

### 12.8.7 Rapid Response Plans

Developing rapid response plans outlining steps to be taken in case of thermal incidents is crucial. These plans should include procedures for containment, evacuation, and handling emergency situations to minimize risks.

### 12.8.8 Regulatory Compliance

Adhering to industry safety standards and regulations ensures that the design, manufacturing, and operation of LIBs meet safety requirements and guidelines.

A comprehensive safety management strategy involves a multi-layered approach, integrating preventive measures, monitoring systems, fail-safe mechanisms, and emergency response protocols. By implementing these strategies, manufacturers, operators, and users can enhance the safety, durability, and reliability of LIBs in diverse applications while mitigating thermal-related risks.

## 12.9 Thermal Analysis of Lithium-Ion Batteries

Thermal analysis of LIBs serves as a crucial cornerstone in understanding their behavior, performance, and durability, particularly concerning temperature-related aspects. This comprehensive analysis involves a range of techniques aimed at characterizing the thermal properties and behavior of these energy storage systems:

### 12.9.1 Calorimetry Techniques

Differential scanning calorimetry (DSC) and ARC are used to measure heat flow and thermal properties during different battery states. These techniques provide insights into the battery's thermal stability, heat generation, and reactions under various conditions.

### 12.9.2 Thermal Imaging

Infrared (IR) thermography or thermal imaging visualizes and maps temperature distributions across the battery surface. Real-time monitoring helps identify hotspots, thermal gradients, and irregularities, aiding in diagnosing heat-related issues within the battery.

### 12.9.3 Thermal Modeling

Computational models simulate the heat generation, transfer, and dissipation processes within LIBs. These multiphysics simulations integrate electrochemical reactions, thermal behavior, and material properties, offering predictive insights into temperature dynamics under diverse operating scenarios.

### 12.9.4 Electrochemical Impedance Spectroscopy (EIS)

Electrochemical impedance spectroscopy (EIS) analyzes changes in impedance across different frequencies to understand the electrochemical and thermal characteristics of batteries. It provides valuable information on how temperature variations affect battery performance and internal resistance.

### 12.9.5 Thermal Conductivity Measurements

Techniques assessing thermal conductivity in battery components, such as electrodes, electrolytes, and separators, offer critical data for comprehending heat transfer mechanisms within the battery assembly.

### 12.9.6 Cycling and Aging Studies

Controlled cycling and aging tests conducted under varied temperature conditions help assess battery performance, capacity retention, and degradation rates over time. These analyses reveal temperature impacts on battery lifespan and durability.

### 12.9.7 Thermal Runaway Assessment

Through abuse tests, researchers evaluate the conditions that lead to thermal runaway, a critical safety concern. Understanding temperature thresholds and runaway triggers aids in designing safety protocols and preventative measures.

By employing these diverse techniques, researchers gain comprehensive insights into how temperature affects LIBs' performance, safety, and longevity. This multifaceted thermal analysis is integral to optimizing battery designs, devising effective thermal management strategies, and ensuring the reliability and safety of LIBs across various applications [19–21].

## 12.10 Failures in Lithium-Ion Batteries Pack

Failures in LIB packs can manifest in various forms, often stemming from thermal issues and other factors. Understanding these failures is crucial in improving durability and safety:

### 12.10.1 Thermal Runaway

This catastrophic failure occurs when a battery rapidly heats up uncontrollably, leading to a chain reaction of increased temperature, further heat generation, and potentially resulting in fire or explosion. Internal short circuits, manufacturing defects, or abuse can trigger thermal runaway.

### 12.10.2 Capacity Fade

Over time, LIBs experience a decrease in their maximum charge-holding capacity. Factors like electrode degradation, electrolyte decomposition, and formation of solid-electrolyte interface (SEI) layers contribute to capacity fade.

### 12.10.3 Internal Resistance Increase

Elevated internal resistance within the battery reduces its efficiency and power output. This can result from electrode degradation, separator breakdown, or electrolyte depletion due to high temperatures or cycling stress.

### 12.10.4 Electrode Degradation

Continuous charge–discharge cycles cause physical and chemical changes in electrodes, leading to cracking, loss of active material, or changes in electrode morphology. These degrade performance and contribute to capacity loss.

### 12.10.5 Electrolyte Breakdown

High temperatures or overcharging can cause electrolyte decomposition, leading to the formation of gas, heat, and potential fire hazards. Degradation of the electrolyte affects battery stability and can lead to internal shorts.

### 12.10.6 Thermal Stress

Large fluctuations in temperature, especially extreme heat or cold, induce mechanical stress within the battery, leading to structural damage, electrode expansion, or separator deformation. This stress accelerates aging and can cause premature failure.

### 12.10.7 External Damage

Physical damage due to impact, vibration, or improper handling can compromise the integrity of the battery pack, leading to internal short circuits, leakage, or thermal issues.

Mitigating these failures requires a holistic approach involving improved materials, advanced manufacturing processes, better thermal management, and robust safety mechanisms. Comprehensive thermal characterization and understanding failure modes are pivotal in enhancing the durability and safety of LIB packs across various applications [22–25].

## 12.11 Conclusion and Suggestions

In conclusion, the thermal characterization of LIBs stands as a pivotal endeavor in ensuring their durability, reliability, and safety across diverse applications. The intricate interplay between temperature, performance, and longevity necessitates a multifaceted approach to understanding and mitigating thermal-related challenges.

Throughout this chapter, we've explored various facets of thermal characterization, including techniques like calorimetry, thermal imaging, modeling, and safety assessments. These methodologies offer invaluable insights into heat generation, distribution, and dissipation within battery systems.

To enhance the durability of LIBs:

### 12.11.1 Advanced Thermal Analysis

Continual advancements in thermal analysis techniques are crucial. Further research into precise thermal modeling and characterization methods will aid in accurately predicting and managing temperature-related issues.

### 12.11.2 Materials Innovation

R&D efforts should focus on developing novel materials and electrode designs with improved thermal properties. Nanostructured materials, conductive additives, and high thermal conductivity components can mitigate heat generation and enhance heat dissipation.

### 12.11.3 Optimized Thermal Management

Innovative exterior thermal management systems, including liquid or air cooling, phase change materials, and thermal insulation, must be optimized to efficiently regulate battery temperatures.

### 12.11.4 Safety Protocols and Training

Robust safety management strategies, rapid response plans, and adherence to stringent regulatory standards are imperative to ensure safety during battery operation and handling.

### 12.11.5 Integrated Design Approach

A holistic approach involving collaboration among researchers, engineers, and manufacturers is essential. Integrating thermal considerations into battery design from inception is crucial for achieving durable and safe LIBs.

By embracing these suggestions and continuing to push the boundaries of thermal characterization, the field can stride toward more durable, reliable, and safer LIBs. These advancements will not only bolster the performance of current applications but also pave the way for the emergence of novel and transformative uses for energy storage systems in the future.

## References

1. Crabtree, G., Kócs, E., Trahey, L., The energy-storage frontier: Lithium-ion batteries and beyond. *MRS Bull.*, *40*, 12, 1067–1078, 2015.
2. Bashir, T., Ismail, S.A., Song, Y., Irfan, R.M., Yang, S., Zhou, S., Gao, L., A review of the energy storage aspects of chemical elements for lithium-ion based batteries. *Energy Mater.*, *1*, 100019, 2021.
3. Pirsaheb, M., Seifi, H., Gholami, T., Ganduh, S.H., Jasim, L.S., Mahdi, M.A., Salavati-Niasari, M., Thermal analysis techniques for evaluating the thermal stability of battery materials: A comprehensive review. *J. Anal. Appl. Pyrolysis*, 106136, 2023.
4. Guo, W., Sun, Z., Vilsen, S.B., Meng, J., Stroe, D., II, Review of "grey box" lifetime modeling for lithium-ion battery: Combining physics and data-driven methods. *J. Energy Storage*, *56*, 105992, 2022.
5. Narayan, R., Laberty-Robert, C., Pelta, J., Tarascon, J.M., Dominko, R., Self-healing: an emerging technology for next-generation smart batteries. *Adv. Energy Mater.*, *12*, 17, 2102652, 2022.
6. Raijmakers, L.H.J., Danilov, D.L., Eichel, R.A., Notten, P.H.L., A review on various temperature-indication methods for Li-ion batteries. *Appl. Energy*, *240*, 918–945, 2019.
7. Gandoman, F.H., Jaguemont, J., Goutam, S., Gopalakrishnan, R., Firouz, Y., Kalogiannis, T., Van Mierlo, J., Concept of reliability and safety assessment of lithium-ion batteries in electric vehicles: Basics, progress, and challenges. *Appl. Energy*, *251*, 113343, 2019.
8. Tahir, M.W. and Merten, C., Multi-scale thermal modeling, experimental validation, and thermal characterization of a high-power lithium-ion cell for automobile application. *Energy Convers. Manage.*, *258*, 115490, 2022.
9. Ling, Z., Lin, W., Zhang, Z., Fang, X., Computationally efficient thermal network model and its application in optimization of battery thermal management system with phase change materials and long-term performance assessment. *Appl. Energy*, *259*, 114120, 2020.

10. Shabani, B. and Biju, M., Theoretical modelling methods for thermal management of batteries. *Energies*, *8*, 9, 10153–10177, 2015.
11. Akinlabi, A.H. and Solyali, D., Configuration, design, and optimization of air-cooled battery thermal management system for electric vehicles: A review. *Renew. Sustain. Energy Rev.*, *125*, 109815, 2020.
12. Zhao, R., Zhang, S., Liu, J., Gu, J., A review of thermal performance improving methods of lithium ion battery: Electrode modification and thermal management system. *J. Power Sources*, *299*, 557–577, 2015.
13. Wang, Y., Gao, Q., Wang, G., Lu, P., Zhao, M., Bao, W., A review on research status and key technologies of battery thermal management and its enhanced safety. *Int. J. Energy Res.*, *42*, 13, 4008–4033, 2018.
14. Zhang, X., Li, Z., Luo, L., Fan, Y., Du, Z., A review on thermal management of lithium-ion batteries for electric vehicles. *Energy*, *238*, 121652, 2022.
15. Zhang, J., Zhang, L., Sun, F., Wang, Z., An overview on thermal safety issues of lithium-ion batteries for electric vehicle application. *IEEE Access*, *6*, 23848–23863, 2018.
16. Masias, A., Marcicki, J., Paxton, W.A., Opportunities and challenges of lithium ion batteries in automotive applications. *ACS Energy Lett.*, *6*, 2, 621–630, 2021.
17. Armand, M., Axmann, P., Bresser, D., Copley, M., Edström, K., Ekberg, C., Zhang, H., Lithium-ion batteries–Current state of the art and anticipated developments. *J. Power Sources*, *479*, 228708, 2020.
18. Pender, J.P., Jha, G., Youn, D.H., Ziegler, J.M., Andoni, I., Choi, E.J., Mullins, C.B., Electrode degradation in lithium-ion batteries. *ACS Nano*, *14*, 2, 1243–1295, 2020.
19. Lyu, P., Liu, X., Qu, J., Zhao, J., Huo, Y., Qu, Z., Rao, Z., Recent advances of thermal safety of lithium ion battery for energy storage. *Energy Storage Mater.*, *31*, 195–220, 2020.
20. Abada, S., Marlair, G., Lecocq, A., Petit, M., Sauvant-Moynot, V., Huet, F., Safety focused modeling of lithium-ion batteries: A review. *J. Power Sources*, *306*, 178–192, 2016.
21. Tian, H., Qin, P., Li, K., Zhao, Z., A review of the state of health for lithium-ion batteries: Research status and suggestions. *J. Clean. Prod.*, *261*, 120813, 2020.
22. Blakey-Milner, B., Gradl, P., Snedden, G., Brooks, M., Pitot, J., Lopez, E., Du Plessis, A., Metal additive manufacturing in aerospace: A review. *Mater. Des.*, *209*, 110008, 2021.
23. Benedetti, M., Du Plessis, A., Ritchie, R.O., Dallago, M., Razavi, S.M.J., Berto, F., Architected cellular materials: A review on their mechanical properties towards fatigue-tolerant design and fabrication. *Mater. Sci. Eng.: R: Rep.*, *144*, 100606, 2021.
24. Rissman, J., Bataille, C., Masanet, E., Aden, N., Morrow III, W.R., Zhou, N., Helseth, J., Technologies and policies to decarbonize global industry: Review and assessment of mitigation drivers through 2070. *Appl. Energy*, *266*, 114848, 2020.

25. Hannan, M.A., Hoque, M.M., Hussain, A., Yusof, Y., Ker, P.J., State-of-the-art and energy management system of lithium-ion batteries in electric vehicle applications: Issues and recommendations. *IEEE Access*, *6*, 19362–19378, 2018.
26. Samanta, A. and Williamson, S.S., A Comprehensive Review of Lithium-Ion Cell Temperature Estimation Techniques Applicable to Health-Conscious Fast Charging and Smart Battery Management Systems. *Energies*, 14, 18, 5960, 2021.
27. Cicconi, P., Kumar, P., Varshney, P., A support approach for the modular design of li-ion batteries: A test case with PCM. *J. Energy Storage*, *31*, 101684, 2020.
28. https://www.libertytrike.com/blogs/liberty-trike/lithium-ion-battery-safety-tips

# 13

# An In-Depth Introduction to State of Health Estimation Methods of Li-Ion Batteries

**Prateek Verma**

*National Taiwan University of Science and Technology, Taipei, Taiwan*

***Abstract***

Lithium-ion batteries (Li-ion) are essential in modern energy storage, powering a wide range of applications due to their high energy density, long lifespan, and versatility. Understanding the state of health (SOH) of these batteries, which reflects their capacity, internal resistance, and chemical stability, is crucial for optimizing performance and extending battery life. Accurate SOH prediction ensures reliability, safety, and economic viability, particularly in electric vehicles and renewable energy storage. Traditional SOH assessment methods, which rely on voltage, current, and temperature, have limitations in detecting subtle degradation mechanisms, such as electrode wear and electrolyte degradation. These methods often lack the sensitivity to capture nuanced changes within the battery. To address these limitations, contemporary research has introduced multidimensional health indicators that encompass a wider array of physical, chemical, and electrochemical parameters, offering a more comprehensive and detailed view of battery health. Advanced diagnostic techniques, including electrochemical impedance spectroscopy (EIS), Kalman filter prediction methods, and machine learning, leverage these indicators for more accurate and real-time SOH predictions. These advancements enable proactive maintenance strategies and optimize battery performance. This chapter delves into methodologies for estimating SOH in batteries, covering fundamental principles to advanced techniques, and provides a comprehensive overview of the diverse methods used in battery health prediction.

***Keywords*:** Li-ion batteries, SOH, multidimensional health indicator, model-based estimation, data-driven method

*Email*: prateekch20@gmail.com; ORCID: 0000-0002-5025-1831

Ashwani Kumar, Nitesh Dutt and Mukesh Kumar Awasthi (eds.) Heat Transfer Enhancement Techniques: Thermal Performance, Optimization and Applications, (291–310) © 2025 Scrivener Publishing LLC

## 13.1 Introduction

Lithium-ion batteries (Li-ion) stand as the cornerstone of modern energy storage, powering a diverse spectrum of applications from portable electronic devices to electric vehicles. Their widespread adoption is attributed to their high energy density, long lifespans, and versatility [1]. The pivotal role played by Li-ion batteries in our daily lives and the global transition toward sustainable energy solutions underscores the need for a nuanced understanding of their state of health (SOH).

The SOH encapsulates the batteries' overall condition, reflecting factors such as capacity, internal resistance, and chemical stability. Accurate SOH prediction is not merely a technical consideration but a linchpin for performance optimization and the extension of battery longevity. In the context of electric vehicles, renewable energy storage, and critical electronics, predicting SOH is paramount for ensuring reliability, safety, and economic viability.

However, conventional methods relying on parameters like voltage, current, and temperature have intrinsic limitations in capturing subtle degradation mechanisms. While these parameters offer valuable insights, they often lack the ability to detect subtle changes within the battery at an early stage. The intricacies of electrode wear, electrolyte degradation, and impedance variations that profoundly influence SOH are often elusive to conventional monitoring techniques. Also the limitations of current commercial sensors limit the direct measurement of the battery SOH. Rather, it can only be approximated through the analysis of measured parameters such as current, temperature, and voltage [2, 3].

To address these limitations, contemporary research has introduced multidimensional health indicators. These indicators encompass a broader spectrum of physical, chemical, and electrochemical parameters, granting a more comprehensive perspective on battery health. By considering parameters such as electrode thickness, internal resistance, AC impedance, self-discharge rate, chemical composition, and more, multidimensional health indicators offer a holistic view of the battery's condition. The adoption of advanced diagnostic techniques, including the electrochemical impedance spectroscopy (EIS), Kalman filter prediction methods, and advanced machine learning techniques, has opened new horizons in battery health assessment. Some of these techniques leverage multidimensional health indicators to provide more accurate and real-time SOH predictions, paving the way for proactive maintenance strategies and the optimization of battery performance.

In this chapter, we explore various methodologies used for estimating the State of Health (SOH) in batteries. The focus extends from fundamental principles to advanced techniques, providing a comprehensive overview of the diverse methods employed in battery health prediction.

## 13.2 State of Health

Let us begin by knowing the basic definition of the SOH of a battery. The SOH of an Li-ion battery is a measure of its overall health and performance relative to its original or designed specifications. It provides an assessment of the battery's capacity to store and deliver energy compared to the brand new cell. State of health is expressed as a percentage, with 100% indicating that the battery is operating at its full, original capacity, and a lower percentage indicating a reduction in capacity due to aging or other factors. The SOH can be derived by capacity and internal resistance of the battery.

$$SOH = C_{actual} / C_{rated} \times 100\%$$

$$SOH = (R_{io} - R_{current}) / (R_{io} - R_{new}) \times 100\%$$

where $C_{rated}$ and $C_{actual}$ represent the rated and actual capacity, respectively. The parameter $R_{current}$ signifies the present internal ohmic resistance following multiple charging/discharging cycles. $R_{new}$ represents the initial internal ohmic resistance of a new battery with 100% SOH, while $R_{io}$ represents the internal ohmic resistance at the end of the battery lifespan [3].

The Li-ion battery consists of four fundamental components: anode, cathode, separator, and electrolyte. Consequently, the battery performance hinges on the characteristics of these constituents. Repeated cycling of Li-ion batteries induces inevitable degradation, encompassing both chemical and mechanical effects. Mechanical effects include altering the cell shape due to the insertion and extraction of lithium ions during charging or discharging. This process can induce stress release, leading to fractures or cracks in both the cathode and anode, ultimately diminishing the battery's performance [4]. Under the chemical degradations, the electrolyte breakdowns and form the solid electrolyte interphase (SEI) layer on the anode surface. These irreversible occurrences deplete electrolytes and active lithium ions, causing capacity degradation and an increase in battery resistance [5, 6]. Beyond the impacts of cycling, battery degradation can also result from calendar ageing behaviors. However, the ageing mechanism at the anode and cathode of Li-ion batteries manifests distinctive characteristics, driven by the dynamic interplay of electrochemical processes and material interactions within each electrode.

In a broad sense, the health of a battery is primarily influenced by the reduction in capacity and the gradual decline in power. These problems do not come from just one reason but result from various processes and how they interact with each other. As shown in Figure 13.1, understanding the aging of Li-ion batteries is a complex task involving many factors, as briefly described in [7]. Accurately estimating their health adds an additional layer of complication to the process. At present, the SOH estimation methods can be classified into two major sections based on methods used, such as (Figure 13.2):

- Direct measurement
- Indirect measurement

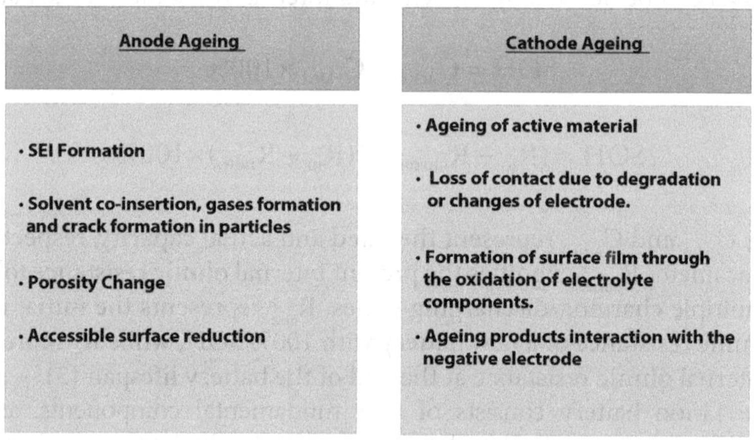


**Figure 13.1** Factors affecting the performance of Li-ion battery based on electrodes.

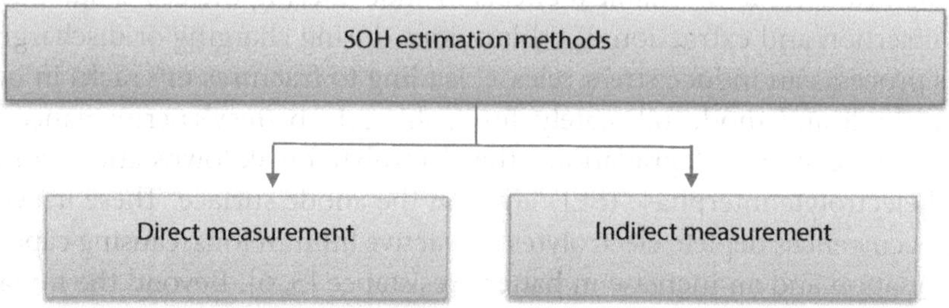


**Figure 13.2** Classifications of SOH estimation methods.

### 13.2.1 Direct Measurement

The direct measurement method centers around comprehensive battery tests, such as internal resistance measurement, impedance measurement, and capacity measurement. Generally, this method utilizes the percentage of remaining battery capacity relative to the rated capacity as a key metric for evaluating battery life. This approach provides a precise assessment of the actual SOH of the battery. However, its implementation involves a meticulous process for both charging and discharging, making it a time-intensive procedure [8]. Determining the maximum capacity through a complete discharging experiment is feasible in theory. However, practical application becomes challenging due to the stringent operating conditions mandated by the experiment. As a result, this method is predominantly utilized for offline applications, where the primary objective is to calibrate and establish the accurate SOH of the battery.

### 13.2.2 Indirect Measurement

#### *13.2.2.1 Model-Based Estimation Method*

Model-based methods have been extensively developed and proven effective in accurately assessing the SOH of energy storage devices in real-time [9, 10]. These methods require replicating the internal dynamics of batteries, employing models such as the electrical equivalent circuit model (EECM) [10], the electrochemical model (ECM) [11], and the electrochemical impedance model (ECIM) [12]. These methods are designed to extract life-related characteristics crucial for assessing the battery's state of health. For instance, the EECM models can effectively capture the internal ohmic resistance and capacitance [3] serving as key parameters for the evaluation of the battery SOH. ECIM employ a system of interconnected non-linear differential equations to depict the various transport, thermodynamic, and kinetic processes within the cell [13]. However, these models commonly rely on partial differential equations (PDEs) containing numerous unknown parameters. This characteristic often results in substantial memory requirements and significant computational loads, rendering electrochemical battery models less practical for real battery management systems (BMS) [14]. In contrast, the EECM can strike a more favorable balance between estimation accuracy and computational efficiency. In Reference [13], various types of battery equivalent circuit models with their corresponding equations were explored, illustrated in the accompanying Figure 13.3.

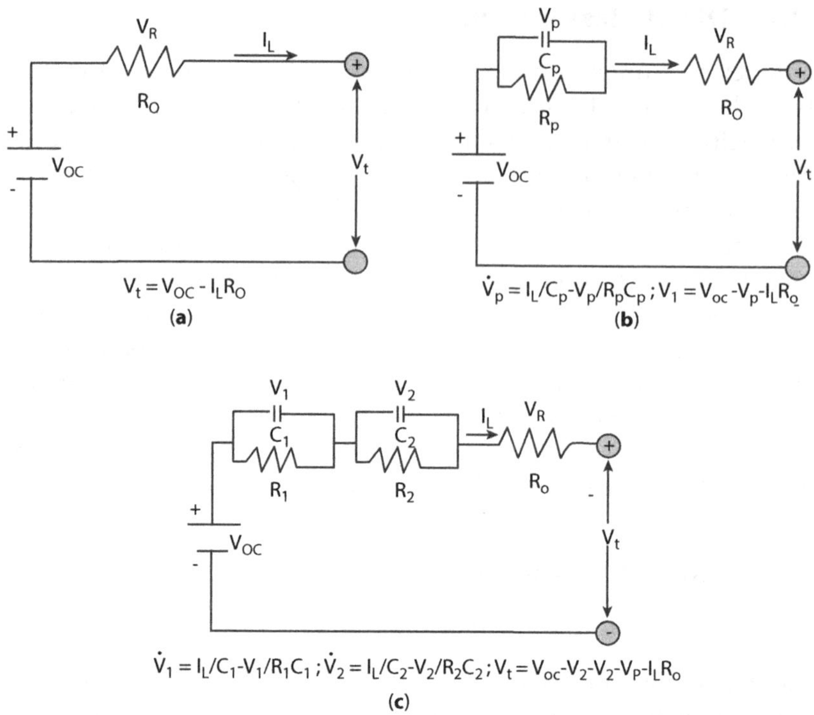


**Figure 13.3** The major types of EECM: (a) the *Rint* model, (b) the *Thevenin* model, and (c) the *DP* model [13].

The terminal voltage ($V_t$) in the model is expressed as a function involving constants ($K_0$, $K_1$, $K_2$, $K_3$, $K_4$) tailored to optimize the model's fitting to data. The equation incorporates parameters such as load current ($I_L$), internal resistance ($R_0$), state of charge ($z_s$), open-circuit voltage ($V_{oc}$), equivalent polarization resistance ($R_p$), and equivalent polarization capacitance ($C_p$) to account for battery relaxation effects during charging and discharging [13]. Additionally, the model includes variables $V_p$ (voltage across $C_p$) and $I_p$ (outflow current of $C_p$). Further details are captured by effective resistances ($R_1$, $R_2$) and effective capacitances ($C_1$, $C_2$), which offer a more intricate representation of the polarization characteristic. The voltages across $C_1$ and $C_2$ are denoted as $V_1$ and $V_2$, respectively.

The model-based estimation methods for battery SOH include:

- **Model Development**
  Creating or selecting a mathematical model that represents the electrochemical and electrical behavior of the battery is

the initial step in the model-based estimation method. This model needs to be appropriate for simulating the electrochemical dynamics of lithium-ion batteries (LIBs) and typically includes equations describing the battery's response to charging, discharging, and aging. Electrochemical impedance spectroscopy (EIS) is widely adopted for developing equivalent circuit models [15] and involves measuring the electrical impedance of the battery across a spectrum of frequencies. Electrochemical impedance spectroscopy (EIS) helps in understanding the electrochemical processes occurring within the battery by examining its response to applied alternating current at different frequencies, providing critical information about internal resistance, capacitance, inductance, and state of charge (SOC) that is directly correlated to the impedance spectrum [16]. The EIS spectrum is typically depicted through a Nyquist plot or a Bode plot. Analyzing measured EIS data involves fitting it with suitable equivalent circuit topologies and relating it to different segments of the plot, each corresponding to various phenomena occurring inside the cell. These phenomena include charge transfer, double-layer capacitance, SEI growth, and diffusion processes [17]. An example of impedance spectra of an aged single cell is shown in Figure 13.4. The cell was subjected to high voltage (3.6 V) and temperature (65°C) conditions for 58 d to understand the changes in the impedance. This

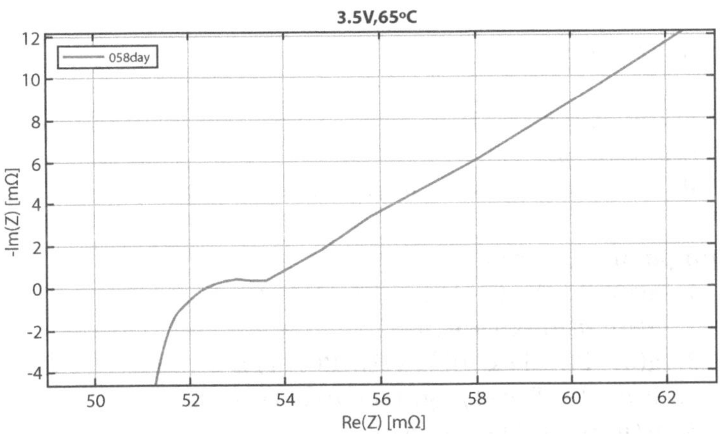


**Figure 13.4** Impedance spectra of aged NCR18650B battery obtained from EIS test under high voltage and high temperature conditions.

information proves valuable for developing models to predict the behavior of aged batteries.

- **Parameter Identification**
  Once we develop the battery model, then the next step is to determine or identify the parameters of the model based on experimental data or known physical properties of the battery. In practical, due to variable usage battery operating conditions, parameter such as internal resistance, capacitance, SOC in the model has irregular values which influences the health estimation results. Offline estimation methods are less complex but incompatible to meet the challenges facing during variable driving conditions because of their dependency on the look-up tables. Several techniques documented in the literature for the online recursive estimation of battery ECM parameters include methods like recursive least squares (RLS) [18–20] and adaptive filter approaches [21], particle swarm optimization (PSO) [22], and H-infinity filter [23]. For real-time parameter identification, RLS method is commonly used due to the fast calculation, better convergence, and accuracy [3]. The RLS algorithm for a p-th order RLS filter can be succinctly summarized as follows [24]:

**Parameters:**

$p$: Filter order.
$\lambda$: Forgetting factor.
$\delta$: Value to initialize $\mathbf{P}$ (0).

**Initialization:**

$\mathbf{w}(0) = 0$,
$\mathbf{x}(k)=0,\ k=-p,\ldots,-1$,
$(k)=0,\ k=-p,\ldots,-1$,
$\mathbf{P}(0)=\delta \boldsymbol{I}$, where $\boldsymbol{I}$ is the identity matrix of rank $p+1$.

**Computation:** For $n=1,2,\ldots$:

1. $\mathbf{x}(n)=[x(n), x(n-1),\ldots,x(n-p)]^T$,
2. $\alpha(n)=d(n)-\mathbf{x}^T(n)\mathbf{w}(n-1)$,
3. $\mathbf{g}(n)=\mathbf{P}(n-1)\,\mathbf{x}(n)[\lambda+\mathbf{x}T(n)\mathbf{P}(n-1)\,\mathbf{x}(n)]\ -1$,
4. $\mathbf{P}(n)=\lambda-1\mathbf{P}(n-1)\ -\mathbf{g}(n)\mathbf{x}T(n)\lambda-1\mathbf{P}(n-1)$,
5. $\mathbf{w}(n)=\mathbf{w}(n-1)\ +\alpha(n)\mathbf{g}(n)$.

The recursion for $\mathbf{P}$ follows an algebraic Riccati equation.

- **State Estimation**
  The EECM falls short of fully reproducing the battery's dynamic characteristics. Hence, it is crucial to choose appropriate filtering algorithms to minimize discrepancies between actual and observed data for precise state estimation. Battery state estimation involves combining available measurements, such as current, temperature, and voltage, with mathematical models and algorithms to infer the current state of the battery. This may include estimating the remaining capacity, internal resistance, lithium-ion concentration, or other relevant parameters that indicate the health of the battery. This process is essential for accurate battery management, determining the available energy, preventing overcharging or over discharging, and extending the overall lifespan of the battery. To achieve online estimation, a combination of the selected EECM and filtering method involves two essential steps. The initial step involves deriving discretized model equations based on the battery EECM. Subsequently, the second step involves establishing a state-space representation using model parameters identified through specific parameter identification methods. One commonly employed approach for state estimation is the utilization of Kalman filters (KF) [25], particularly the extended Kalman filter (EKF) [26] and the unscented Kalman filter (UKF) [27]. These filters have found extensive application either independently or in conjunction with various other filtering techniques, particularly in the context of nonlinear battery state estimation [26], particle filter (PF) [28], and adaptive extended Kalman filter (AEKF) [29].
- **Model Validation**
  Validation is needed to verify the accuracy of the model-based estimates by comparing them to real-world measurements. It ensures that the mathematical representation aligns closely with the actual behavior of the battery. The model can be validated under variable current, voltage, temperature, and aging conditions. Validation is generally carried out by using various current profiles such as such as constant current discharge (CCD) and periodic current profile [9], specifically for dynamic changes while using electric vehicles (EVs), the transient cycles are used such as dynamic stress test (DST) [30], urban dynamometer driving schedule (UDDS) [31], and New York City Cycle (NYCC) driving cycle [9],

some examples has shown in Figure 13.5. NYCC and UDDS describe the urban route through New York and urban route with cold start, respectively. To evaluate the accuracy and robustness of the proposed battery model under complex working conditions, these tests can be useful. ADVISOR is a good simulation tool to generate real time current profiles and working conditions to validate the battery model.

Model-based estimation methods offer several advantages, including the ability to provide detailed insights into the internal state of the battery, such as the health of individual components and dynamic behaviors. These methods can be particularly useful for predictive maintenance, where the goal is to anticipate battery degradation and take corrective actions before significant performance loss occurs.

However, model-based methods also come with challenges, such as the need for accurate models and parameters, sensitivity to variations in operating conditions, and computational complexity. Despite these challenges, ongoing research aims to enhance the accuracy and applicability of model-based estimation methods for battery health monitoring.

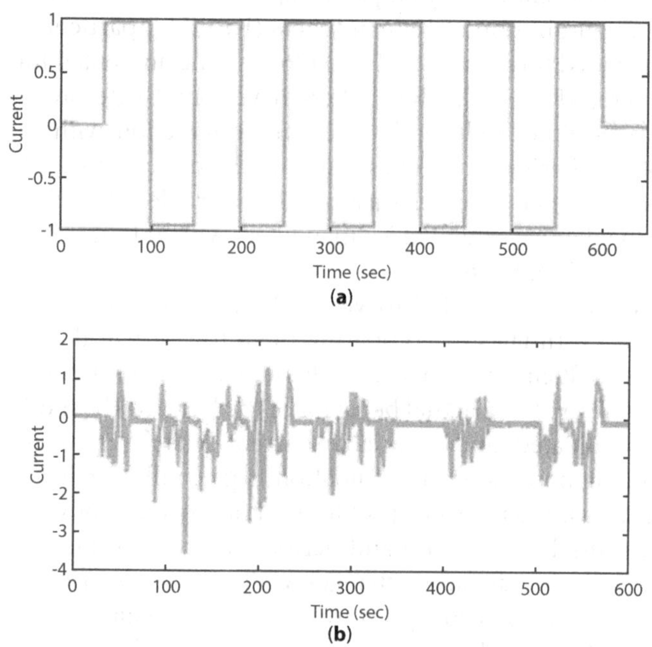


**Figure 13.5** Charging/discharging current profile: (a) periodic. (b) NYCC.

### *13.2.2.2 Data-Driven Model*

A data-driven method in the context of battery health or SOH estimation relies on analyzing and interpreting experimental data directly, without explicit reliance on a predefined mathematical model. Instead of using a theoretical model to represent the underlying processes, data-driven methods extract patterns, relationships, and trends directly from the available data. Furthermore, data-driven methods concentrate on capturing the nonlinear mapping between health features and the SOH of the battery. The data-driven method for SOH estimation of batteries involves a systematic four-step process [3] as shown in Figure 13.6. Initially, relevant data is collected from the battery system, encompassing variables such as voltage, temperature and current, acquired under diverse operational conditions and over the battery's lifecycle. Subsequently, meaningful features indicative of the battery's health is extracted from the collected data, employing statistical, frequency domain, or time-domain analyses. The third step encompasses training a data-driven model using the extracted features and known SOH values, utilizing machine learning (ML) techniques like regression, or neural networks. Finally, the trained model is applied to new, unseen data to predict the current SOH of the battery, providing a comprehensive approach to estimating and monitoring the battery's health in real-world applications.

Here are the key aspects of data-driven methods:

- **Data Collection**
  The method begins with collecting relevant battery data for data-driven methods to estimate SOH under various charging and discharging profiles and operating conditions. This includes employing various charging rates, discharge depths, and cycling frequencies to comprehensively assess the batteries' behavior across different scenarios. Different

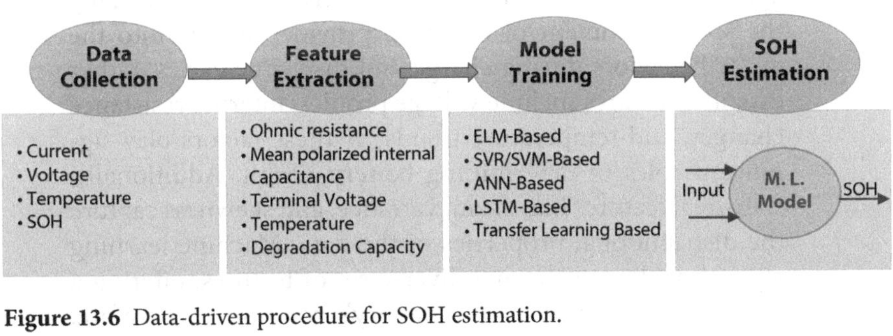


**Figure 13.6** Data-driven procedure for SOH estimation.

temperature conditions should be considered, as thermal variations significantly impact battery performance and aging. Both constant current and constant voltage charging profiles, along with intermittent charging and discharging cycles, should be included in the data collection process to simulate real-world usage scenarios. Moreover, capturing data under extreme conditions, such as high temperatures or rapid charge–discharge cycles, provides insights into the battery's resilience and durability. Integrating a wide range of operating conditions ensures that the data-driven methods have a robust foundation for accurate SOH estimation, allowing for a more thorough understanding of the battery's health under various circumstances. There are some battery data, such as NASA battery dataset [32, 33], MIT dataset, and CALCE battery dataset [34], which can be found online and has been used widely in the literature [35].

- **Feature Extraction**
  Feature extraction is a process in which relevant information or features are identified and extracted from raw data. These features may include statistical properties, trends, or other patterns that can be indicative of the battery's health or performance. In the context of battery health monitoring or any data-driven analysis, feature extraction methods play a crucial role in identifying key characteristics or patterns that can be used for analysis, classification, or prediction. The goal is to identify and quantify relevant characteristics from the collected battery data that directly influence the health status. Features should encapsulate key information about the battery's performance and degradation patterns. Time-domain features, such as cycle count and cumulative capacity loss, are fundamental indicators of a battery's aging. Frequency-domain features, derived from analyzing the spectral content of the data, provide insights into the cyclic behaviors during charge and discharge cycles. Other essential features include voltage profiles, internal resistance changes, and temperature trends, as these factors play significant roles in determining battery health. Additionally, statistical features like mean, variance, and skewness capture the distributional properties of the data. Machine learning algorithms benefit from a diverse set of features, offering a comprehensive view of the battery's condition. Therefore,

a judicious selection of features that encapsulate various aspects of the battery's behavior is crucial for accurate and robust SOH estimation in data-driven approaches. Here are some common feature extraction methods such as statistical features, time or frequency domain features (RMS, FFT), waveform shape analysis, principal component analysis (PCA), independent component analysis (ICA), and auto-aggressive model coefficients. Selecting high-quality input characteristics related to aging mechanisms is crucial for accurate SOH estimation, while simultaneously considering the operational complexity during parameter extraction in online estimation processes.

- **Machine Learning Techniques**
  Data-driven methods often involve the application of ML techniques, where algorithms learn from the data patterns to make predictions or classifications. These are computational algorithms and models that enable computers to learn patterns, make predictions, and improve their performance over time without being explicitly programmed. This phase requires selecting and applying model technique and algorithms while setting and adjusting hyper parameters if needed. Common ML algorithms include extreme learning machine (ELM), artificial neural networks (ANN), support vector regression (SVR), support vector machines (SVM), long short-term memory neural network (LSTM), and regression models. As an example, in Reference [26], a method based on SVR is proposed to find the relationship between the resistor–capacitor (RC) parameters and SOH. Recently, in Reference [45], an ELM prediction method with beetle antenna search (BAS) optimization algorithm was proposed for the online prediction of SOH. Some of the ML methods and their corresponding input features are detailed in Table 13.1, as referenced in [3].
- **Training and Testing**
  The ML algorithm is trained on a subset of the collected data, known as the training set, to teach the model by adjusting its parameters to minimize the difference between predicted and actual labels. Optionally, a validation set aids in fine-tuning. Following training, the model's performance is evaluated on a separate testing set, unseen during training, to assess its ability to generalize to new data. This process helps identify

**Table 13.1** Input feature used in various ML methods for SOH estimation.

| Methods based on | Input features | References |
|---|---|---|
| ELM | Terminal voltage, temperature, current, ohmic polarization resistance, and capacitance | [36, 37] |
| SVR/SVM | Charging profile, current /voltage from current pulse test, charging capacity in constant current/voltage phase, average temperature, polarization resistance/capacity, ohmic resistance, and SOC | [38–41] |
| ANN | Charging capacity, current, voltage and temperature | [42, 43] |
| LSTM | Current, voltage, temperature, and sampling time | [44] |

potential issues such as overfitting or under fitting. Cross-validation and careful consideration of model complexity are common practices for robust evaluation. The goal is to ensure the model not only fits the training data well but also performs effectively on previously unseen instances, a crucial aspect of ML model development. A commonly used ratio of training and testing data is 80:20, respectively, based on the rule of thumb and used in many literatures. Other ratios such as 70:30, 60:40, and even 50:50 are also used in practice [46].

- **Prediction or Classification**
  Once the model is trained, it can be used to predict or classify certain aspects of the battery SOH. For example, it may estimate remaining capacity, internal resistance, or classify the battery into different health states. In Reference [40], the authors predict the battery SOH across diverse operational conditions using the least squares support vector machine (LSSVM) in conjunction with a series of health features derived from charging curves. The utilization of LSSVM, coupled with intricate health features extracted from charging profiles, showcased the effectiveness of this

methodology in capturing the nuanced dynamics of battery health under varying working conditions. Prediction methods utilize historical data and mathematical models to anticipate how a battery's health may evolve over time. On the other hand, classification involves sorting batteries into distinct health states based on identified patterns or features. The ultimate goal is to provide actionable insights into a battery's condition, enabling proactive maintenance and optimized performance in diverse applications ranging from electric vehicles to renewable energy systems.

## 13.3 Conclusion

Data-driven methods are advantageous in situations where the underlying physical processes are complex, not fully understood, or difficult to model accurately. They are particularly effective when dealing with real-world variations and uncertainties. Additionally, data-driven approaches can adapt to changing conditions and do not necessarily rely on a fixed set of assumptions. They also used to develop the correlation between multiple health indicators and battery health conditions [37]. However, these methods also have limitations, such as the need for large and diverse datasets, potential challenges in handling outliers or noisy data, and the risk of overfitting the model to the training data. Striking a balance between the complexity of the model and its generalizability is crucial in achieving reliable results with data-driven methods. While model-based methods and data-driven approaches involve high implementation complexity, they will remain applicable for practical online SOH estimation, unlike the direct method. The practical application of direct methods becomes challenging due to the stringent operating conditions mandated by the experiment. As a result, this method is predominantly utilized for offline applications. In the future, with advanced sensors, hardware, and more real-time data, combining intelligent methods with battery models could be a beneficial practice. This approach can integrate the smart techniques with the battery model, enhancing the coverage of battery dynamics and, as a result, improving the accuracy of health estimations.

## References

1. Guo, Y., Yang, Z., Liu, K., Zhang, Y., Feng, W., A compact and optimized neural network approach for battery state-of-charge estimation of energy storage system. *Energy*, 219, 119529, 2021, http://dx.doi.org/10.1016/j.energy.2020.119529.
2. Zhao, X., Hu, J., Hu, G., Qiu, H., A state of health estimation framework based on real world electric vehicles operating data. *J. Storage Mater.*, 63, 107031, 2023, https:// doi.org/10.1016/j.est.2023.107031.
3. Wang, Z., Feng, G., Zhen, D., Gu, F., Ball, A., A review on online state of charge and state of health estimation for lithium-ion batteries in electric vehicles. *Energy Rep.*, 7, 5141–5161, 2021, https://doi.org/10.1016/j.egyr.2021.08.113.
4. Vennam, G., Sahoo, A., Ahmed, S., A survey on lithium-ion battery internal and external degradation modeling and state of health estimation. *J. Storage Mater.*, 52, 104720, 2022.
5. Kabir, M.M. and Demirocak, D.E., Degradation mechanisms in Li-ion batteries: a state-ofthe-art review. *Int. J. Energy Res.*, 41, 14, 1963–86, 2017.
6. Xu, R., Sun, H., de Vasconcelos, L.S., Zhao, K., Mechanical and structural degradation of LiNixMnyCozO2 cathode in Li-ion batteries: an experimental study. *J. Electrochem. Soc.*, 164, 13, A3333, 2017.
7. Vetter, J., Novák, P., Wagner, M.R., Veit, C., Möller, K.-C., Besenhard, J.O., Winter, M., Wohlfahrt-Mehrens, M., Vogler, C., Hammouche, A., Ageing mechanisms in lithium-ion batteries. *J. Power Sources*, 147, 1–2, 269–281, 2005, ISSN 0378-7753, https://doi.org/10.1016/j.jpowsour.2005.01.006.
8. Peng, K., Deng, Z., Bao, Z., Hu, X., Data-driven battery capacity estimation based on partial discharging capacity curve for lithium-ion batteries. *J. Storage Mater.*, 67, 107549, 2023, https://doi.org/10.1016/j.est.2023.107549.
9. Chiang, C.-J., Verma, P., Huang, W.-R., A lifetime prediction model for the thermo-electric behaviors of ultracapacitors. *J. Power Sources*, 590, 233790, 2024, ISSN 0378-7753, https://doi.org/10.1016/j.jpowsour.2023.233790.
10. Real-time fractional-order equivalent circuit model for internal resistance estimation of lithium-ion cells. *J. Storage Mater.*, 67, 107516, 2023, https://doi.org/10.1016/j.est.2023.107516.].
11. Sadabadi, K.K., Jin, X., Rizzoni, G., Prediction of remaining useful life for a composite electrode lithium ion battery cell using an electrochemical model to estimate the state of health. *J. Power Sources*, 481, 228861, 2021.
12. Mu, H., Xiong, R., Zheng, H., Chang, Y., Chen, Z., A novel fractional order model based state-of-charge estimation method for lithium-ion battery. *Appl. Energy*, 207, 384–393, 2017, http://dx.doi.org/10.1016/j.apenergy.2017.07.003.
13. He, H., Xiong, R., Guo, H., Li, S., Comparison study on the battery models used for the energy management of batteries in electric vehicles. *Energy Convers. Manage.*, 64, 113–121, 2012, ISSN 0196-8904, https://doi.org/10.1016/j.enconman.2012.04.014.

14. Smith, K., Rahn, C.D., Wang, C., Model-based electrochemical estimation and constraint management for pulse operation of lithium ion batteries. *IEEE Trans. Control Syst. Technol.*, 18, 654–63, 2010.
15. Buller, S., Thele, M., DeDoncker, R.W.A.A., Karden, E., Impedance-based simulation models of supercapacitors and Li-ion batteries for power electronic applications. *IEEE Trans. Ind. Appl.*, 41, 742–7, 2005.
16. Westerhoff, U., Kurbach, K., Lienesch, F., Kurrat, M., Analysis of lithium-ion battery models based on electrochemical impedance spectroscopy. *Energy Technol.*, 4, 1620–30, 2016.
17. Kallel, A.Y., Petrychenko, V., Kanoun, O., State-of-Health of Li-Ion Battery Estimation Based on the Efficiency of the Charge Transfer Extracted from Impedance Spectra. *Appl. Sci.*, *12*, 885, 2022, https://doi.org/10.3390/app12020885.
18. Zhu, K., Wan, Y., Li, C., Luo, X., Battery Parameter Identification Using Recursive Least Squares with Variable Directional Forgetting. *2020 IEEE 16th International Conference on Control & Automation (ICCA)*, Singapore, pp. 755–760, 2020, doi: 10.1109/ICCA51439.2020.9264444.
19. Verbrugge, M., Adaptive, multi-parameter battery state estimator with optimized time-weighting factors. *J. Appl. Electrochem.*, 37, 5, 605–16, 2007.
20. Zhang, C., Allafi, W., Dinh, Q., Ascencio, P., Marco, J., Online estimation of battery equivalent circuit model parameters and state of charge using decoupled least squares technique. *Energy*, 142, 678–688, 2018, ISSN 0360-5442, https://doi.org/10.1016/j.energy.2017.10.043.
21. Rahimi–Eichi, H., Baronti, F., Chow, M.-Y., Online adaptive parameter identification and state-of-charge coestimation for lithium-polymer battery cells. *IEEE Trans. Ind. Electron.*, 61, 4, 2053–61, 2014.
22. Li, S., Li, Y., Zhao, D., Zhang, C., Adaptive state of charge estimation for lithium-ion batteries based on implementable fractional-order technology. *J. Energy Storage*, 32, 101838, 2020b.
23. Shu, X., Li, G., Shen, J., Yan, W., Chen, Z., Liu, Y., An adaptive fusion estimation algorithm for state of charge of lithium-ion batteries considering wide operating temperature and degradation. *J. Power Sources*, 462, 228132, 2020c.
24. https://en.wikipedia.org/wiki/Recursive_least_squares_filter#:~:text=Recursive%20least%20squares%20(RLS)%20is,reduce%20the%20mean%20square%20error.
25. Topan, P.A., Ramadan, M.N., Fathoni, G., Cahyadi, A.I., Wahyunggoro, O., State of charge (SOC) and state of health (SOH) estimation on lithium polymer battery via Kalman filter. In. *2016 2nd International Conference on Science and Technology-Computer (ICST)*, pp. 93–96, 2016.
26. Tan, X., Zhan, D., Lyu, Rao, J., Fan, Y., Online state-of-health estimation of lithium-ion battery based on dynamic parameter identification at multi timescale and support vector regression. *J. Power Sources*, 484, 229233, 2021.

27. Cai, Y., Wang, Q., Qi, W., D-UKF based state of health estimation for 18650 type lithium battery. *2016 IEEE International Conference on Mechatronics and Automation*, Harbin, China, pp. 754–758, 2016, doi: 10.1109/ICMA.2016.7558657.
28. Xiong, R., Zhang, Y., He, H., Zhou, X., Pecht, M.G., A double scale, particle filtering, 2017. Energy state prediction algorithm for lithium-ion batteries. *IEEE Trans. Ind. Electron.*, 65, 2, 1526–1538.
29. He, H., Xiong, R., Guo, H., Online estimation of model parameters and state-of-charge of LiFePO4 batteries in electric vehicles. *Appl. Energy*, 89, 1, 413–420, 2012, http://dx.doi.org/10.1016/j.apenergy.2011.08.005.
30. Idaho National Engineering & Environmental Laboratory, *Electric vehicle battery test procedures manual; assistant secretary for energy efficiency and renewable energy (EE)*, Idaho Operations Office, Idaho Falls, ID, USA, 1996.
31. Duong, V.-H., Bastawrous, H.A., See, K.W., Accurate approach to the temperature effect on state of charge estimation in the LiFePO4 battery under dynamic load operation. *Appl. Energy*, 204, 560–571, 2017.
32. Chen, L., Bao, X., Lopes, A.M., Xu, C., Wu, X., Kong, H., Ge, S., Huang, J., State of health estimation of lithium-ion batteries based on equivalent circuit model and data-driven method. *J. Energy Storage*, 73, Part D, 109195, 2023, ISSN 2352-152X, https://doi.org/10.1016/j.est.2023.109195.
33. https://www.kaggle.com/datasets/patrickfleith/nasa-battery-dataset
34. Xing, Y., Ma, E.W.M., Tsui, K.-L., Pecht, M., An ensemble model for predicting the remaining useful performance of lithium-ion batteries. *Microelectron. Reliab.*, 53, 6, 811–820, 2013, ISSN 0026-2714, https://doi.org/10.1016/j.microrel.2012.12.003.
35. Jorge, I., Mesbahi, T., Samet, A., Boné, R., Time Series Feature extraction for Lithium-Ion batteries State-Of-Health prediction. *J. Energy Storage*, 59, 106436, 2023, ISSN 2352-152X, https://doi.org/10.1016/j.est.2022.106436.
36. Chen, L., Wang, H., Liu, B., Wang, Y., Ding, Y., Pan, H., Battery state-ofhealth estimation based on a metabolic extreme learning machine combining degradation state model and error compensation. *Energy*, 215, 119078, 2021a.
37. Pan, H., Lu, Z., Wang, H., Wei, H., Chen, L., Novel battery state-of-health online estimation method using multiple health indicators and an extreme learning machine. *Energy*, 160, 466–477, 2018.
38. Yang, D., Wang, Y., Pan, R., Chen, R., Chen, Z., State-of-health estimation for the lithium-ion battery based on support vector regression. *Appl. Energy*, 227, 273–283, 2018b.
39. Cai, L., Meng, J., Stroe, D.-I., Luo, G., Teodorescu, R., An evolutionary framework for lithium-ion battery state of health estimation. *J. Power Sources*, 412, 615–622, 2019.
40. Deng, Y. *et al.*, Feature parameter extraction and intelligent estimation of the state-of-health of lithium-ion batteries. *Energy*, 176, 91–102, 2019.

41. Shu, X., Li, G., Zhang, Y., Shen, J., Chen, Z., Liu, Y., Online diagnosis of state of health for lithium-ion batteries based on short-term charging profiles. *J. Power Sources*, 471, 228478, 2020d.
42. Song, L., Zhang, K., Liang, T., Han, X., Zhang, Y., Intelligent state of health estimation for lithium-ion battery pack based on big data analysis. *J. Energy Storage*, 32, 101836, 2020b.
43. Fan, Y., Xiao, F., Li, C., Yang, G., Tang, X., A novel deep learning framework for state of health estimation of lithium-ion battery. *J. Energy Storage*, 32, 101741, 2020.
44. Li, P. *et al.*, State-of-health estimation and remaining useful life prediction for the lithium-ion battery based on a variant long short term memory neural network. *J. Power Sources*, 459, 228069, 2020f, http://dx.doi.org/10.1016/j.jpowsour.2020.228069.
45. Hou, X., Guo, X., Yuan, Y., Zhao, K., Tong, L., Yuan, C., Teng, L., The state of health prediction of Li-ion batteries based on an improved extreme learning machine. *J. Energy Storage*, 70, 108044, 2023, ISSN 2352-152X, https://doi.org/10.1016/j.est.2023.108044.
46. https://onlinelibrary.wiley.com/doi/epdf/10.1002/sam.11583

# 14

# Heat and Mass Transportation Enhancement of Casson Cu-AA7075-AA7072/Methanol Tri-Hybrid Nanofluid Flow Past A Porous Spinning Disk: A Computational Assessment

**Bhagyashri Patgiri*** **and Ashish Paul**

*Department of Mathematics, Cotton University, Guwahati, Assam, India*

### *Abstract*

This study focuses on the computational analysation of mass and heat relocation upon the flow of Casson Cu-AA7075-AA7072/methanol tri-hybrid nanofluids across a spinning disk under thermal radiation, slip condition, and chemical reaction. The obtained PDEs are transposed into non-linear ordinary differential equations (ODEs) by adapting the appropriate similarity substitution. Bvp4c method is opted to derive the numerical outcomes. Implications of important physical factors like porosity, magnetic, Casson, radiation parameters and nanoparticle's volume fractions, Schmidt number, and Prandtl number are displayed and discussed via graphs. Methanol can serve as a fuel source in cell batteries as it can reserve an enormous amount of energy while occupying a minuscule amount of space. These cell batteries generally exist in electronic devices like phones and laptops. Furthermore, Cu nanoparticles have many advantageous thermophysical features which enable Cu nanoparticles to be utilized as coolant in disk drives, computer monitors, televisions, and in many other electronic equipment. Our investigation unveils that sturdy values of Cu nanoparticles lessens the temperature profile and so it can be utilized as a coolant. Besides, our acquired results explain that sturdy values of the Casson parameter enhance both the mass and heat transmission rates while reverse tendency is depicted for

**Corresponding author*: iambhagyashripatgiri@gmail.com
Bhagyashri Patgiri: ORCID: 0009-0008-5455-841X
Ashish Paul: ORCID: 0000-0001-7542-4025

Ashwani Kumar, Nitesh Dutt and Mukesh Kumar Awasthi (eds.) Heat Transfer Enhancement Techniques: Thermal Performance, Optimization and Applications, (311–330) © 2025 Scrivener Publishing LLC

the skin friction. Also, chemical reaction parameter's bigger values amplify both mass and heat transmission rates.

***Keywords*:** Casson model, methanol base fluid, tri-hybrid nanofluid, AA7075 and AA7072 nanoparticles, heat and mass transmission

## Nomenclature

| | |
|---|---|
| $(\mathbb{r},\phi,\mathbb{z})$: | cylindrical coordinates |
| $(\mathbb{u},\mathbb{v},\mathbb{w})$: | velocity components |
| $\mathfrak{L}$: | angular speed |
| $\mathcal{K}$: | thermal conductivity |
| $\mathfrak{J}$: | electrical conductivity |
| $\mathfrak{m}$: | dynamic viscosity |
| $\vartheta$: | kinematic viscosity |
| $\mathfrak{p}$: | density |
| $\mathfrak{D}$: | mass diffusivity |
| $c_p$: | heat capacity |
| $\mathbb{B}$: | magnetic coefficient |
| $\mathfrak{X}$: | slip coefficient |
| $\mathbb{T}$: | temperature |
| $\mathfrak{R}^*$: | chemical reaction coefficient |
| $\mathbb{C}$: | concentration |
| $(\mathbb{T}_W,\mathbb{C}_W)$: | wall temperature and concentration |
| $k^*$: | mean absorption coefficient |
| $(\mathbb{T}_\infty,\mathbb{C}_\infty)$: | ambient temperature and concentration |
| $\mathfrak{G}^*$: | Stefan Boltzmann constant |
| $\mathcal{P}$: | porosity coefficient |
| $\delta_m$: | magnetic parameter |
| $\delta_c$: | Casson parameter |
| $P_r$: | Prandtl number |
| $\delta_p$: | porosity parameter |
| $\delta_s$: | slip parameter |
| $S_c$: | Schmidt number |
| $R_o$: | radiation parameter |
| $c_r$: | chemical reaction parameter |
| Re: | Reynold number |
| $\eta$: | dimensionless similarity variable |
| $p'$, q: | dimensionless velocity profiles |

| | |
|---|---|
| r: | dimensionless temperature |
| s: | dimensionless concentration |
| THNF: | tri-hybrid nanofluid |

## 14.1 Introduction

Tri-hybrid nanofluids are the newest genre of nanofluids produced by mixing three separate nanoparticles. The thermophysical assets of this genre of nanofluids are superior to hybrid and mono nanoparticles. Therefore, many researchers seeks to investigate this particular genre of nanofluids. Sundar *et al.* [1] describes rGO+$TiO_2$+$Fe_3O_4$ /EG THNF (ethylene glycol tri-hybrid nanofluid) flow on a circular tube. Elnaqeeb *et al.* [2] addressed the flow due to two water-based THNFs with varying densities and shapes of the nanoparticles. Kumar *et al.* [3] studied the heat transmission analysis of (carbon nanotube) CNT+graphene+ZnO/water-based THNF flow with shape effects. Mahmood *et al.* [4] reported Cu+$TiO_2$+$Al_2O_3$/polymer THNF flow through a shrinking sheet accompanied by slip impacts. Khan *et al.* [5] provided the flow of a water-based THNF across a rotating sphere. Mahmood *et al.* [6] investigated Cu+ $Fe_3O_4$+$SiO_2$/polymer THNF flow due to a heated cylinder with shape impacts. Bilal *et al.* [7] analysed $TiO_2$+$Fe_3O_4$+MgO/$H_2O$ THNF flow across three surfaces: plate, wedge, and cone. Xiu *et al.* [8] elaborated the forced convective THNF flow across wedge surfaces. Animasaun *et al.* [9] elucidated the Darcy flow of CNT+graphene+$Al_2O_3$/water-based THNF on horizontal surfaces. Elsebaee *et al.* [10] addressed MgO+$TiO_2$+ $Fe_3O_4$/$H_2O$ THNF flow past a slender stretching sheet.

The flow induced by Casson fluid is of tremendous industrial importance because of its marvellous applications in coating, drilling processes, lubrication, biological treatments, etc. A large number of surveillance have been performed on flows due to Casson fluids. One may notice the associated studies in References [11–17].

The inspection of thermal radiation impacts in fluid flow systems has grabbed the attention of academicians and researchers owing to its several real life utilizations such as extraction, crude oil, nuclear power reactors, and geothermal systems. A list of research articles on fluid flow due with thermal radiation impacts can be perceived in References [18–24].

The investigation of mass and heat relocation with chemical reactions has become a relevant significant topic to scientists and engineers because of its incidence in several areas in engineering and science. This phenomenon illustrates an important role in the cooling industry, evaporation, chemical industry, cooling tower, desert cooler, etc. One may perceive the connected articles in References [25–27].

The major motive of the current surveillance is to analyse the mass and heat relocation of Casson Cu-AA7075-AA7072/methanol tri-hybrid nanofluid flow in occurrence of thermal radiation and chemical reaction past a porous spinning disk in attendance of slip effects and by applying bvp4c procedure. As far authors' awareness, no such inspection has been conducted yet. Tabular and graphical illustrations for several flow variables are scrutinized in detail.

## 14.2 Problem Formulation

In this research, the influence of magnetohydrodynamics (MHD), thermal radiation, nanoparticle volume fraction, and chemical reaction on the 3D flow of the micropolar Cu–$H_2O$ nanofluid across one porous rotatory disk is examined. A demonstration of this is presented in Figure 14.1. The cylindrical coordinates that are applied are applied are $(\mathbb{r},\phi,\mathbb{z})$, and $(\mathbb{u},\mathbb{v},\mathbb{w})$ are the corresponding velocities. $(\mathbb{T}_W, \mathbb{C}_W)$ mean wall temperature and concentration and $(\mathbb{T}_\infty, \mathbb{C}_\infty)$ mean ambient temperature and concentration, and $\mathbb{B}$ is the magnetic field. Also, the spinning disk rotates around $\mathbb{z} = 0$ and $\mathfrak{J}$ corresponds to the angular speed.

By keeping the stated presumptions in mind, the flow equations are provided as (Rehman *et al.* [28] and Shoaib *et al.* [29]):

$$\frac{\partial \mathbb{u}}{\partial \mathbb{u}} + \frac{\mathbb{u}}{\mathbb{r}} + \frac{\partial \mathbb{w}}{\partial \mathbb{z}} = 0 \tag{14.1}$$

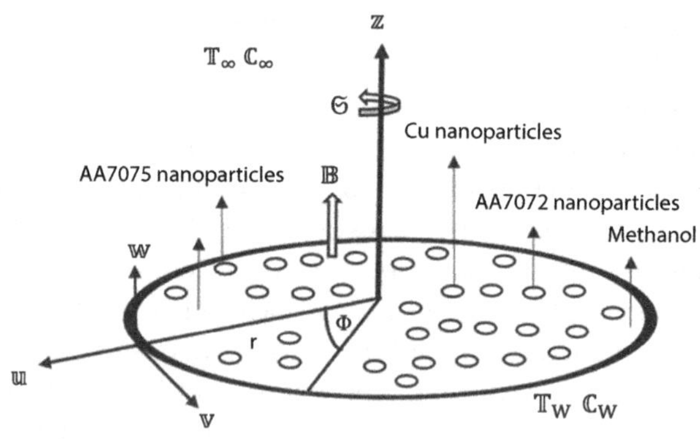


**Figure 14.1** Physical model.

$$\mathbb{u}\frac{\partial \mathbb{u}}{\partial \mathbb{r}}+\mathbb{w}\frac{\partial \mathbb{u}}{\partial \mathbb{z}}-\frac{\mathbb{v}^2}{\mathbb{r}}=\frac{\mathfrak{m}_{thnf}}{\mathfrak{p}_{thnf}}\left(1+\frac{1}{\delta}\right)\left(\frac{\partial^2 \mathbb{u}}{\partial \mathbb{r}^2}+\frac{1}{\mathbb{r}}\frac{\partial \mathbb{u}}{\partial \mathbb{r}}-\frac{\mathbb{u}}{\mathbb{r}^2}+\frac{\partial^2 \mathbb{u}}{\partial \mathbb{z}^2}\right)-\frac{\mathfrak{m}_{thnf}}{\mathfrak{p}_{thnf}}\left(\frac{\mathbb{u}}{\mathcal{P}}\right)-\frac{\mathfrak{L}_{thnf}}{\mathfrak{p}_{thnf}}\cdot\mathbb{u}\cdot\mathbb{B}^2 \tag{14.2}$$

$$\mathbb{u}\frac{\partial \mathbb{v}}{\partial \mathbb{r}}+\mathbb{w}\frac{\partial \mathbb{v}}{\partial \mathbb{z}}-\frac{\mathbb{u}\mathbb{v}}{\mathbb{r}}=\frac{\mathfrak{m}_{thnf}}{\mathfrak{p}_{thnf}}\left(1+\frac{1}{\delta_c}\right)\left(\frac{\partial^2 \mathbb{v}}{\partial \mathbb{r}^2}+\frac{1}{\mathbb{r}}\frac{\partial \mathbb{v}}{\partial \mathbb{r}}-\frac{\mathbb{v}}{\mathbb{r}^2}+\frac{\partial^2 \mathbb{v}}{\partial \mathbb{z}^2}\right)-\frac{\mathfrak{m}_{thnf}}{\mathfrak{p}_{thnf}}\left(\frac{\mathbb{v}}{\mathcal{P}}\right)-\frac{\mathfrak{L}_{thnf}}{\mathfrak{p}_{thnf}}\cdot\mathbb{u}\cdot\mathbb{B}^2 \tag{14.3}$$

$$\mathbb{u}\frac{\partial \mathbb{w}}{\partial \mathbb{r}}+\mathbb{w}\frac{\partial \mathbb{w}}{\partial \mathbb{z}}=\frac{\mathfrak{m}_{thnf}}{\mathfrak{p}_{thnf}}\left(1+\frac{1}{\delta_c}\right)\left(\frac{\partial^2 \mathbb{w}}{\partial \mathbb{r}^2}+\frac{1}{\mathbb{r}}\frac{\partial \mathbb{w}}{\partial \mathbb{r}}+\frac{\partial^2 \mathbb{w}}{\partial \mathbb{z}^2}\right)-\frac{\mathfrak{m}_{thnf}}{\mathfrak{p}_{thnf}}\left(\frac{\mathbb{u}}{\mathcal{P}}\right) \tag{14.4}$$

$$\mathbb{u}\frac{\partial \mathbb{T}}{\partial \mathbb{r}}+\mathbb{w}\frac{\partial \mathbb{T}}{\partial \mathbb{z}}=\left[\frac{\mathcal{K}_{thnf}}{(\mathfrak{p}c_p)_{thnf}}+\frac{1}{(\mathfrak{p}c_p)_{thnf}}\left(\frac{16\mathfrak{G}^*\,\mathbb{T}_\infty{}^3}{3k^*}\right)\right]\left(\frac{\partial^2 \mathbb{T}}{\partial \mathbb{r}^2}+\frac{1}{\mathbb{r}}\frac{\partial \mathbb{T}}{\partial \mathbb{r}}+\frac{\partial^2 \mathbb{T}}{\partial \mathbb{z}^2}\right) \tag{14.5}$$

$$\mathbb{u}\frac{\partial \mathbb{C}}{\partial \mathbb{r}}+\mathbb{w}\frac{\partial \mathbb{C}}{\partial \mathbb{z}}=\mathfrak{D}_{thnf}\left(\frac{\partial^2 \mathbb{C}}{\partial \mathbb{r}^2}+\frac{1}{\mathbb{r}}\frac{\partial \mathbb{C}}{\partial \mathbb{r}}+\frac{\partial^2 \mathbb{C}}{\partial \mathbb{z}^2}\right)-\mathfrak{R}^*(\mathbb{C}-\mathbb{C}_\infty) \tag{14.6}$$

And the decided boundary assumptions are:

$$\left.\begin{aligned}&\mathbb{u}=\mathfrak{X}\left(\frac{\partial \mathbb{u}}{\partial \mathbb{r}}\right),\ \mathbb{v}=\mathbb{r}\mathfrak{J}+\mathfrak{X}\frac{\partial \mathbb{v}}{\partial \mathbb{z}},\ \mathbb{w}=0,\ \mathbb{T}=\mathbb{T}_W \text{ and } \mathbb{C}=\mathbb{C}_W \text{ at } \mathbb{z}=0\\ &\mathbb{u}\to 0,\mathbb{v}\to 0,\mathbb{w}\to 0,\mathbb{T}\to\mathbb{T}_\infty \text{ and } \mathbb{C}\to\mathbb{C}_\infty \text{ at } \mathbb{z}\to\infty\end{aligned}\right\} \tag{14.7}$$

Here, subscripts f, nf, hnf, and thnf correspond to working fluid, nanofluid, hybrid nanofluid, and tri-hybrid nanofluid, respectively. Besides, the symbols $\delta_c, \mathfrak{m}, \mathfrak{p}, \mathcal{P}, \mathfrak{L}, \mathcal{K}, \mathfrak{p}c_p, \mathfrak{G}^*, \mathbb{T}_\infty, k^*, \mathfrak{D}, \mathfrak{R}^*, \mathfrak{X}, \mathfrak{J}$, and $\mathbb{C}_\infty$ are depicted in the nomenclature segment.

**Table 14.1** Physical properties of methanol, AA7075, AA7072, and Cu (Tlili *et al.* [31] and Vishalakshi *et al.* [32]).

| Name | ρ (kg/m³) | $c_p$ (J/kgK) | k (W/mK) | σ (S/m) |
|---|---|---|---|---|
| Methanol | 792 | 2545 | 0.2035 | $0.5 \times 10^{-6}$ |
| AA7075 | 2810 | 960 | 173 | $26.77 \times 10^{6}$ |
| AA7072 | 2270 | 893 | 222 | $34.83 \times 10^{6}$ |
| Cu | 8933 | 385 | 401 | $59.7 \times 10^{6}$ |

The physical traits of Cu, AA7072, AA7075, and methanol are stated in Table 14.1. Furthermore, thermophysical correspondence relating base fluid (methanol), AA7075/methanol (nanofluid), AA7072-AA7075/methanol (hybrid nanofluid), and Cu-AA7072-AA7075/methanol (tri-hybrid nanofluid) are outlined as follows:

$$\mathfrak{m}_{nf} = \frac{\mathfrak{m}_f}{(1-n_1)^{2.5}}, \mathfrak{m}_{hnf} = \frac{\mathfrak{m}_f}{(1-n_1)^{2.5}(1-n_2)^{2.5}},$$

$$\mathfrak{m}_{thnf} = \frac{\mathfrak{m}_f}{(1-n_1)^{2.5}(1-n_2)^{2.5}(1-n_3)^{2.5}},$$

$$\mathfrak{p}_{nf} = (1-n_1)\mathfrak{p}_f + n_1\mathfrak{p}_1, \ \mathfrak{p}_{hnf} = (1-n_2)\{(1-n_1)\mathfrak{p}_f + n_1\mathfrak{p}_1\} + n_2\mathfrak{p}_2,$$

$$\mathfrak{p}_{thnf} = (1-n_3)\left[(1-n_2)\{(1-n_1)\mathfrak{p}_f + n_1\mathfrak{p}_1\} + n_2\mathfrak{p}_2\right] + n_3\mathfrak{p}_3,$$

$$(\mathfrak{p}c_p)_{nf} = (1-n_1)(\mathfrak{p}c_p)_f + n_1(\mathfrak{p}c_p)_1,$$
$$(\mathfrak{p}c_p)_{hnf} = (1-n_2)\{(1-n_1)(\mathfrak{p}c_p)_f + n_1(\mathfrak{p}c_p)_1\} + n_2(\mathfrak{p}c_p)_2,$$

$$(\mathfrak{p}c_p)_{thnf} = (1-n_3)\left[(1-n_2)\{(1-n_1)(\mathfrak{p}c_p)_f + n_1(\mathfrak{p}c_p)_1\} + n_2(\mathfrak{p}c_p)_2\right] + n_3(\mathfrak{p}c_p)_3,$$

$$\mathcal{K}_{nf} = \mathcal{K}_f \left\{ \frac{\mathcal{K}_1 + 2\mathcal{K}_f - 2n_1(\mathcal{K}_f - \mathcal{K}_1)}{\mathcal{K}_1 + 2\mathcal{K}_f + n_1(\mathcal{K}_f - \mathcal{K}_1)} \right\},$$

$$\mathcal{K}_{hnf} = \mathcal{K}_f \left\{ \frac{\mathcal{K}_2 + 2\mathcal{K}_{nf} - 2n_2(\mathcal{K}_{nf} - \mathcal{K}_2)}{\mathcal{K}_2 + 2\mathcal{K}_{nf} + n_2(\mathcal{K}_{nf} - \mathcal{K}_2)} \right\},$$

$$\mathcal{K}_{thnf} = \mathcal{K}_f \left\{ \frac{\mathcal{K}_3 + 2\mathcal{K}_{hnf} - 2n_3\left(\mathcal{K}_{hnf} - \mathcal{K}_3\right)}{\mathcal{K}_3 + 2\mathcal{K}_{hnf} + n_3\left(\mathcal{K}_{hnf} - \mathcal{K}_3\right)} \right\},$$

$$\mathfrak{L}_{nf} = \mathfrak{L}_f \left[ \frac{\mathfrak{L}_1 + 2\mathfrak{L}_f - 2n_1\left(\mathfrak{L}_f - \mathfrak{L}_1\right)}{\mathfrak{L}_1 + 2\mathfrak{L}_f + n_1\left(\mathfrak{L}_f - \mathfrak{L}_1\right)} \right],$$

$$\mathfrak{L}_{hnf} = \mathfrak{L}_f \left[ \frac{\mathfrak{L}_2 + 2\mathfrak{L}_{nf} - 2n_2(\mathfrak{L}_{nf} - \mathfrak{L}_2)}{\mathfrak{L}_2 + 2\mathfrak{L}_{nf} + n_2(\mathfrak{L}_{nf} - \mathfrak{L}_2)} \right],$$

$$\mathfrak{L}_{thnf} = \mathfrak{L}_f \left[ \frac{\mathfrak{L}_3 + 2\mathfrak{L}_{hnf} - 2n_3(\mathfrak{L}_{hnf} - \mathfrak{L}_3)}{\mathfrak{L}_3 + 2\mathfrak{L}_{hnf} + n_3(\mathfrak{L}_{hnf} - \mathfrak{L}_3)} \right]$$

Now, we alter Equations 14.1–14.7 into a dimensionless set by using the following (Shafiq *et al.* [30]) transformations:

$$\eta = \left(\sqrt{\frac{2\mathfrak{I}}{x}}\right) \mathbb{z}, \mathbb{u} = \mathbb{r}\, \mathfrak{I} p'(\eta), \mathbb{v} = \mathbb{r}\, \mathfrak{I} p(\eta), \mathbb{w} = -\sqrt{2\mathfrak{I}x}\, p(\eta),$$

$$\left. r(\eta), \frac{(\mathbb{T} - \mathbb{T}_\infty)}{(\mathbb{T}_W - \mathbb{T}_\infty)}, s(\eta) = \frac{(\mathbb{C} - \mathbb{C}_\infty)}{(\mathbb{C}_W - \mathbb{C}_\infty)} \right\} \tag{14.8}$$

Using Equation 14.8 in Equations 14.1 to 14.7, we procure the following equations:

$$\frac{2\left(1 + \frac{1}{\delta_c}\right)}{\mathbb{M}_1 \mathbb{M}_2} p''' - \left[ (p')^2 - 2pp'' - q^2 \right]$$
$$- \frac{\left(1 + \frac{1}{\delta_c}\right)}{\mathbb{M}_1 \mathbb{M}_5} \cdot \delta_p \cdot p' - \frac{2\mathbb{M}_5}{\mathbb{M}_2} \cdot \delta_m \cdot p' = 0 \tag{14.9}$$

$$\frac{2\left(1+\frac{1}{\delta_c}\right)}{\mathbb{M}_1\mathbb{M}_2} q'' - \left[2p'q - 2pq'\right] - \frac{2\left(1+\frac{1}{\delta_c}\right)}{\mathbb{M}_1\mathbb{M}_2} \cdot \delta_p \cdot q - \frac{2\mathbb{M}_5}{\mathbb{M}_2} \cdot \delta_m \cdot q = 0 \tag{14.10}$$

$$\left[\mathbb{M}_4 + \frac{4}{3}R_o\right] r'' + \mathbb{M}_3 \cdot P_r \cdot p \cdot r' = 0 \tag{14.11}$$

$$\left(\frac{1}{S_c}\right) s'' + ps' - c_r \cdot s = 0 \tag{14.12}$$

And

$$\left.\begin{array}{l} p'(0) = \delta_s \cdot p''(0), q(0) = 1 + s \cdot q'(0), p(0) = 0, r(0) = 1, s(0) = 1 \text{ at } \eta = 0 \\ p(\infty) \to 0, p'(\infty) \to 0, q(\infty) \to 0, r(\infty) \to 0, s(\infty) \to 0 \text{ as } \eta \to \infty \end{array}\right\} \tag{14.13}$$

Here,

$$\mathbb{M}_1 = \frac{\mathfrak{m}_{thnf}}{\mathfrak{m}_f}, \mathbb{M}_2 = \frac{\mathfrak{p}_{thnf}}{\mathfrak{p}_f}, \mathbb{M}_3 = \frac{(\mathfrak{p}c_p)_{thnf}}{(\mathfrak{p}c_p)_f}, \mathbb{M}_4 = \frac{\mathcal{K}_{thnf}}{\mathcal{K}_f}, \mathbb{M}_5 = \frac{\mathfrak{L}_{thnf}}{\mathfrak{L}_f}$$

Also,

$$\vartheta_f = \frac{\mathfrak{m}_f}{\mathfrak{p}_f}\left(\text{kinematic viscosity}\right), \quad \delta_m = \frac{\mathfrak{L}_f \mathbb{B}^2}{2\mathfrak{J}\mathfrak{p}_f}(\text{magnetic parameter}),$$

$$Re = \frac{\mathfrak{J}\mathbb{r}^2}{\vartheta_f} \text{ (Reynold's number)}, P_r = \frac{\vartheta_f (\mathfrak{p}c_p)_f}{\mathcal{K}_f} \text{ (Prandtl number)},$$

$$\delta_s = \mathfrak{X}\left(\sqrt{\frac{2\mathfrak{J}}{\vartheta_f}}\right)(\text{velocity slip parameter}), \ S_c = \frac{\vartheta_f}{\mathfrak{D}_f}(\text{Schimdt number}),$$

$$R_O = \frac{4\mathfrak{G}^*\mathbb{T}_\infty{}^3}{3k^*\mathcal{K}} \text{ (radiation parameter),}$$

$$c_r = \frac{\mathfrak{R}^*}{2\mathfrak{J}} \text{ (chemical reaction parameter)}$$

and $\delta_p = \dfrac{\vartheta_f}{\mathfrak{J}\cdot\mathcal{P}}$ (porosity parameter)

The dimensionless physical quantities are:
Skin friction coefficient:

$$Ce_0Re^{0.5} = \left(1+\frac{1}{\delta_c}\right)\cdot\frac{\sqrt{\left(p''(0)\right)^2+\left(q(0)\right)^2}}{(1-n_1)^{2.5}(1-n_2)^{2.5}(1-n_3)^{2.5}}$$

Local Nusselt number (heat relocation rate):

$$Ne_0Re^{-0.5} = -\left(\frac{\mathcal{K}_{thnf}}{\mathcal{K}_f}\right)r'(0) = -\mathbb{M}_4\cdot r'(0)$$

Local Sherwood number (mass relocation rate):

$$Se_0Re^{-0.5} = -s'(0)$$

## 14.3 Numerical Method and Validation

Equations 14.9 to 14.13 are treated numerically operating the bvp4c methodology. In this technique, we alter Equations 14.9 to 14.13 into a first order ODE set. The mesh size, $\eta_\infty$ and is perceived as 0.01 and the boundary layer thicknesses, $\Delta\eta$, is taken as 20. Also, Table 14.2 reports the comparison outputs that we procured, and it is displayed that all outcomes are within justifiable range.

## 14.4 Results and Discussion

Here we concentrate on the implications of various crucial parameters against radial velocity, $p'(\eta)$, tangential velocity, $q(\eta)$, temperature, $r(\eta)$, and concentration, $s(\eta)$ profiles and skin friction $Ce_0Re^{0.5}$, local Nusselt, $Ne_0Re^{-0.5}$, and local Sherwood, $Se_0Re^{-0.5}$ numbers.

Figures 14.2 and 14.3 deliberate that both the velocities, $p'(\eta)$ and $q(\eta)$, are depleting capacities of the porosity parameter, $\delta_p$. As values of $\delta_p$ upsurge, the hindrance in fluid's movement also grows, resulting in abatement of both $p'(\eta)$ and $q(\eta)$.

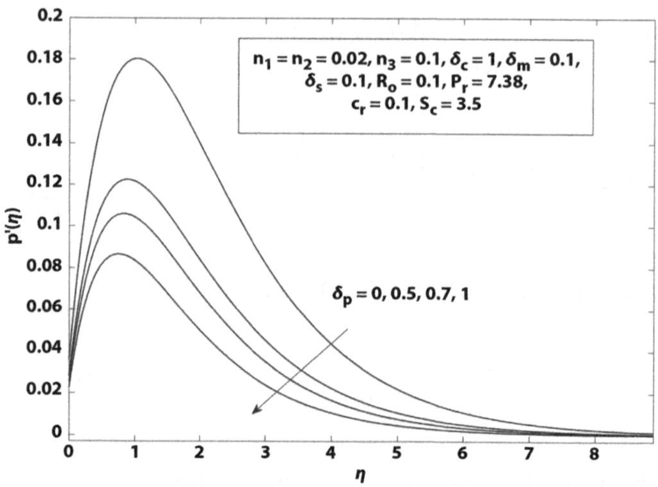


**Figure 14.2** Radial velocity, $p'(\eta)$ via porous parameter, $\delta_p$.

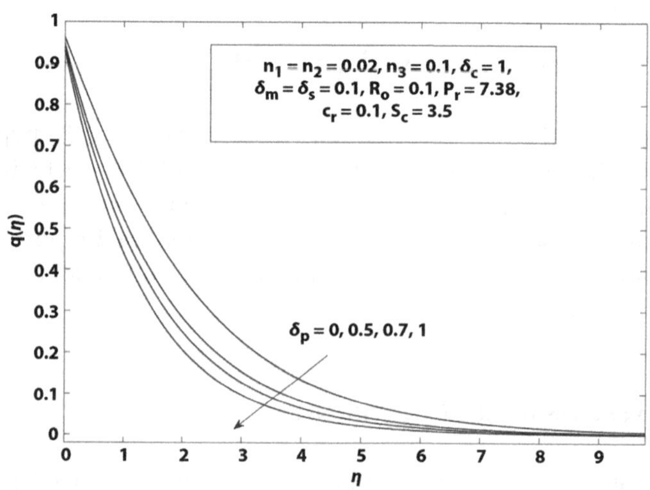


**Figure 14.3** Tangential velocity, $q(\eta)$ via porous parameter, $\delta_p$.

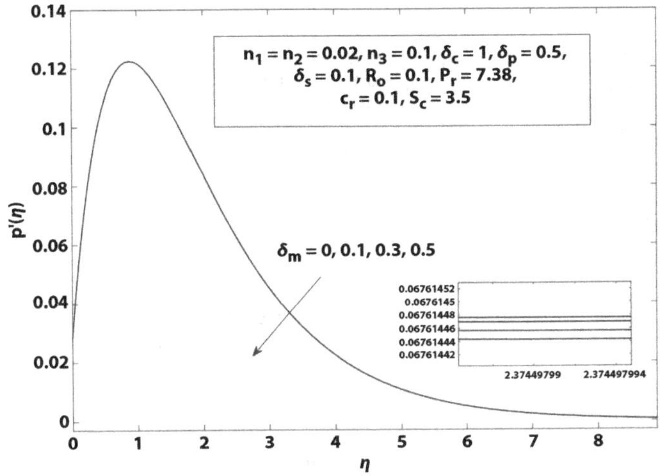


**Figure 14.4** Radial velocity, $p'(\eta)$ via magnetic parameter, $\delta_m$.

**Table 14.2** Comparison of $p''(0)$ for dissimilar $\delta_m$'s values with Mustafa *et al.* [33].

| $\delta_m$ | $\delta_s$ | $p''(0)$ (Mustafa *et al.* [33]) | $p''(0)$ (Our output) |
|---|---|---|---|
| 0 | 0.25 | 0.259534 | 0.064833 |
| 0.6 | 0.25 | 0.146057 | 0.029175 |

Figure 14.4 describes that the magnetic parameter, $\delta_m$, detoriates the radial velocity, $p'(\eta)$. As M augments, the impelled Lorentz force upsurges, restricting the fluid flow and, eventually, decreasing $p'(\eta)$. The result is shown in Table 14.2.

Figure 14.5 confirms that the tangential velocity, $q(\eta)$, faces declination for the slip parameter, $\delta_s$'s mounting values. As the $\delta_s$'s values upsurge, the flow opposition also augments, so $q(\eta)$ decelerates.

Figures 14.6 and 14.7 report that Casson parameter, $\delta_c$'s mounting values abate both the tangential velocity, $q(\eta)$ and temperature, $r(\eta)$. Casson parameter, $\delta_c$ succumbs to the fluid motion by weakening the yield stress's strength and thus $q(\eta)$ decreases as $\delta_s$ increases. Furthermore, mounting values of $\delta_c$ causes a curtailment of the thermal boundary layer's thickness, and correspondingly temperature, $r(\eta)$, falls off, too.

Figures 14.8 and 14.9 explains that Cu nanoparticle's volume fraction, $n_3$ aggrandizes the tangential velocity, $q(\eta)$, but curtails the temperature, $r(\eta)$. It happens because as the volume fraction of Cu nanoparticles aggrandizes,

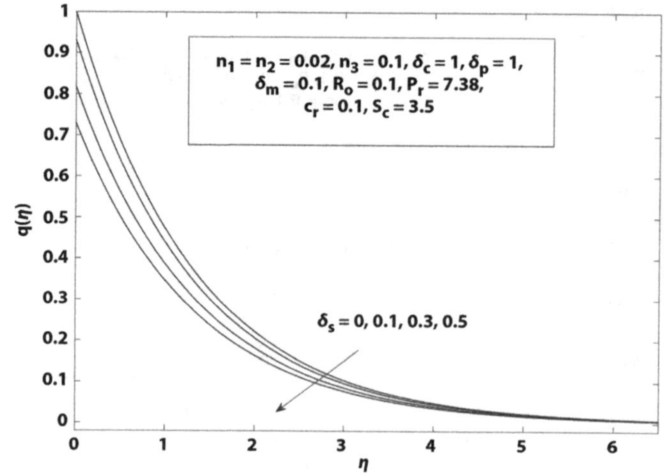


**Figure 14.5** Tangential velocity, $q(\eta)$ via slip parameter, $\delta_s$.

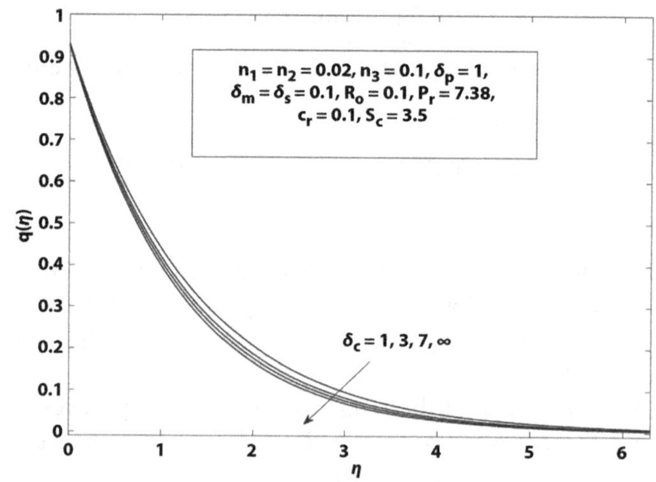


**Figure 14.6** Tangential velocity, $q(\eta)$ via Casson parameter, $\delta_c$.

the fluid turns to be thicker, and a retarding force arises that causes abatement in velocity. However, an increment in Cu nanoparticles volume fraction causes a contractionof the thermal boundary layer's thickness, and correspondingly temperature, $r(\eta)$ deteriorates, too. Cu nanoparticles have many advantageous thermophysical features which enable Cu nanoparticles to be utilized as coolant in disk drives, computer monitors, televisions, and in many other electronic equipment.

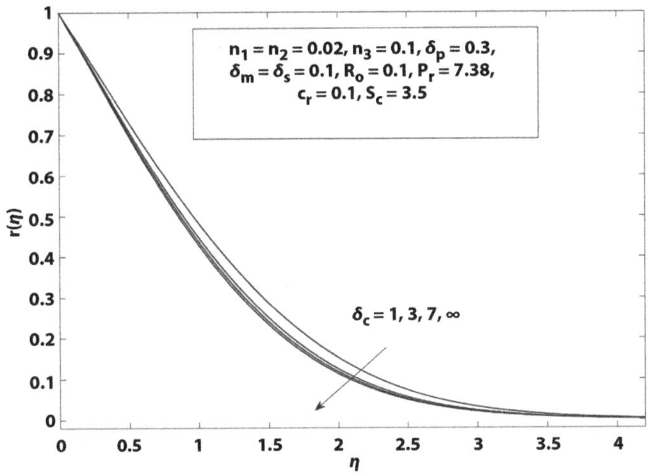


**Figure 14.7** Temperature, $r(\eta)$ via Casson parameter, $\delta_c$.

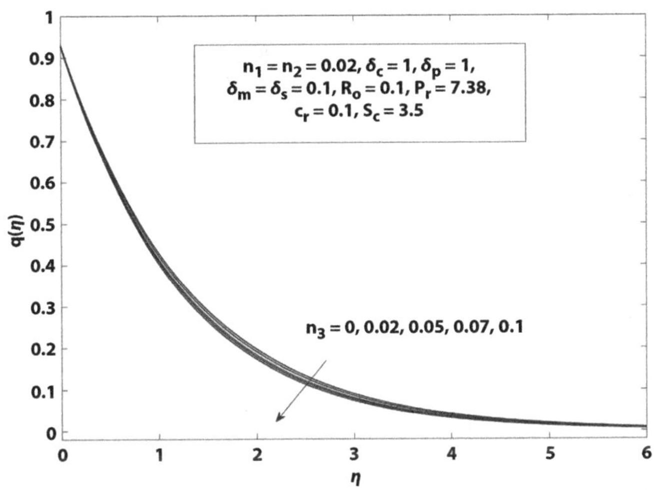


**Figure 14.8** Tangential velocity, $q(\eta)$ via nanoparticles volume fraction, $n_3$.

Figure 14.10 depicts that the temperature, r(η) diminishes for the uplifting estimates of the Prandtl number, $P_r$. Sturdy $P_r$ values cause curtailment into the fluid's thermal diffusivity, further leading to an abatement of the thermal boundary layer's thickness, and consequently, temperature depletes.

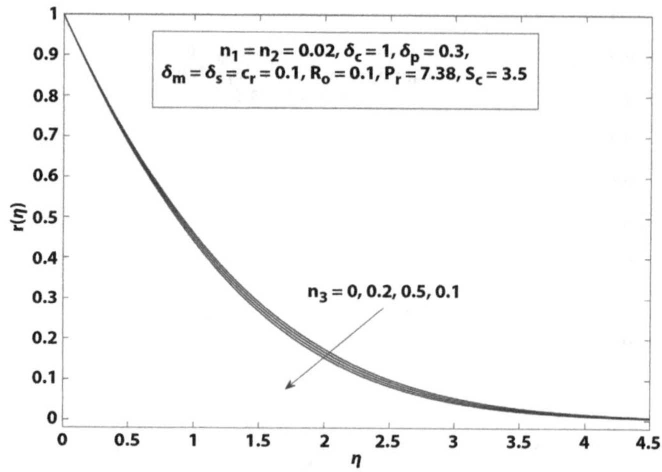


**Figure 14.9** Temperature, r(η) via nanoparticles volume fraction, $n_3$.

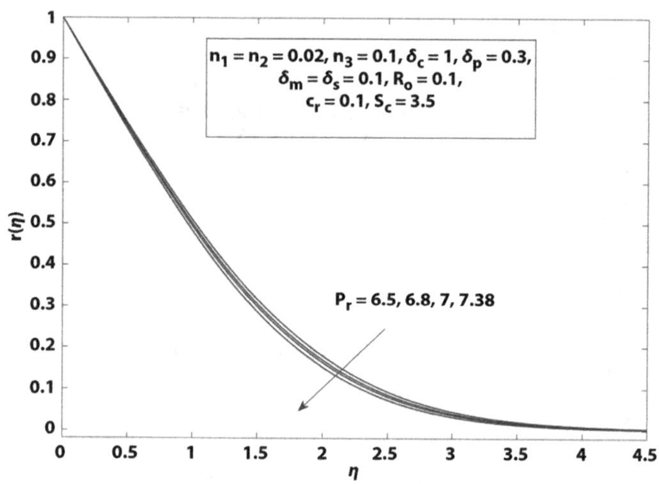


**Figure 14.10** Temperature, r(η) via Prandtl number, $P_r$.

Figure 14.11 expounds that growing values of the radiation parameter, $R_o$ escalates the temperature, r(η). Observingly, due to radiation fine quantity of energy is delivered in the system. Therefore, r(η) increases as $R_o$ increases, too.

Figure 14.12 expresses that concentration, s(η) is a reducing capacity of the chemical reaction parameter, $c_r$. An increment in $c_r$ translates into an

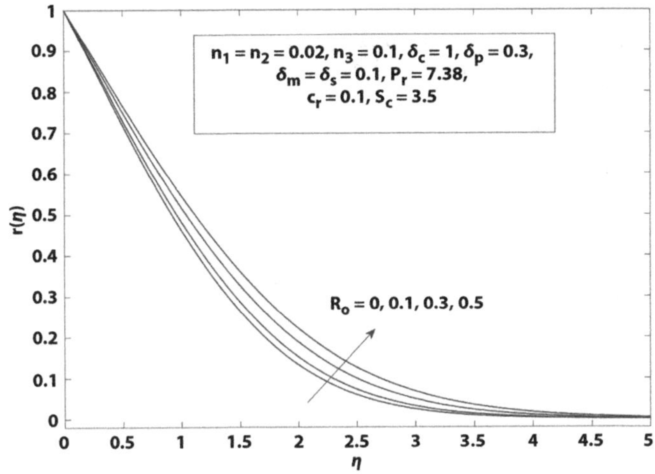


**Figure 14.11** Temperature, $r(\eta)$ via radiation parameter, $R_o$.

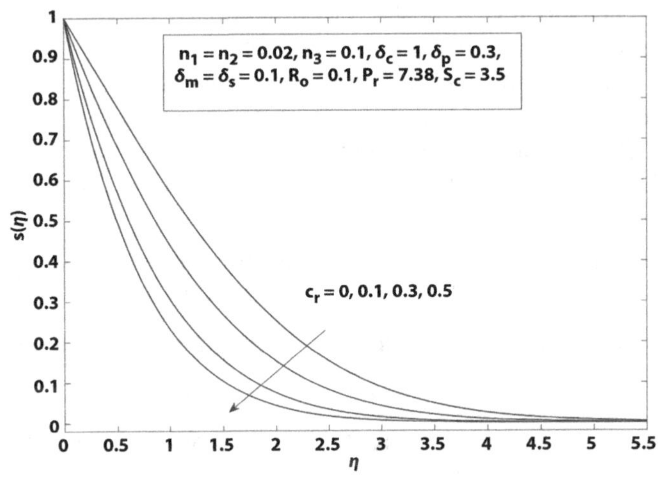


**Figure 14.12** Concentration, $s(\eta)$ via chemical reaction parame,ter $c_r$.

augmentation in the number of fluid molecules engaging in chemical reactions, conducing to a diminution in the concentration, s(η).

Figure 14.13 delineates that the concentration, s(η) declines for the rising estimates of the Schmidt number, $S_c$. As estimates of $S_c$ boost up, the molecular diffusivity drops down, so s(η) decelerates.

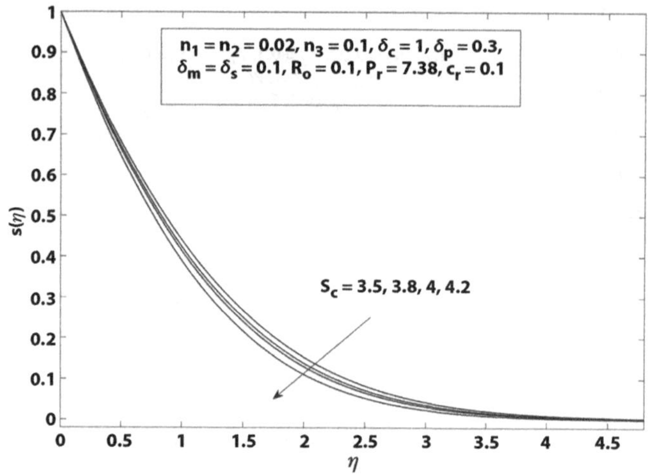


**Figure 14.13** Concentration $s(\eta)$ via Schmidt number $S_c$.

Table 14.3 elucidates the upshot of Cu nanoparticle's volume fraction $n_3$, Casson, $\delta_c$ and chemical reaction, $c_r$ parameters on $Ce_0Re^{0.5}$ (skin friction), $Ne_0Re^{-0.5}$ (heat relocate rate), and $Se_0Re^{-0.5}$ (mass relocate rate). It is vivid from Table 14.3 that $n_3$ enhances $Ce_0Re^{0.5}$, $Ne_0Re^{-0.5}$, and $Se_0Re^{-0.5}$. Besides, $\delta_c$ reduces $Ce_0Re^{0.5}$ but increases both $Ne_0Re^{-0.5}$ and $Se_0Re^{-0.5}$. Moreover, $c_r$ amplifies both $Ne_0Re^{-0.5}$ and $Se_0Re^{-0.5}$ but $c_r$ has no worthy impact on $Ce_0Re^{0.5}$.

**Table 14.3** Repercussions of pertinent parameters on $Ce_0Re^{0.5}$, $Ne_0Re^{-0.5}$, and $Se_0Re^{-0.5}$ when $n_1 = n_2 = 0.02$, $\delta_m = \delta_s = \delta_p = R_o = 0.1$, $P_r = 6.5$, $S_c = 4$.

| $n_3$ | $\delta_c$ | $c_r$ | $Ce_0Re^{0.5}$ | $Ne_0Re^{-0.5}$ | $Se_0Re^{-0.5}$ |
|---|---|---|---|---|---|
| 0 | 1 | 0.1 | 0.889651 | 0.667665 | 0.776217 |
| 0.02 | | | 0.984404 | 0.705694 | 0.782358 |
| 0.04 | | | 1.079113 | 0.741662 | 0.787055 |
| | 3 | | 0.859830 | 0.782710 | 0.805928 |
| | 7 | | 0.789537 | 0.796241 | 0.812363 |
| | 10 | | 0.773057 | 0.799403 | 0.813885 |
| | | 0.2 | 0.773057 | 0.799451 | 1.018440 |
| | | 0.3 | 0.773057 | 0.799496 | 1.191881 |
| | | 0.4 | 0.773057 | 0.799501 | 1.344795 |

## 14.5 Conclusions

Here, 3D MHD flow of Casson Cu-AA7075-AA7072/methanol tri-hybrid nanofluid flow with chemical reaction and thermal radiation past a porous spinning disk in attendance of slip effects is examined. The principal outcomes of this inspection are as follows:

- Velocities $p'(\eta)$ and $q(\eta)$ are deteriorating capacities of the porosity parameter, $\delta_p$.
- Both temperature, $r(\eta)$ and transversal velocity, $q(\eta)$ are depleting capacities of the Casson parameter, $\delta_c$.
- Thermal radiation, $R_o$ enhances the temperature, $r(\eta)$.
- Heat and mass relocation rates aggrandize for higher estimates of Cu nanoparticles' volume fraction, $n_3$.
- Casson parameter, $\delta_c$ reduces the skin friction while aggrandizes both mass and heat relocation rates.
- Chemical reaction parameter, $c_r$ amplifies the mass relocation rate.

**Data availability statement:** Our manuscript does not contain any associated data.

**Conflict of interests:** The authors announce no conflict of interests.

**Funding statement:** Authors declare no funding involved.

## References

1. Sundar, L.S., Chandra Mouli, K.V., Said, Z., Sousa, A.C., Heat transfer and second law analysis of ethylene glycol-based ternary hybrid nanofluid under laminar flow. *J. Therm. Sci. Eng. Appl.*, 13, 5, 051021, 2021, https://doi.org/10.1115/1.4050228.
2. Elnaqeeb, T., Animasaun, I.L., Shah, N.A., Ternary-hybrid nanofluids: significance of suction and dual-stretching on three-dimensional flow of water conveying nanoparticles with various shapes and densities. *Z. Naturforsch. A*, 76, 3, 231–243, 2021, https://doi.org/10.1515/zna-2020-0317.
3. Kumar, R.N., Gamaoun, F., Abdulrahman, A., Chohan, J.S., Gowda, R.P., Heat transfer analysis in three-dimensional unsteady magnetic fluid flow of water-based ternary hybrid nanofluid conveying three various shaped nanoparticles: A comparative study. *Int. J. Mod. Phys. B*, 36, 25, 2250170, 2022, https://doi.org/10.1142/S0217979222501703.

4. Mahmood, Z., Ahammad, N.A., Alhazmi, S.E., Khan, U., Bani-Fwaz, M.Z., Ternary hybrid nanofluid near a stretching/shrinking sheet with heat generation/absorption and velocity slip on unsteady stagnation point flow. *Int. J. Mod. Phys. B*, 36, 29, 2250209, 2022, https://doi.org/10.1142/S0217979222502095.
5. Khan, U., Kumar, R.N., Zaib, A., Prasannakumara, B.C., Ishak, A., Galal, A.M., Gowda, R.P., Time-dependent flow of water-based ternary hybrid nanoparticles over a radially contracting/expanding and rotating permeable stretching sphere. *Therm. Sci. Eng. Prog.*, 36, 101521, 2022, https://doi.org/10.1016/j.tsep.2022.101521.
6. Mahmood, Z., Iqbal, Z., Alyami, M.A., Alqahtani, B., Yassen, M.F., Khan, U., Influence of suction and heat source on MHD stagnation point flow of ternary hybrid nanofluid over convectively heated stretching/shrinking cylinder. *Adv. Mech. Eng.*, 14, 9, 16878132221126278, 2022, https://doi.org/10.1177/16878132221126278.
7. Bilal, M., Ullah, I., Alam, M.M., Weera, W., Galal, A.M., Numerical simulations through PCM for the dynamics of thermal enhancement in ternary MHD hybrid nanofluid flow over plane sheet, cone, and wedge. *Symmetry*, 14, 11, 20222419, https://doi.org/10.3390/sym14112419.
8. Xiu, W., Animasaun, I.L., Al-Mdallal, Q.M., Alzahrani, A.K., Muhammad, T., Dynamics of ternary-hybrid nanofluids due to dual stretching on wedge surfaces when volume of nanoparticles is small and large: forced convection of water at different temperatures. *Int. Commun. Heat Mass Transfer*, 137, 106241, 2022, https://doi.org/10.1016/j.icheatmasstransfer.2022.106241.
9. Animasaun, I.L., Kumar, T.K., Noah, F.A., Okoya, S.S., Al-Mdallal, Q.M., Bhatti, M.M., Insight into Darcy flow of ternary-hybrid nanofluid on horizontal surfaces: Exploration of the effects of convective and unsteady acceleration. *ZAMM-J. Appl. Math. Mech./Z. Angew. Math. Mech.*, 103, 5, e202200197, 2023, https://doi.org/10.1002/zamm.202200197.
10. Elsebaee, F.A.A., Bilal, M., Mahmoud, S.R., Balubaid, M., Shuaib, M., Asamoah, J.K., Ali, A., Motile micro-organism based trihybrid nanofluid flow with an application of magnetic effect across a slender stretching sheet: Numerical approach. *AIP Adv.*, 13, 3, 2023, https://doi.org/10.1063/5.0144191.
11. Ramesh, G.K., Roopa, G.S., Rauf, A., Shehzad, S.A., Abbasi, F.M., Time-dependent squeezing flow of Casson-micropolar nanofluid with injection/suction and slip effects. *Int. Commun. Heat Mass Transfer*, 126, 105470, 2021, https://doi.org/10.1016/j.icheatmasstransfer.2021.105470.
12. Sandeep, N., Sulochana, C., Ashwinkumar, G.P., Understanding the dynamics of chemically reactive Casson liquid flow above a convectively heated curved expanse. *Proc. Inst. Mech. Eng., Part C: J. Mech. Eng. Sci.*, 236, 24, 11420–11430, 2022, https://doi.org/10.1177/09544062221115084.
13. Mushtaq, T., Rauf, A., Shehzad, S.A., Mustafa, F., Hanif, M., Abbas, Z., Numerical and statistical approach for Casson-Maxwell nanofluid flow

with Cattaneo-Christov theory. *Appl. Math. Mech.*, 42, 7, 1063–1076, 2021, https://doi.org/10.3390/sym14122658.

14. Raza, A., Almusawa, M.Y., Ali, Q., Haq, A.U., Al-Khaled, K., Sarris, I.E., Solution of water and sodium alginate-based casson type hybrid nanofluid with slip and sinusoidal heat conditions: A prabhakar fractional derivative approach. *Symmetry*, 14, 12, 2658, 2022, https://doi.org/10.3390/sym14122658.
15. Alqarni, M.M., Bilal, M., Allogmany, R., Tag-Eldin, E., Ghoneim, M.E., Yassen, M.F., Mathematical analysis of casson fluid flow with energy and mass transfer under the influence of activation energy from a non-coaxially spinning disc. *Front. Energy Res.*, 10, 986284, 2022, https://doi.org/10.3389/fenrg.2022.986284.
16. Goud, B.S., Reddy, Y.D., Asogwa, K.K., Chemical reaction, Soret and Dufour impacts on magnetohydrodynamic heat transfer Casson fluid over an exponentially permeable stretching surface with slip effects. *Int. J. Mod. Phys. B*, 37, 13, 2350124, 2023, https://doi.org/10.1142/S0217979223501242.
17. Asogwa, K.K., Alsulami, M.D., Prasannakumara, B.C., Muhammad, T., Double diffusive convection and cross diffusion effects on Casson fluid over a Lorentz force driven Riga plate in a porous medium with heat sink: An analytical approach. *Int. Commun. Heat Mass Transfer*, 131, 105761, 2022, https://doi.org/10.1016/j.icheatmasstransfer.2021.105761.
18. Megahed, A.M., Williamson fluid flow due to a nonlinearly stretching sheet with viscous dissipation and thermal radiation. *J. Egypt. Math. Soc.*, 27, 1, 1–10, 2019, https://doi.org/10.1186/s42787-019-0016-y.
19. Mandal, I.C. and Mukhopadhyay, S., Nonlinear convection in micropolar fluid flow past an exponentially stretching sheet in an exponentially moving stream with thermal radiation. *Mech. Adv. Mater. Struct.*, 26, 24, 2040–2046, 2019, https://doi.org/10.1080/15376494.2018.1472325.
20. Hosseinzadeh, K., Gholinia, M., Jafari, B., Ghanbarpour, A., Olfian, H., Ganji, D.D., Nonlinear thermal radiation and chemical reaction effects on Maxwell fluid flow with convectively heated plate in a porous medium. *Heat Transfer—Asian Res.*, 48, 2, 744–759, 2019, https://doi.org/10.1002/htj.21404.
21. Ramesh, K., Khan, S.U., Jameel, M., Khan, M.I., Chu, Y.M., Kadry, S., Bioconvection assessment in Maxwell nanofluid configured by a Riga surface with nonlinear thermal radiation and activation energy. *Surf. Interfaces*, 21, 100749, 2020, https://doi.org/10.1016/j.surfin.2020.100749.
22. Ahmad, S., Ashraf, M., Ali, K., Simulation of thermal radiation in a micropolar fluid flow through a porous medium between channel walls. *J. Therm. Anal. Calorim.*, 144, 941–953, 2021, https://doi.org/10.1007/s10973-020-09542-w.
23. Rehman, K.U., Shatanawi, W., Al-Mdallal, Q.M., A comparative remark on heat transfer in thermally stratified MHD Jeffrey fluid flow with thermal radiations subject to cylindrical/plane surfaces. *Case Stud. Therm. Eng.*, 32, 101913, 2022, https://doi.org/10.1016/j.csite.2022.101913.

24. Mohanty, D., Mahanta, G., Shaw, S., Analysis of irreversibility for 3-D MHD convective Darcy–Forchheimer Casson hybrid nanofluid flow due to a rotating disk with Cattaneo–Christov heat flux, Joule heating, and nonlinear thermal radiation. *Numer. Heat Transf., Part B: Fundam.*, 84, 2, 115–142, 2023, https://doi.org/10.1080/10407790.2023.2189644.
25. Mehmood, T., Ramzan, M., Howari, F., Kadry, S., Chu, Y.-M., Application of response surface methodology on the nanofluid flow over a rotating disk with autocatalytic chemical reaction and entropy generation optimization. *Sci. Rep.*, 11, 1, 4021, 2021, https://doi.org/10.1038/s41598-021-81755-x.
26. Mahesh, A., Varma, S.V.K., Raju, C.S.K., Babu, M.J., Vajravelu, K., Al-Kouz, W., Significance of non-Fourier heat flux and radiation on PEG–water based hybrid nanofluid flow among revolving disks with chemical reaction and entropy generation optimization. *Int. Commun. Heat Mass Transfer*, 127, 105572, 2021, https://doi.org/10.1016/j.icheatmasstransfer.2021.105572.
27. Sharma, B.K., Khanduri, U., Mishra, N.K., Mekheimer, K.S., Combined effect of thermophoresis and Brownian motion on MHD mixed convective flow over an inclined stretching surface with radiation and chemical reaction. *Int. J. Mod. Phys. B*, 37, 10, 2350095, 2023, https://doi.org/10.1142/S0217979223500959.
28. Rehman, K.U., Malik, M.Y., Zahri, M., Tahir, M., Numerical analysis of MHD Casson Navier's slip nanofluid flow yield by rigid rotating disk. *Results Phys.*, 8, 744–751, 2018. 10.1016/j.rinp.2018.01.017.
29. Shoaib, M., Raja, M.A.Z., Sabir, M.T., Awais, M., Islam, S., Shah, Z., Kumam, P., Numerical analysis of 3-D MHD hybrid nanofluid over a rotational disk in presence of thermal radiation with Joule heating and viscous dissipation effects using Lobatto IIIA technique. *Alexandria Eng. J.*, 60, 4, 3605–3619, 2021, 10.1016/j.aej.2021.02.015.
30. Shafiq, A., Rasool, G., Alotaibi, H., Aljohani, H.M., Wakif, A., Khan, I., Akram, S., Thermally enhanced Darcy-Forchheimer Casson-water/glycerine rotating nanofluid flow with uniform magnetic field. *Micromachines*, 12, 6, 605, 2021, 10.3390/mi12060605.
31. Tlili, I., Nabwey, H.A., Ashwinkumar, G.P., Sandeep, N., 3-D magnetohydrodynamic AA7072-AA7075/methanol hybrid nanofluid flow above an uneven thickness surface with slip effect. *Sci. Rep.*, 10, 1, 4265, 2020, 10.1038/s41598-020-61215-8.
32. Vishalakshi, A.B., Mahesh, R., Mahabaleshwar, U.S., Rao, A.K., Pérez, L.M., Laroze, D., MHD Hybrid Nanofluid Flow over a Stretching/Shrinking Sheet with Skin Friction: Effects of Radiation and Mass Transpiration. *Magnetochemistry*, 9, 5, 118, 2023, 10.3390/magnetochemistry9050118.
33. Mustafa, M., MHD nanofluid flow over a rotating disk with partial slip effects: Buongiorno model. *Int. J. Heat Mass Transfer*, 108, 1910–1916, 2017, https://doi.org/10.1016/j.ijheatmasstransfer.2017.01.064.

# 15

# Thermal Performance of MXene ($Ti_3C_2$) Nanoparticles in Blood Flow Over a Curved Region: A Biomedical Application

**Niraj Rathore and N. Sandeep***

*Department of Mathematics, Central University of Karnataka, Kalaburagi, India*

***Abstract***

MXenes are a family of two-dimensional compounds that consist of transition metal carbides, nitrides, and carbonatites. MXene gained noteworthy values in various fields, including biomedicine, due to its unique properties. The nano-sized MXene particles have several potential biomedical applications, such as drug delivery, thermal therapy, imaging, and antimicrobial applications. This study investigates the thermal conducting and rate of heat transmission due to MXene ($Ti_3C_2$) nanoparticles flowing in blood arteries. For this purpose, mathematical modeling for blood flow on the curved region of the artery is developed using the Navier-Stokes equation. To predict the physical phenomena of the flow, the momentum equation is fortified with the magnetic effect and buoyancy effect. The equation for energy is fulfilled with nonlinear thermal radiation conditions and the Joule heating effect. Since the fluid is blood, it is considered non-Newtonian, and the boundary conditions are assumed based on geometry. The established system of nonlinear partial differential equation (PDE) is solved numerically by means of the Bvp5c MATLAB package. Outcomes of the study are drawn to analyze thermal performance due to MXene ($Ti_3C_2$) nanoparticles for different thermal conductivity models. It is noticed that thermal performance in the blood flow is enhanced after mixing MXene nanoparticles. The outcomes of this work are helpful in thermal ablation, photothermal therapy, and biomedical applications wherever heat transfer significantly plays a role.

***Keywords:*** MHD, curved artery, MXene ($Ti_3C_2$), heat transfer, Darcy-Forchheimer, nonlinear thermal radiation

**Corresponding author*: nsandeep@cuk.ac.in

Ashwani Kumar, Nitesh Dutt and Mukesh Kumar Awasthi (eds.) Heat Transfer Enhancement Techniques: Thermal Performance, Optimization and Applications, (331–358) © 2025 Scrivener Publishing LLC

## 15.1 Introduction

Problems of heart disease are due to pulmonary physiology pressure where all deoxygenated blood flows from right ventricle to the lungs with the help of pulmonary arteries and oxygenated blood returns from the lungs to the left part of the heart by the pulmonary vein. The shape and curvature of the artery differ at different body positions. Heat transfer enhancement in biofluids is an essential part of biomedical engineering. During the last decade, applications of nanoparticles have improved, and a new kind of solid-liquid, called nanofluid, has been prepared. Nanoparticles are very useful for nano-drug delivery. It helps to deal with drug resistance problems. Nanoscience and nano-drug carriage systems supply helpful agents to the targeted portions of the body.

The valuable results of gold nanoparticles merged with blood flowing through an artery explored by authors in Reference [1]. It is perceived that the energy and shear strain of the gold bloodstream can be improved by increasing particle concentration. Also, nanoparticles can help overcome blood clotting problems. Flow behavior of copper nanoparticles merged in the bloodstream in the ciliated tube was studied by the authors in Reference [2]. They discerned that the magnetic field effect and nanoparticles are beneficial in breaking blood clots. Magneto-hydro-dynamics (MHD) ferro-nanofluid movement through a bent tube in the occurrence of a Hall and joule heating effect is observed in Reference [3]. This study revealed that the energy transfer rate is enriched, and surface shear stress is reduced because of ferro particles. Since some fluids have less thermal conductivity, nanoparticles significantly enhance blood's thermal conductivity. Biomedical applications of nanoparticles are studied by the authors in References [4–6]. These studies show that the heat transfer rate is enhanced for solid suspension in blood flow.

The study of blood flow is vital to understanding pulmonary vascular physiology pressure and overcoming hypertension. The blood flow against a wedge region by assuming blood as a Williamson fluid [7]. It is noticed that the flow pattern varies for different wedge angle parameters $\beta$. That is, if wedge angle parameters $\beta = 0$ flow is above a flat plate, and $\beta = 0.5$ flow is adjacent to a stagnation point. Biomagnetic flow over a wedge; it has been noticed that by improving the values of the wedge angle, there is a rise in velocity [8]. Also, unsteadiness parameters lead to reducing the fluid temperature. The authors also took flow in curved shape geometry and analyzed the flow pattern numerically [9–11].

The influence of thermophysical effects is essential to analyzing the flow pressure in curved-shaped geometry. A comparison of "Cattaneo–Christov heat flux on radiative Oldroyd-B flow in cone and wedge" is made [12]. It is observed that the physical constraints are highly effective in a cone when compared to a wedge. MHD flow flowing is considered under viscous dissipation effect together with convective boundary conditions in Reference [13]. It has shown that the thermal boundary layer is affected by slip parameters, *i.e.*, the heat is reduced due to growing values of slip parameters. The electrically conducting nanofluid in a temperature-dependent isothermal is analyzed by Mahanthesh *et al.* [14]. Here, the rising importance of the porosity parameter enhances resistance to the flow; due to that, there is a decline in velocity. The study of bioconvective thermal viscoelastic nanoparticles moving in wedge-shaped geometry has been explored [15]. This study reveals that the rise in Biot number augments temperature distribution.

Nanofluids prepared by emerging nanoparticles into base fluids exhibit a change in thermal conductivity. The nanoparticles' nanolayers play essential roles in effective thermal conductivity, confirmed by experimental data [16]. Unsteady second-order fluids flow in a wedge with velocity slip, and thermal jump and nanolayer concepts are considered [17]. It is perceived that the growing diameter of the nanolayer energy and momentum transmission rate is enhanced, but the wall temperature declined. The two-dimensional (2D) steady incompressible flow of $Al_2O_3$-water flow in the sloping magnetic field with the nanolayer concept [18]. The heat transfer is enhanced due to the higher nanolayer value and increasing nanoparticle volume fraction values. An experimental investigation of the significance of the nanolayer of CuO nanoparticles that emerged in water was investigated by the authors in Reference [19]. This study found that the interfacial layer highly impacts the nanofluids' thermal conductivity. Therefore, different thermal models play a crucial role in enhancing thermal conductivity. Further study about nanofluids and the nanolayer concept is observed by References [20, 21].

Physical impact on fluid flow impacts the thermal performance and momentum activities. The magnetized flow of Newtonian liquid with viscous degeneracy and Joule heating effect above the extending piece is considered by Reference [22]. The heat transmission frequency is reduced because of sizeable viscous dissipation and Joule heating constraints. $Al_2O_3$-water nano liquid in a wince tube using Buongiorno's model together with the nanolayer approach and nanoparticles diameter was explored by Reference [23]. The size of nanoparticles is one of the reasons for the variation in heat allocation rate; that is, small particles enhance the heat transfer rate significantly when compared with large nanoparticles. The importance of the artery's curvature on blood flow, wall shear, flow rate profile, and velocity have been observed [24]. The momentum and energy decline by improving the curvature parameter. The magnetic treatment has grown in recent years for that purpose; the observation of magnetic impact is observed using mathematical modeling of the flow on the curved-shaped artery [25]. The shear stress and pressure for deoxygenated blood have been increased under a magnetic field, and it affects shear stress significantly more than pressure. Moreover, it was noticed that the change in the blood velocity is directly correlated to the strength of the magnetic field. Flow over the curved shirking/stretching surface has been studied to analyze the couple stress, force drag, and heat transport rate in the presence of gold nanoparticles [26]. The results showed that the velocity of the fluid is enhanced due to suction in the first region and decreases in the second region also, thermal radiation improves the temperature of both fluids.

Nanomedicine applications and drug agent properties of MXene nanoparticles encourage the exploration of biofluids. MXene nanoparticle is made of combining the 'M', 'A', and 'X' elements of familiar MAX phases. It is a 2D material showing prominent behavior favoring biomedical applications. The applications of Mxenes and their amalgams for anticancer, drug distribution, antimicrobial, smart devices, biosensors, and tissue engineering are explored in the study [27]. Also, it explores the drug loading capacity, efficiency, payload, and antimicrobial activity of MXene nanoparticles. The important fact about MXene is that its biocompatibility can be improved by coating it with a suitable polymer. The study of MXene nanoparticles is improving each year, as shown by researchers in Reference [28]. Various benefits, applications, and developments have been discussed by authors in References [29–31].

Blood flows in the human body with different arteries and shows diverse flow behavior. The study of curved arterial flow is essential to understanding pulmonary vascular physiology pressure and overcoming hypertension, including large arteries. This work can assist in understanding the nature of the bloodstream in curved arteries and heat transmission variation due to MXene nanoparticles.

## 15.2 Characteristics of MXene Nanomaterials

MXenes, a class of 2D materials with unique properties, are rapidly gaining traction in the biomedical field. Their exceptional characteristics include high surface area, tolerable surface chemistry, biocompatibility, biodegradability, electrical conductivity, and optical properties [32]. MXene tiny particles have grown substantially in biomedicine because of their unique properties and valuable applications. MXenes are a class of 2D transition metal carbides, nitrides, or carbonitrides with excellent electronic, thermal, and mechanical properties. Here are some biomedical uses of MXene nanoparticles [27]:

### 15.2.1 Drug Delivery

MXene nanoparticles can be functionalized and loaded with therapeutic agents such as drugs or nucleic acids. Their large surface area and the ability to carry multiple payloads make them promising candidates for targeted drug conveyance. The surface interaction of MXenes permits easy modification, enabling the controlled release of drugs at specific sites within the body.

### 15.2.2 Imaging

MXenes can be used as contrast mediators in numerous imaging practices, including magnetic resonance imaging (MRI) and computed tomography (CT). Their ability to absorb and scatter X-rays and their magnetic properties make them suitable for enhancing imaging contrast, enabling better visualization of tissues and organs.

### 15.2.3 Photothermal Therapy

MXene nanoparticles have excellent photothermal conversion properties, which can efficiently convert light into heat. This property makes them suitable for photothermal therapy, where they can selectively destroy cancer cells when illuminated with near-infrared (NIR) light. This approach minimizes damage to surrounding healthy tissues.

### 15.2.4 Biosensing

MXene-based sensors can be designed for the detection of various biomolecules. The unique electronic assets of MXenes allow for the growth of highly sensitive and selective biosensors. These sensors could be employed in the early detection of diseases, monitoring of biomarkers, and point-of-care diagnostics.

### 15.2.5 Antibacterial Applications

MXene nanoparticles have shown antibacterial properties, which can be harnessed for wound healing and infection prevention. The surface charge and chemistry of MXenes can be modified to enhance their antibacterial activity.

### 15.2.6 Tissue Engineering

MXenes can be incorporated into scaffolds for tissue engineering applications. The biocompatibility and mechanical properties of MXenes make them suitable for enhancing the structural integrity of tissue scaffolds. Additionally, they can support cell growth and differentiation.

### 15.2.7 Neural Interfaces

MXenes have been explored for their potential use in neural interfaces, such as neural electrodes. Their electrical conductivity and compatibility with neural tissues make them promising materials for improving the performance and biocompatibility of microelectrodes.

### 15.2.8 Anti-Inflammatory Applications

MXene nanoparticles have been investigated for their anti-inflammatory possessions, which could be helpful in treating circumstances characterized by inflammation.

It's important to note that while MXene nanoparticles hold great promise in various biomedical applications, ongoing research is necessary to address issues such as toxicity, long-term effects, and scalability before widespread clinical use.

The properties and characters depend on the synthetization methods of nanoparticles. Wang *et al.* [33] discussed the electric conductivity of MXene nanoparticles and showed that it can be improved up to 2140 S/cm by calcination of $Ti_3C_2$ at 600°C (Table 15.1). Therefore, synthetization approaches also play a vibrant role in operative conductivity, biotoxicity,

**Table 15.1** Physical assets of blood and nanomaterial [33–35].

| Type of particle | $\rho(Kgm^{-3})$ | $C_p(J/KgK)$ | $K(W/mK)$ | $\sigma(Sm^{-1})$ |
|---|---|---|---|---|
| $Ti_3C_2$ | 3700 | 2800 | 55.8 | $2.6\times10^5$ |
| Blood | 1050 | 3617 | 0.52 | 1.25 |

photothermal treatment, photoacoustic imaging abilities, etc. Hence, the appropriate handling of $Ti_3C_2$ helps to enhance electrical conductivity.

## 15.3 Problem Description

According to the human arteries, the shape of the geometry in this study is considered to be curved. The lower wall of the curved shape geometry is supposed to be an elongating layer of radius, $\hat{R}$. The curvature of the model depends on the radius, $\hat{R}$. The curvilinear coordinates are used to show flow direction. The fluid is moving significantly in the *z*-direction, and the *r*-axis is upright to it. The establishment is made to demonstrate blood flow in the curved artery [36]. Various thermal engineering relations between fluid and nanofluid are given in Table 15.2. Let us assume the nanofluid (blood+$Ti_3C_2$) flows along the *z*-direction with velocity, $U = bz^n$ (b > 0), and the functional magnetic effect of strength, $M_0$, vertical to the fluid stream, as shown in Figure 15.1. The *r* directional velocity is assumed to be negligible in comparison to the others.

The main equations for the fluid flow are given below:

$$\hat{R}\frac{\partial}{\partial z}(rU)+\frac{\partial}{\partial r}\left(V\left(r+\hat{R}\right)\right)=0, \tag{15.1}$$

$$\rho_f\frac{U^2}{r+\hat{R}}=\frac{\partial P}{\partial r}, \tag{15.2}$$

$$\rho_{nf}\begin{pmatrix} VU_r+\dfrac{\hat{R}U}{r+\hat{R}}U_z \\ +\dfrac{UV}{r+\hat{R}} \end{pmatrix}=\left(1+\gamma^{-1}\right)\mu_{nf}\begin{pmatrix} U_{rr}-\dfrac{U}{(r+\hat{R})^2} \\ +\dfrac{1}{r+\hat{R}}U_r \end{pmatrix}$$
$$-\frac{\hat{R}}{r+\hat{R}}P_z-\sigma_{nf}M_0^2U+g\left(\rho\beta\right)_{nf}\left(T-T_\infty\right), \tag{15.3}$$

**Table 15.2** Thermal engineering relation between fluid and nanofluid [36].

| **Features** | **Hybrid nanofluid parameter** | |
|---|---|---|
| Density | $\Gamma_1 = \frac{\rho_{nf}}{\rho_f} = (1-\phi) + \frac{\phi\rho_{sp}}{\rho_f}$ | |
| Viscosity | $\Gamma_2 = \frac{\mu_{nf}}{\mu_f} = (1-\phi)^{-2.5}$ | |
| Electrical conductivity | $\Gamma_3 = \frac{\sigma_{nf}}{\sigma_f} = 1 + \frac{3\left(\frac{\sigma_{sp}}{\sigma_f} - 1\right)\phi}{\left(\frac{\sigma_{sp}}{\sigma_f} + 2\right) - \left(\frac{\sigma_{sp}}{\sigma_f} - 1\right)\phi}$ | |
| Thermal expansion | $\Gamma_4 = \frac{(\rho\beta)_{nf}}{(\rho\beta)_f} = (1-\phi_1) + \phi_1 \frac{(\rho\beta)_{sp}}{(\rho\beta)_f}$ | |
| Heat capacity | $\Gamma_5 = \frac{(\rho C_p)_{nf}}{(\rho C_p)_f} = (1-\phi) + \frac{\phi(\rho C_p)_{sp}}{(\rho C_p)_f}$ | |
| Thermal conductivity | Hamilton and Crosser model | $\Gamma_6 = \frac{k_{nf}}{k_f} = \frac{k_{sp} + k_f(m-1) + (k_{sp} - k_f)(m-1)\phi}{k_{sp} + k_f(m-1) + \phi(k_f - k_{sp})}$ |
| | Maxwell model | $\Gamma_6 = \frac{k_{nf}}{k_f} = \frac{k_{sp} + 2k_f + 2(k_{sp} - k_f)\phi}{k_{sp} + 2k_f + \phi(k_f - k_{sp})}$ |

$$(\rho c_p)_{nf}\left(UT_r + \frac{\hat{R}U}{r+\hat{R}}T_z\right) = k_{nf}\left(T_{rr} + \frac{1}{r+\hat{R}}T_r\right) - \frac{1}{r+\hat{R}}\left(\frac{\partial(r+\hat{R})q_r}{\partial r}\right) + \sigma_{nf}M_0^2U^2, \tag{15.4}$$

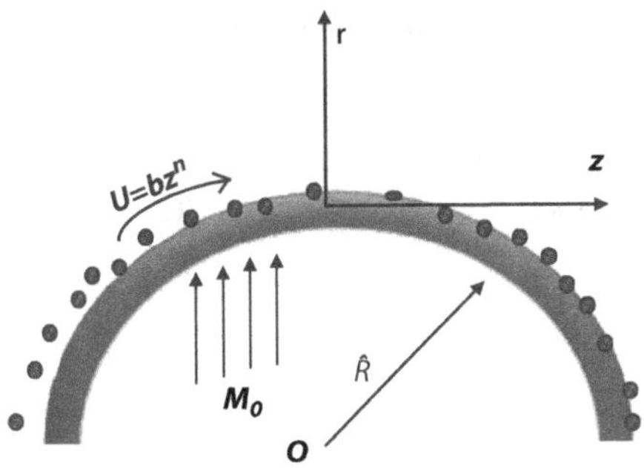


**Figure 15.1** Flow of nanofluid on curved shape region.

Where $q_r = -\frac{4\sigma^*}{3k^*}\frac{\partial T^4}{\partial r}$ is the variable thermal radiation.

Boundary situations are:

$$\left.\begin{array}{l} U = bz^n,\ V = 0,\ -k_{nf}\dfrac{\partial T}{\partial r} = h_f\left(T_f - T\right),\ \text{at } r = 0, \\ U \to 0,\ \dfrac{\partial U}{\partial r} \to 0,\ T \to T_\infty,\ \text{as } r \to \infty \end{array}\right\} \tag{15.5}$$

Using the below similarity transformation [36].

$$\left.\begin{array}{l} U = bz^n f'(\xi),\ V = \dfrac{n+1}{2}\dfrac{-\hat{R}}{r+\hat{R}}\sqrt{b\upsilon_f z^{n-1}}\left(f(\xi) + \dfrac{n-1}{n+1}\xi f'(\xi)\right), \\ \xi = r\sqrt{\dfrac{bz^{n-1}}{\upsilon_f}},\ P = \rho_f b^2 z^{2n} p(\xi), \theta = \dfrac{T - T_\infty}{T_w - T_\infty} \end{array}\right\} \tag{15.6}$$

Using the above similarity transformations, Equations 15.2–15.5 translate in the below form:

Dimensionless form of motion equation,

$$\left.\begin{aligned}&\Gamma_1\left(\frac{(n+1)\xi}{2(\xi+k)}ff''' - \frac{n+1}{2(\xi+k)^2}ff'' - \frac{n+1}{2(\xi+k)^3}ff' + \frac{3n-1}{2(\xi+k)}kf'f'' - \frac{(3n-1)k}{2(\xi+k)}f'^2\right)\\&+\left(1+\frac{1}{\gamma}\right)\Gamma_2\left(f''+\frac{2}{\xi+k}f''' - \frac{f''}{(\xi+k)^2}+\frac{f'}{(\xi+k)^3}\right)\\&-M\Gamma_3\left(\frac{f'}{\xi+k}+f''\right)+\Gamma_4 Gr\theta' = 0,\end{aligned}\right\} \tag{15.7}$$

Dimensionless form of the heat equation,

$$\begin{aligned}&Ra\left(\begin{aligned}&3(\theta_w-1)\left(1+(\theta_w-1)\theta\right)^2\theta'^2\\&+\frac{\left(1+(\theta_w-1)\theta\right)^3\left(\theta'+(\xi+k)\theta''\right)}{k+\xi}\end{aligned}\right)\\&-\Gamma_5\,\mathrm{Pr}\frac{n+1}{2}\frac{k}{\xi+k}f\theta'+\Gamma_6\left(\theta''+\frac{\theta'}{\xi+k}\right)+\Gamma_3 M\,Brf'^2=0,\end{aligned} \tag{15.8}$$

The dimensionless boundary conditions (BCs) are:

$$\begin{aligned}&f(0)=0, f'(0)=1, \frac{k_{nf}}{k_f}\theta'(0)=Bi\left(\theta(0)-1\right), at\,\xi=0,\\&f'(\xi)\to 0, f''\to 0, \theta(\xi)\to 0, \text{as } \xi\to\infty\end{aligned} \tag{15.9}$$

Where $\Gamma_i$, $(i = 1, 2, 3, 4, 5)$ are non-dimensional constraints explained in the following table.

Here, *nf*, *f*, and *sp* indicate the nanofluid, fluid, and solid nanoparticles, respectively.

The dimensionless parameters that appeared in the above equations are given as follows:

| | |
|---|---|
| $Gr = \frac{\rho_f \beta_f (T_w - T_\infty)}{a^2 z^n}$ Grashoff number | $\Pr = \frac{\upsilon_f (\rho Cp)_{nf}}{k_{nf}}$ Prandtl number |
| $M = \frac{M_0^2 \sigma_f}{b \rho_f}$ Hartmann number | $Ec = \frac{\rho_f U^2 (T_w - T_\infty)^{-1}}{(\rho c_p)_f}$ Eckert number |
| $Ra = \frac{16\sigma^* T_0^3}{3k^* k_f}$ Thermal radiation parameter | $Bi = \frac{h_f}{k_f}\sqrt{\frac{\upsilon_f}{bz^{n-1}}}$ Biot number |
| $Br = \Pr Ec = \frac{\mu_f U^2}{k_f (T_w - T_\infty)}$ Brinkman number | $k = \frac{h_f}{k_f Bi}\hat{R}$ is the curvature parameter. |

The boundary region resistance, $C_{fz}$, and Nusselt numeral, $Nu_z$, flow are [37, 38]:

$$\mathrm{Re}_z^{0.5} C_{fz} = (1+\gamma^{-1})\frac{\mu_{nf}}{\mu_f}\left( f''(0) - \frac{f'(0)}{k} \right), \tag{15.10}$$

$$\mathrm{Re}_z^{-0.5} Nu_z = -\frac{k_{nf}}{k_f}\left(1 + Ra\left(1 + (\theta_w - 1)\theta(0)\right)^3\right)\theta'(0), \tag{15.11}$$

Where $\mathrm{Re}_z = \frac{bz^{n+1}}{\upsilon_f}$ is the local Reynolds number.

## 15.4 Flow Nature for Different Thermal Conductivity Models

Here, we have compared velocity, temperature, heat transfer rate, and drag using Hamilton-Crosser and Maxwell flow models. Figures 15.2 and 15.3 represent the change in velocity and temperature in both heat transfer models. Here, we have noticed that due to changes in magnetic numbers, $M$ velocity decreases and temperature increases. Here, the velocity and temperature are more in the case of the Hamilton-Crosser model when compared with

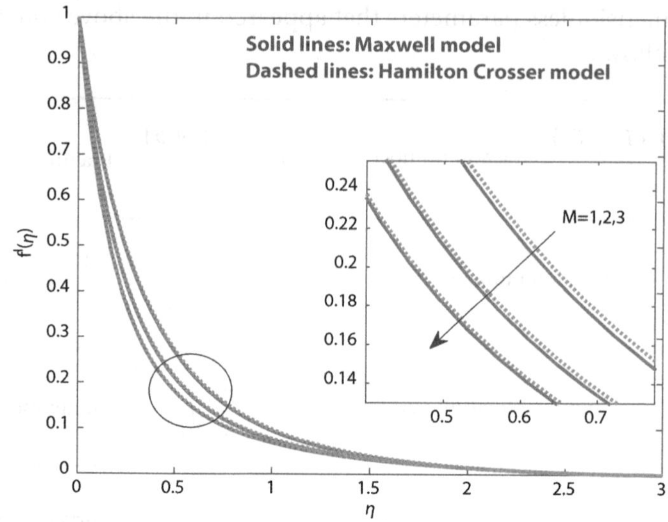


**Figure 15.2** Velocity change in both models due to increasing $M$ values.

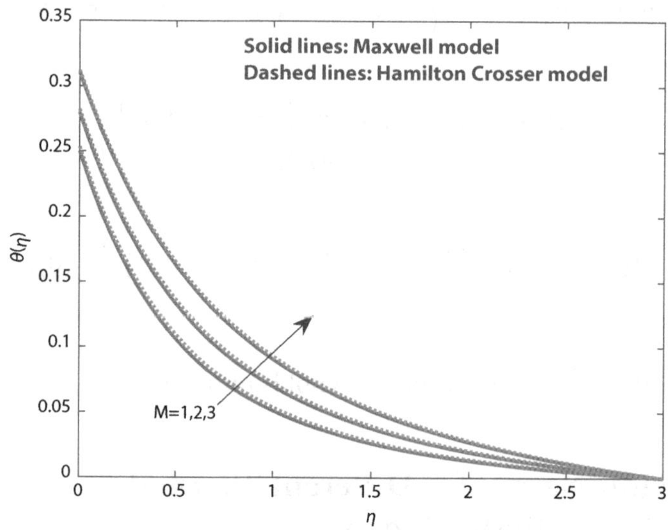


**Figure 15.3** Temperature change in both models due to increasing $M$ values.

the Maxwell model. Figures 15.4 and 15.5 display the change in velocity and heat due to changes in thermic radiation and drawn for both thermal conductivity models. Here, it is seen that by improving the values of radiation parameters, both velocity and heat upsurge. Effect on the velocity and temperature is more for the Hamilton-Crosser model than the Maxwell model.

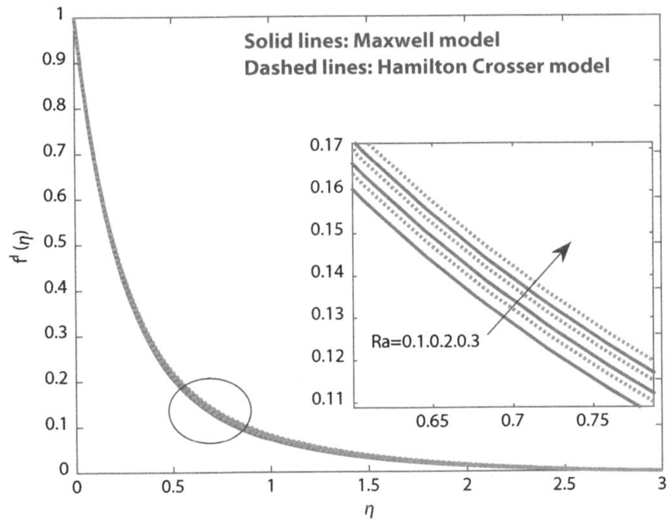


**Figure 15.4** Velocity change in both models due to increasing *Ra* values.

The heat transmission rate for both models is compared in Figures 15.6 and 15.7. For increasing values of magnetic numbers, they decline, but the Hamilton-Crosser model gives a higher heat transfer rate than the Maxwell model, as revealed in Figure 15.6. Growing standards of thermal radiation parameters upsurge the heat transmission amount in both models. Also, it is noticed that the Hamilton-Crosser model helps to obtain a higher heat transfer rate when compared with the Maxwell model (Table 15.3). Hence, the Hamilton-Crosser model plays a significant role in improving the velocity, temperature, and percentage of heat transmission of the flow.

**Table 15.3** Heat transfer rate for Maxwell and Hamiltonian-Crosser models.

| | | **Maxwell model** | | **Hamiltonian-Crosser model** | |
|---|---|---|---|---|---|
| *M* | *Ra* | **Drag values** | **Nusselt number** | **Drag values** | **Nusselt number** |
| 1 | | -5.752363 | 0.621654 | -5.709046 | 0.659556 |
| 2 | | -6.690554 | 0.599104 | -6.657622 | 0.636070 |
| 3 | | -7.412254 | 0.575342 | -7.389765 | 0.611585 |
| | 0.1 | -6.752111 | 0.546071 | -6.715855 | 0.579458 |
| | 0.2 | -6.690554 | 0.599104 | -6.657622 | 0.636070 |
| | 0.3 | -6.633901 | 0.652023 | -6.603890 | 0.692552 |

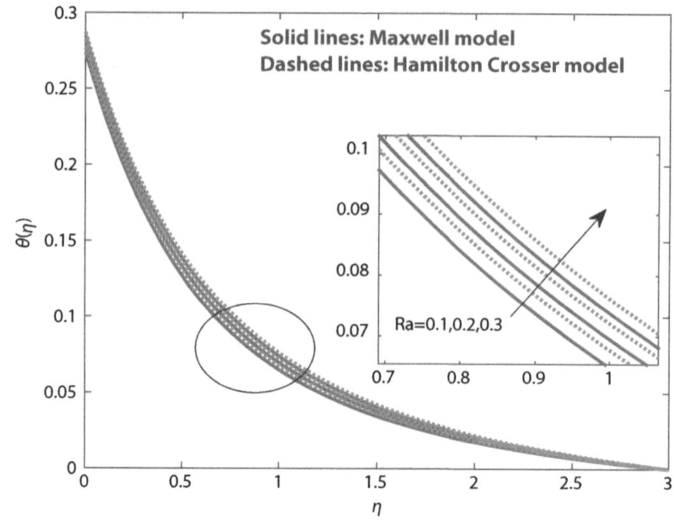


**Figure 15.5** Temperature change in both models due to increasing *Ra* values.

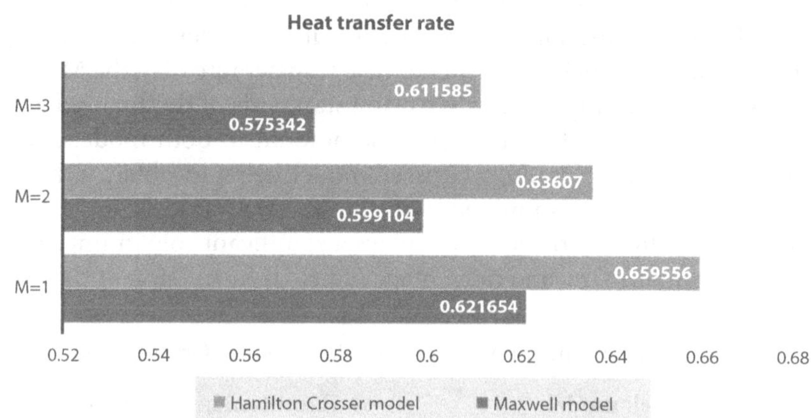


**Figure 15.6** Heat transfer rate for both models due to increasing values of magnetic numbers.

## 15.5 Outcomes and Discourse of Results

A mathematical and computational performance has been done in this chapter to analyze the blood flow on the curved outline region of the blood artery. Consider the 2D blood flow of MXene ($Ti_3C_2$), which is an emerging nanoparticle flowing through the curved-shaped artery. The thermal

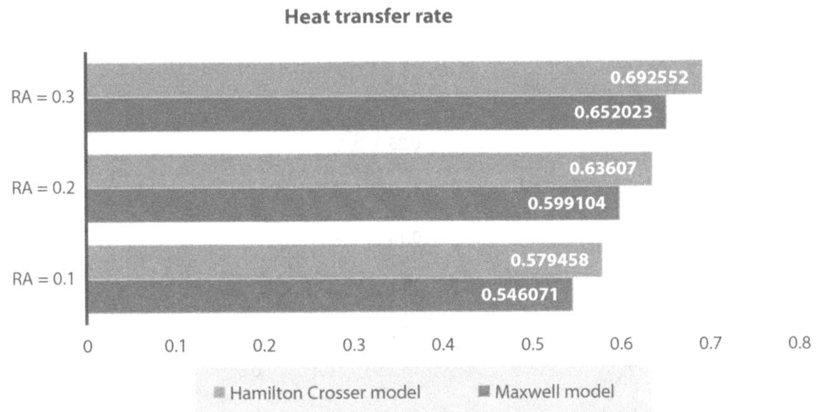


**Figure 15.7** Heat transfer rate for both models due to increasing values of thermal radiation.

and momentum nature of the nanofluids is examined for different values of physical constraints. Also, the heat transmission rate and drag of the pure blood and nanoparticle mixed blood are compared for better understanding and variance for different parametric values is shown with the help of graphs. The physical parametric value has been kept fixed at a time for computational purposes, that is, $M = 2$, $\gamma = 0.2$, $Gr = 2$, $Ra = 0.2$, $Br = .8$, $Bi = 0.6$, $\Pr = 10$, $\theta_w = 1.2$, the particle shape factor, $m = 4.9$, for cylindrical particles. The mathematical model of the flow, together with the boundary condition, is first transformed into ordinary differential equation (ODE) using suitable similarity transformation. Then, it was solved numerically by means of the bvp5c method, and graphs have been drawn using MATLAB software. The results are outlined to associate the momentum, energy, and rate of heat transfer for pure blood and MXene ($Ti_3C_2$) mixed blood for different values of physical parameters. The Hamilton-Crosser model is used to draw Figures 15.8 to 15.20, which represent the velocity and thermal changes for increasing standards of physical constraints involved in this study.

### 15.5.1 Velocity and Temperature Variance for MXene ($Ti_3C_2$) Emerged Blood Flow

This section discusses the change in velocity and temperature due to increasing values of physical parameters. Figures 15.8 and 15.9 display the velocity and thermal changes for the increasing values of the Casson parameter and compare the change with and without MXene ($Ti_3C_2$) nanoparticles. Increasing values of Casson fluid parameters cause the velocity to decline while temperature improves. Figures 15.10 and 15.11

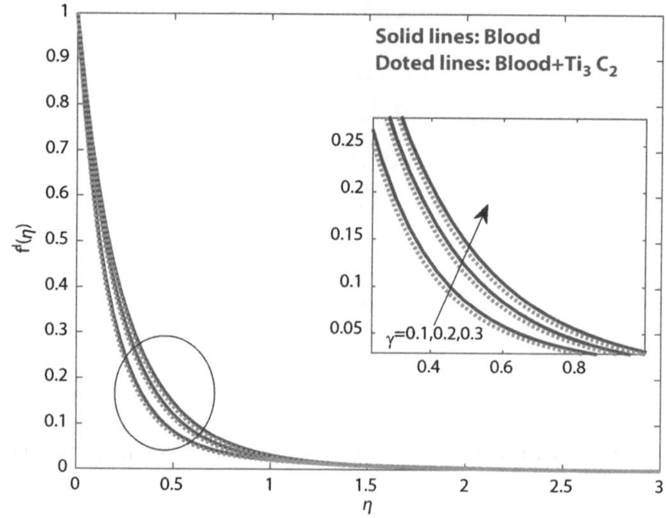


**Figure 15.8** Velocity modification for improving values of Casson constraint, *γ*.

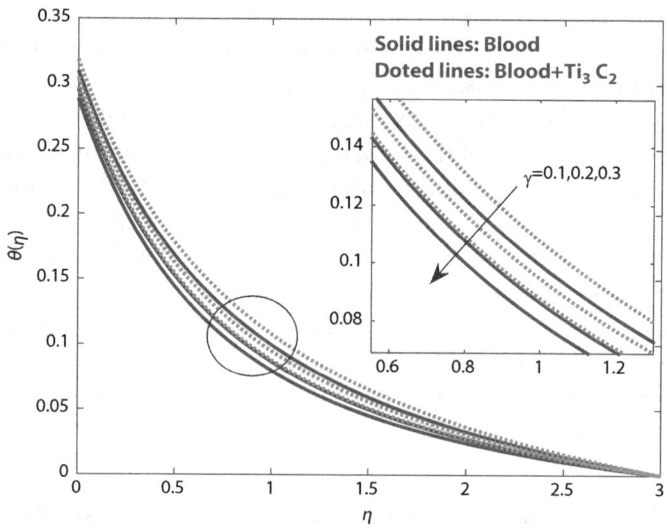


**Figure 15.9** Heat modification for improving values of Casson constraint, *γ*.

show the change in the moment and temperature for growing values of magnetic constraints; here, by improving the magnetic parameters, the velocity declines, but temperature is enhanced in both fluids. The influence of the Grashof numeral on velocity and heat is displayed in Figures 15.12 and 15.13. Here, it can be noticed that the improving standards of Grashof

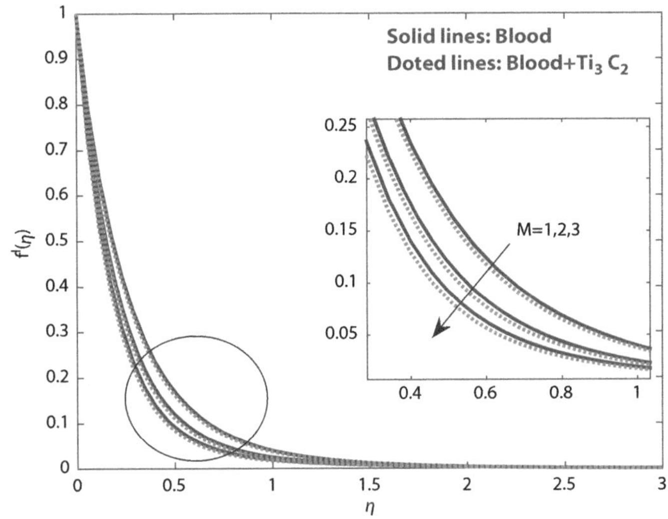


**Figure 15.10** Velocity for different values of *M*.

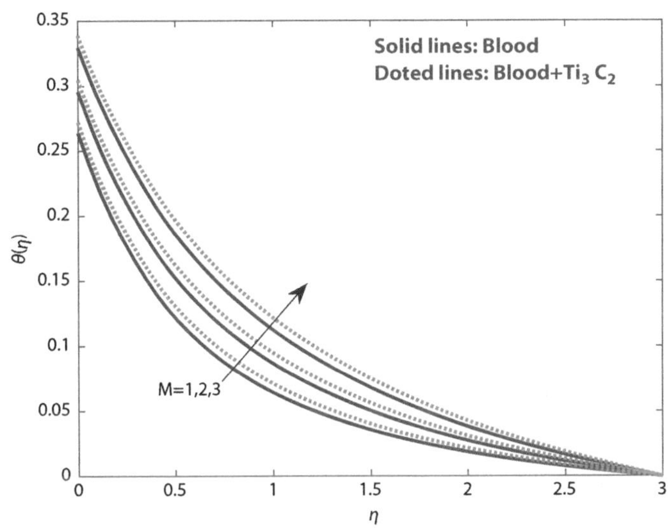


**Figure 15.11** Heat modification for improving values of *M*.

number the velocity and temperature upsurges in both fluids. Similarly, we can notice changes in moment and temperature for mounting standards of radiation factors *Ra*, Brinkman number *Br*, and Biot number *Bi* in Figures 15.14–15.19. Here, it is noticed that the velocity and temperature improve in both fluids by accumulating values of *Ra*, *Br*, and *Bi*.

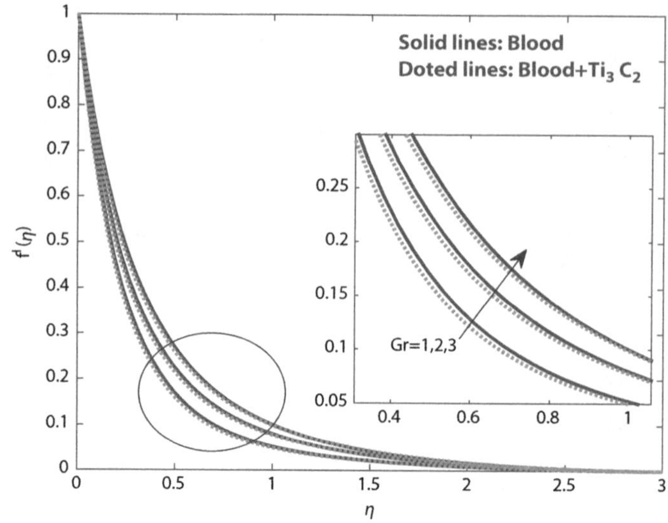


**Figure 15.12** Velocity modification for improving values of *Gr.*

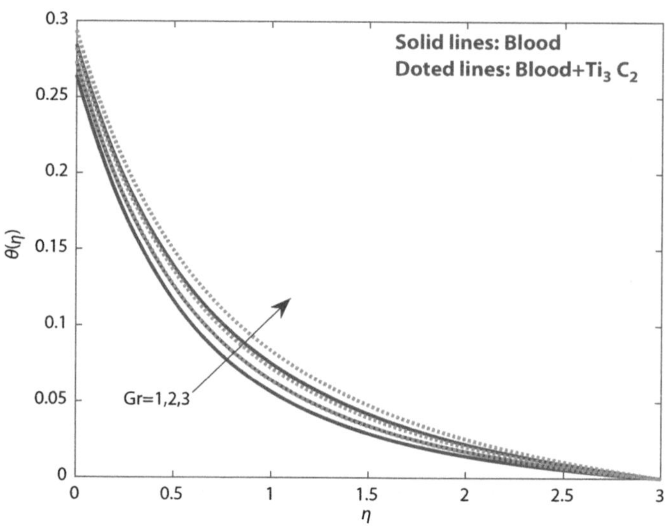


**Figure 15.13** Heat modification for improving values of *Gr*.

The most important thing is to notice that the momentum and heat of the nano bloodstream (blood+$Ti_3C_2$) have higher velocity and temperature than the pure blood flow. This reflects the importance of $Ti_3C_2$ nanoparticles, which help improve the momentum and energy of the blood flow.

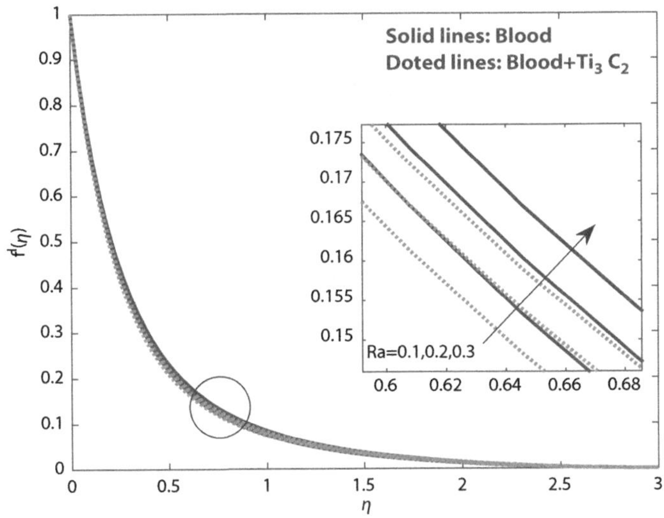


**Figure 15.14** Velocity modification for improving values radiation constraint, *Ra*.

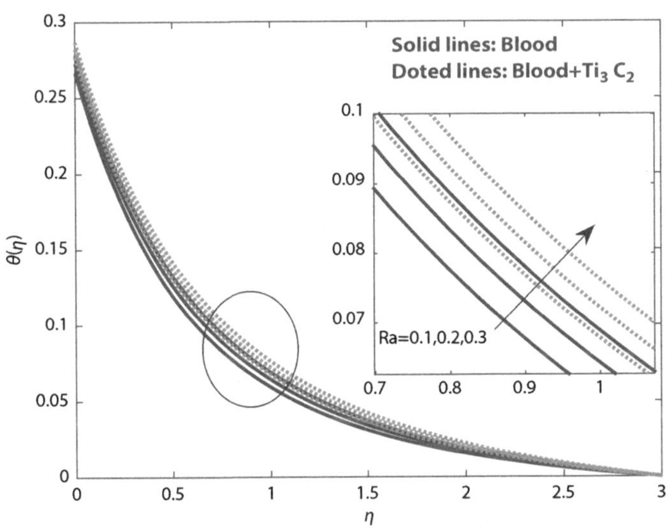


**Figure 15.15** Thermal modification for improving values of radiation constraint, *Ra*.

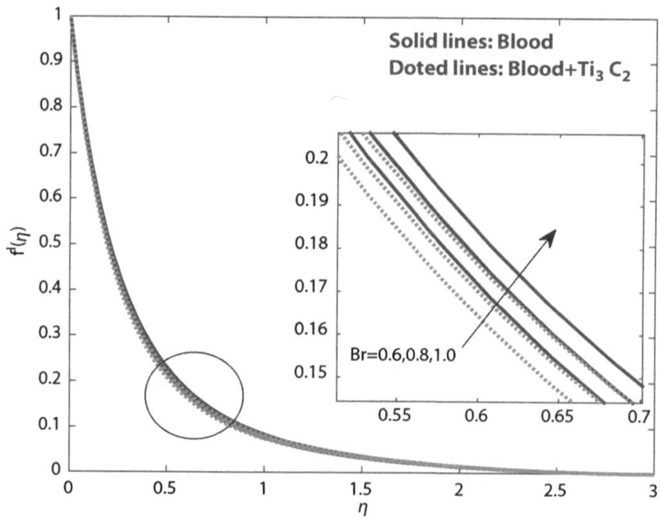


**Figure 15.16** Velocity modification for improving values of *Br*.

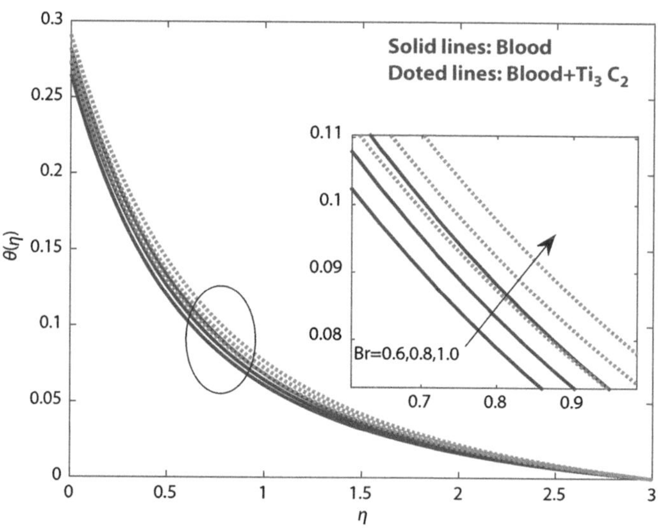


**Figure 15.17** Thermal modification for improving values of *Br*.

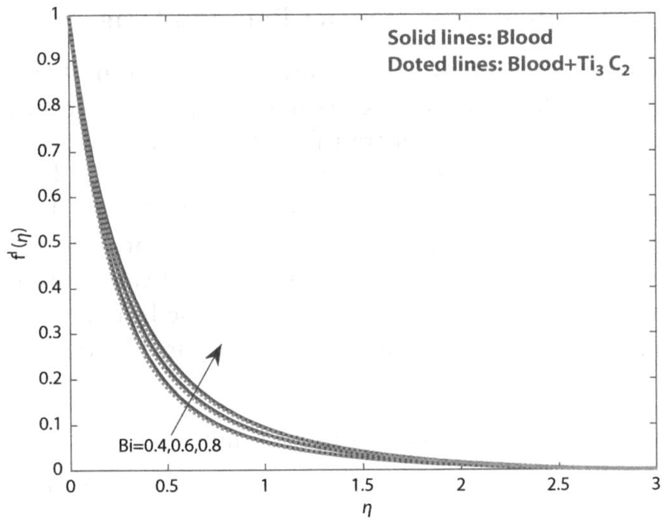


**Figure 15.18** Velocity modification for improving values of Biot number, *Bi*.

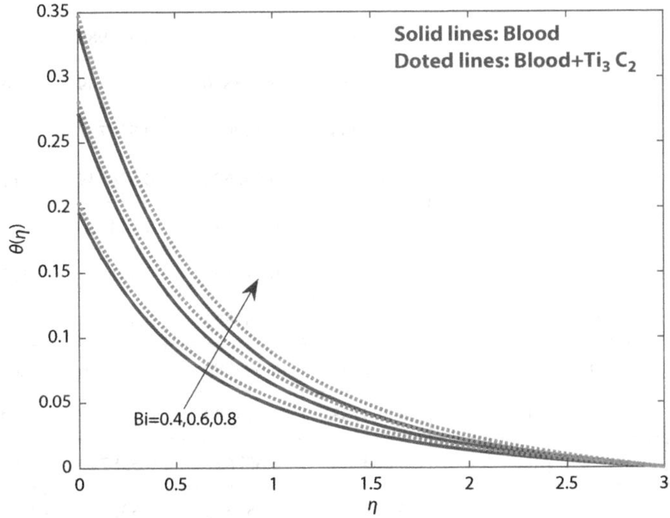


**Figure 15.19** Thermal modification for improving values of Biot number, *Bi*.

### 15.5.2 Variation in Heat Transfer Rate and Drag of the Flow

Table 15.4 shows the output for flow drag and rate of heat transmission for changing standards of parameters and compares the results with or without nanoparticles. The heat transfer rate advances by improving values of *γ*, *Ra*, *Gr*, and *Bi*. Moreover, significant change can be noticed in the case of thermic radiation *Ra* and Biot quantity *Bi*. For the rest of the parameters, *M* and *Br*, the heat transmission percentage declines, and the maximum heat transfer rate can occur at the lowest values of the range of these parameters. Additionally, it is noted that the heat allocation rate is advanced in nano blood flow when linked with pure blood flow for all the

**Table 15.4** Values of drag aspect and Nusselt quantity.

| Parameters | | | | | | Drag coefficient | | Nusselt number | |
|---|---|---|---|---|---|---|---|---|---|
| $\gamma$ | $M$ | $Gr$ | $Ra$ | $Br$ | $Bi$ | Blood | Blood+$Ti_3C_2$ | Blood | Blood+$Ti_3C_2$ |
| 0.1 | | | | | | -8.201213 | -9.352889 | 0.513324 | 0.606031 |
| 0.2 | | | | | | -6.951836 | -7.891940 | 0.522988 | 0.617407 |
| 0.3 | | | | | | -6.430702 | -7.282142 | 0.528021 | 0.623398 |
| | 1 | | | | | -6.128812 | -6.906503 | 0.544804 | 0.644075 |
| | 2 | | | | | -6.951836 | -7.891940 | 0.522988 | 0.617407 |
| | 3 | | | | | -7.620687 | -8.688870 | 0.499689 | 0.589296 |
| | | 1 | | | | -6.449344 | -7.340075 | 0.530792 | 0.626540 |
| | | 2 | | | | -5.833354 | -6.657622 | 0.538749 | 0.636070 |
| | | 3 | | | | -5.293456 | -6.056523 | 0.544616 | 0.643209 |
| | | | 0.1 | | | -5.899704 | -6.715855 | 0.491622 | 0.579458 |
| | | | 0.2 | | | -5.833354 | -6.657622 | 0.538749 | 0.636070 |
| | | | 0.3 | | | -5.772542 | -6.603890 | 0.585796 | 0.692552 |
| | | | | 0.6 | | -5.894258 | -6.724280 | 0.544260 | 0.642837 |
| | | | | 0.8 | | -5.833354 | -6.657622 | 0.538749 | 0.636070 |
| | | | | 1.0 | | -5.767470 | -6.585588 | 0.532907 | 0.628903 |
| | | | | | 0.4 | -6.209072 | -7.068941 | 0.393572 | 0.466231 |
| | | | | | 0.6 | -5.833354 | -6.657622 | 0.538749 | 0.636070 |
| | | | | | 0.8 | -5.514047 | -6.310193 | 0.658825 | 0.775680 |

parameters. At the same time, the number of skin friction is reduced due to Mxene nanoparticles. Hence, the heat allocation percentage is determined to be enhanced, and drag is reduced by mixing the MXene ($Ti_3C_2$) nanoparticles in the blood flow.

Figures 15.20–15.22 show changes in heat transfer rate for magnetic constraint, radiation constraint, and Biot number. The heat allocation rate is increasing for increasing values of physical constraints. Also, it is noticed that the heat allocation percentage is enhanced after mixing MXene nanoparticles.

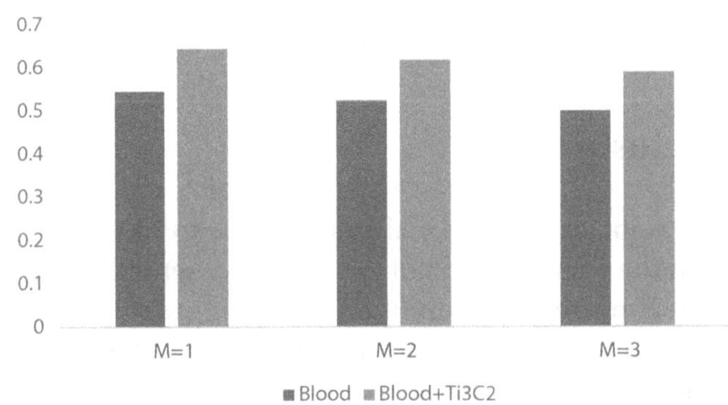


**Figure 15.20** Comparison of heat transfer rate for *M*.

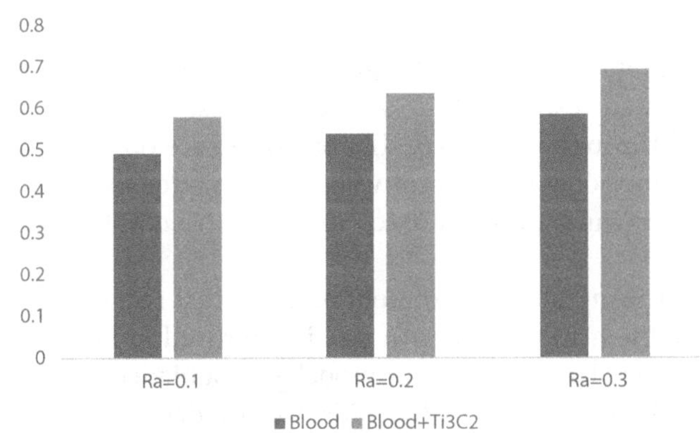


**Figure 15.21** Comparison of heat transfer rate for *Ra*.

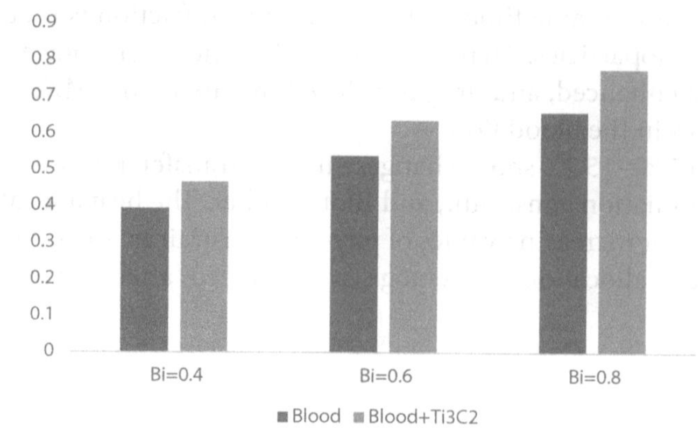


**Figure 15.22** Evaluation of heat transfer rate for *Bi*.

## 15.6 Conclusion

To observe the velocity and thermal changes due to MXene ($Ti_3C_2$) of the bloodstream on the curved region of the artery, we settled the mathematical model together with boundary conditions according to the curved geometry. The momentum equation is influenced by magnetic and buoyance effects to predict the physical phenomena of the flow. The temperature equation is fulfilled with nonlinear thermic radiation and Joule heating conditions. The governing conditions of the flow are transformed to the ODE and solved numerically using the bvp5c method. Here, we have compared the thermal conductivity for different models, namely Maxwell and Hamiltonian-Crosser models, and drawn results graphically and in tabular form. The velocity and thermal changes are noticed for diverse values of physical constraints and exhibited through the graphs. The drag of the flow and heat transmission rate is examined for different values, shown in tabular form, and compiled through bar charts. The outcomes are obtained numerically with the assistance of MATLAB software and exhibited through graphs and tables. The following significant outcomes are noticed in this chapter.

- For growing numerals of nonlinear thermal radiation parameters, the heat transmission rate is improved in both models. Also, the Hamilton-Crosser model gives a higher heat transfer rate when compared with the Maxwell model.
- Velocity is reduced, and temperature is improved by mixing the MXene ($Ti_3C_2$) nanoparticles, which helps in several

medical applications, such as thermal therapies for tumors, imaging, and others.

- MXene ($Ti_3C_2$) nanoparticles help to improve the heat transmission proportion and diminish the drag of the flow.
- Growing Biot quantity and nonlinear thermal radiation parameters upsurge the heat transfer rate in both fluids.
- The ability to manipulate heat allocation rate due to MXene ($Ti_3C_2$) nanoparticles at the nanoscale offers a range of benefits, including targeted hyperthermia for cancer treatment, improved drug delivery, advanced diagnostic imaging, and therapeutic heating for cardiovascular diseases.
- Observations of this study are helpful in enhancing cryopreservation techniques, treating infectious diseases, expediting wound healing, and providing thermotherapy for neurological disorders.

## References


1. Changdar, S. and De, S., Analytical Investigation of Nanoparticle as a Drug Carrier Suspended in a MHD Blood Flowing Through an Irregular Shape Stenosed Artery. *Iran. J. Sci. Technol. Trans. A Sci.*, 43, 1259–1272, 2019, https://doi.org/10.1007/s40995-018-0601-1.
2. Maqbool, K., Shaheen, S., Siddiqui, A.M., Effect of nanoparticles on MHD flow of tangent hyperbolic fluid in a ciliated tube: An application to fallopian tube. *Math. Biosci. Eng.*, 16, 2927–2941, 2019.
3. Abdel-wahed, M.S., Magnetohydrodynamic Ferro-Nano fluid flow in a semi-porous curved tube under the effect of hall current and nonlinear thermal radiative. *J. Magn. Magn. Mater.* [Internet], 474, 347–354, 2019, https://doi.org/10.1016/j.jmmm.2018.11.050.
4. Reddy, P.B.A., Biomedical aspects of entropy generation on electromagnetohydrodynamic blood flow of hybrid nanofluid with nonlinear thermal radiation and non-uniform heat source/sink. *Eur. Phys. J. Plus*, 135, 1–30, 2020, https://doi.org/10.1140/epjp/s13360-020-00825-7.
5. Tripathi, J., Vasu, B., Bég, O.A. *et al.*, Numerical simulation of the transport of nanoparticles as drug carriers in hydromagnetic blood flow through a diseased artery with vessel wall permeability and rheological effects. *Microvasc. Res.*, 139, 2022.
6. Wang, F., Rehman, S., Bouslimi, J. *et al.*, Comparative study of heat and mass transfer of generalized MHD Oldroyd-B bio-nano fluid in a permeable medium with ramped conditions. *Sci. Rep.*, **11**, 23454, 2021, https://doi.org/10.1038/s41598-021-02326-8.

7. Subbarayudu, K., Suneetha, S., Bala Anki Reddy, P., The assessment of time dependent flow of Williamson fluid with radiative blood flow against a wedge. *Propul. Power Res.*, 9, 87–99, 2020.
8. Ayeche, C.M., Kezzar, M., Sari, M.R. *et al.*, Analytical ADM study of time-dependent hydromagnetic flow of biofluid over a wedge. *Indian J. Phys.*, 95, 2769–2784, 2021.
9. Khan, S.A., Hayat, T., Alsaedi, A., Entropy optimization in passive and active flow of liquid hydrogen based nanoliquid transport by a curved stretching sheet. *Int. Commun. Heat Mass Transfer*, 119, 104890, 2020, https://doi.org/10.1016/j.icheatmasstransfer.2020.104890.
10. Ullah, B., Khan, U., Wahab, H.A., Khan, I., Alam, M.N., Entropy Generation Analysis for MHD Flow of Hybrid Nanofluids over a Curved Stretching Surface with Shape Effects. *J. Nanomater.*, 2022, 2022, https://doi.org/10.1155/2022/8929985.
11. Levy, E.Z. and Levy, Y., Comparing the Effect of Stent Geometry on Blood Flow Rate of Curved Coronary Artery Stenosis. *J. Emerging Invest.*, 1–6, 2016, https://doi.org/10.59720/15-047.
12. Reddy, M.G., Rani, MVVNLS, Kumar, K.G. *et al.*, Cattaneo–Christov heat flux and non-uniform heat-source/sink impacts on radiative Oldroyd-B two-phase flow across a cone/wedge. *J. Braz. Soc. Mech. Sci. Eng.*, 40, 1–21, 2018, https://doi.org/10.1007/s40430-018-1033-8.
14. Pandey, A.K. and Kumar, M., Effect of viscous dissipation and suction/injection on MHD nanofluid flow over a curved with porous medium and slip. *Alexandria Eng. J.*, 55, 3115–3123, 2016, Available from: http://dx.doi.org/10.1016/j.aej.2016.08.018.
15. Khan, M.I. and Alzahrani, F., Bio-convective thermal melting applications of viscoelastic nanoparticles due to moving wedge. *Case Stud. Therm. Eng.*, 28, 101493, 2021, Available from: https://doi.org/10.1016/j.csite.2021.101493.
16. Feng, Y., Yu, B., Xu, P. *et al.*, The effective thermal conductivity of nanofluids based on the nanolayer and the aggregation of nanoparticles. *J. Phys. D Appl. Phys.*, 40, 3164–3171, 2007.
17. Zhu, J. and Cao, J., Effects of nanolayer and second order slip on unsteady nanofluid flow past awedge. *Mathematics*, 7, 11, 1043, 2019, https://doi.org/10.3390/math7111043.
18. Rana, P. and Bég, O.A., Mixed convection flow along an inclined permeable plate: Effect of magnetic field, nanolayer conductivity and nanoparticle diameter. *Appl. Nanosci.*, **5**, 569–581, 2015, https://doi.org/10.1007/s13204-014-0352-z.
19. Dehkordi, B.A.F. and Abdollahi, A., Experimental investigation toward obtaining the effect of interfacial solid-liquid interaction and base fluid type on the thermal conductivity of CuO-loaded nanofluids. *Int. Commun. Heat Mass Transfer*, 97, 151–162, 2018.
20. Jiang, H., Xu, Q., Huang, C. *et al.*, The role of interfacial nanolayer in the enhanced thermal conductivity of carbon nanotube-based nanofluids. *Appl. Phys. A Mater. Sci. Process.*, 118, 197–205, 2015.

21. Benos, L.T. and Sarris, I.E., The interfacial nanolayer role on magnetohydrodynamic natural convection of an Al2O3-water nanofluid. *Heat Transfer Eng.*, 42, 89–105, 2021, http://dx.doi.org/10.1080/01457632.2019.1692487.
22. Swain, B.K., Parida, B.C., Kar, S. *et al.*, Viscous dissipation and joule heating effect on MHD flow and heat transfer past a stretching sheet embedded in a porous medium. *Heliyon*, 6, e 05338, 2020, https://doi.org/10.1016/j.heliyon.2020.e05338.
23. Rana, P., Dhanai, R., Kumar, L., MHD slip flow and heat transfer of Al2O3-water nanofluid over a horizontal shrinking cylinder using Buongiorno's model: Effect of nanolayer and nanoparticle diameter. *Adv. Powder Technol.*, 28, 1727–1738, 2017, http://dx.doi.org/10.1016/j.apt.2017.04.010.
24. Sharma, B.K., Poonam, Chamkha, A.J., Effects of heat transfer, body acceleration and hybrid nanoparticles (Au–Al2O3) on MHD blood flow through a curved artery with stenosis and aneurysm using hematocrit-dependent viscosity. *Waves Random Complex Media*, 2022, https://doi.org/10.1080/17455030.2022.2125597.
25. Teimouri, K., Tavakoli, M.R., Ghafari, A., Kim, K.C., Investigation of the plaque morphology effect on changes of pulsatile blood flow in a stenosed curved artery induced by an external magnetic field. *Comput. Biol. Med.*, 135, March, 104600, 2021, https://doi.org/10.1016/j.compbiomed.2021.104600.
26. Nisar, K.S., Khan, U., Zaib, A., Khan, I., Morsy, A., A novel study of radiative flow involving micropolar nanoliquid from a shrinking/stretching curved surface including blood gold nanoparticles. *Eur. Phys. J. Plus*, 135, 10, 1–19, 2020, https://doi.org/10.1140/epjp/s13360-020-00830-w.
27. Koyappayil, A., Chavan, S.G., Roh, Y.G., Lee, M.H., Advances of MXenes; Perspectives on Biomedical Research. *Biosensors*, 12, 7, 2022, https://doi.org/10.3390/bios12070454.
28. Saharudin, M.S., Ayub, A., Hasbi, S., Muhammad-Sukki, F., Shyha, I., Inam, F., Recent advances in MXene composites research, applications and opportunities. *Mater. Today: Proc.*, 0–4, 2023, https://doi.org/10.1016/j.matpr.2023.02.435.
29. Zamhuri, A., Lim, G.P., Ma, N.L., Tee, K.S., Soon, C.F., MXene in the lens of biomedical engineering: synthesis, applications and future outlook. *Biomed. Eng. Online*, 20, 1, 1–24, 2021, https://doi.org/10.1186/s12938-021-00873-9.
30. Xing, C., Chen, S., Liang, X., Liu, Q., Qu, M., Zou, Q., Li, J., Tan, H., Liu, L., Fan, D., Zhang, H., Two-Dimensional MXene (Ti3C2)-Integrated Cellulose Hydrogels: Toward Smart Three-Dimensional Network Nanoplatforms Exhibiting Light-Induced Swelling and Bimodal Photothermal/Chemotherapy Anticancer Activity. *ACS Appl. Mater. Interfaces*, 10, 33, 27631–27643, 2018, https://doi.org/10.1021/acsami.8b08314.
31. Sreenilayam, S.P., Ul Ahad, I., Nicolosi, V., Brabazon, D., MXene materials based printed flexible devices for healthcare, biomedical and energy storage applications. *Mater. Today*, 43, March, 99–131, 2021, https://doi.org/10.1016/j.mattod.2020.10.025.

32. Chao, M., He, L., Gong, M., Li, N., Li, X., Peng, L., Shi, F., Zhang, L., Wan, P., Breathable Ti3C2TxMXene/Protein Nanocomposites for Ultrasensitive Medical Pressure Sensor with Degradability in Solvents. *ACS Nano*, 15, 6, 9746–9758, 2021, https://doi.org/10.1021/acsnano.1c00472.
33. Wang, H., Wu, Y., Zhang, J., Li, G., Huang, H., Zhang, X., Jiang, Q., Enhancement of the electrical properties of MXene Ti3C2 nanosheets by post-treatments of alkalization and calcination. *Mater. Lett.*, 160, 537–540, 2015, https://doi.org/10.1016/j.matlet.2015.08.046.
34. Hussain, A., Sarwar, L., Rehman, A., Al Mdallal, Q., Almaliki, A.H., El-Shafay, A.S., Mathematical analysis of hybrid mediated blood flow in stenosis narrow arteries. *Sci. Rep.*, 12, 1, 2022, https://doi.org/10.1038/s41598-022-15117-6.
35. Ansarpour, M., Aslfattahi, N., Mofarahi, M., Saidur, R., Numerical study on the convective heat transfer performance of a developed MXene IoNanofluid in a horizontal tube by considering temperature-dependent properties. *J. Therm. Anal. Calorim.*, 147, 21, 12067–12078, 2022, https://doi.org/10.1007/s10973-022-11414-4.
36. Hayat, T., Rashid, M., Alsaedi, A., MHD convective flow of magnetite-Fe3O4 nanoparticles by curved stretching sheet. *Results Phys.*, 7, 3107–3115, 2017, https://doi.org/10.1016/j.rinp.2017.08.015.
37. Ahmed, A., Khan, M., Irfan, M., Ahmed, J., Transient MHD flow of Maxwell nanofluid subject to non linear thermal radiation and convective heat transport. *Appl. Nanosci.*, 0123456789, 2020, https://doi.org/10.1007/s13204-020-01375-1.
38. Khan, U., Zaib, A., Khan, I., Sooppy, K., Insight into the dynamics of transient blood conveying gold nanoparticles when entropy generation and Lorentz force are significant. *Int. Commun. Heat Mass Transfer*, 127, July, 105415, 2021, https://doi.org/10.1016/j.icheatmasstransfer.2021.105415.

# 16

# Strong Magnetic Shock Wave Propagation in a Dusty Gas

**Akmal Husain[1]*, S. A. Haider[2], M. K. Shukla[2], Mohd. Miyan[2] and A. Taqvi[2]**

[1]*Department of Mathematics, Applied Science Cluster, SoAE, University of Petroleum and Energy Studies, Dehradun, India*
[2]*Department of Mathematics, Shia P.G. College, Lucknow, India*

***Abstract***

The aim of this study is to create a precise mathematical solution for the movement of a shock wave in a gas that is not perfect and contains tiny solid particles. Additionally, the study will take into account the impact of a magnetic field. In this regard, a relationship has been demonstrated between thermodynamic flow parameters that are affected by Rankine-Hugoniot jump boundary conditions. Through the application of Murata's direct analytical approach, we have obtained a succinct resolution for the physical flow parameters and the radius of the shock front. This solution effectively demonstrates the correlation between time and space. A comprehensive analysis was undertaken to investigate the total energy released during the propagation of a severe discontinuity from the center of the explosion. To perform the present study, the gaseous medium is assumed to comprise a non-ideal gas that is infused with dusty particles. Furthermore, we have examined the impact of dust particles and suboptimal parameters in the presence of a magnetic field (quantified by the Shock-Cowling number) on the characteristics of physical flow profiles and overall energy. The magnetic field has a significant impact on the precise solutions derived for flow-field variables, including density, pressure, and magnetic pressure. Nevertheless, the dimensions of the shock front and the speed distribution remain unaltered in the presence of the magnetic field.

**Corresponding author*: akmal.h04@gmail.com
Akmal Husain: ORCID: 0000-0003-3016-2898
S. A. Haider: ORCID: 0000-0002-0628-6375

Ashwani Kumar, Nitesh Dutt and Mukesh Kumar Awasthi (eds.) Heat Transfer Enhancement Techniques: Thermal Performance, Optimization and Applications, (359–374) © 2025 Scrivener Publishing LLC

***Keywords*:** Shock waves, Rankine-Hugoniot jump boundary conditions, van der Waals gas, magnetic field, dust particles

## 16.1 Introduction

Studying the propagation of waves in different gaseous environments has become highly significant in both mathematical and physical contexts, as well as for their applications in various fields such as nuclear science, aerodynamics, plasma physics, and quantum physics. Blast waves refer to intermittent compression waves that occur in the gaseous medium as a result of the abrupt concentration of a vast quantity of energy inside a relatively confined space. In the realm of mathematics, the phenomenon of blast wave propagation is expressed as a set of conservation laws that exhibit rigorous hyperbolic behavior. Due to the non-linear nature of hyperbolic conservative systems, analyzing these systems analytically is a demanding and difficult endeavor. However, under certain assumptions, it is possible to determine closed-form solutions.

The study of shock wave proliferation in a dust-filled real medium, which involves the movement of non-ideal gas including particles of dust, has become an increasingly important field of research from both theoretical and experimental perspectives in recent years. In a turbulent gas flow with suspended solid particles, there is a sudden exchange of momentum and energy between the particles and the gas. This exchange leads to a major divergence from the behavior of an ideal gas flow. When a shock propagates in a real medium containing small solid particles, the velocity of the gas-particle mixture increases abruptly near the leading wavefront. However, the shock velocity decreases because of the scattering of solid particles. Conversely, the particle velocity eventually converges with the gas velocity behind the shock front as a result of the significantly enormous inertia at a microscopic level. The intricate nature of the system allows for a wide range of research, encompassing natural processes as well as issues pertaining to industrial and astrophysical applications. Examples of such applications include the production of reactive powder materials in industries, ensuring explosion safety in confined buildings and coal mines, studying the motion of supersonic vehicles in multiphase flows, exploring coating technologies, and investigating supernova remnants, among others.

Over the years, numerous authors have conducted extensive research to enhance our comprehension of the robust shock propagation phenomenon in a dusty medium. These investigations include studies by Carrier [1], Soo [2], Marble [3], Pai [4] and Rudinger [5], and the references mentioned in their works. Elperin *et al.* [6] conducted a numerical study on the direct impact of normal shock waves in dusty gases. They used a modified random-choice method and observed that reflected shock waves in dusty gases travel at a slower pace compared to those in a perfect gas. Hamad and El-Fayes [7] theoretically determined the structure of fully dispersed waves in dusty gases by taking into account the molecular relaxation process and macroscopic relaxation process. Nath [8] conducted a study to examine the impact of dust parameters on the flow variables and provided a comparison between the solutions for isothermal and adiabatic flows. With the aid of the power series method, Singh and Arora [9] derived the first and second-order approximate solutions that qualitatively analyze the shock propagation in dust-filled gas.

Most of the studies mentioned above focused on analyzing the behavior of shock waves in a dusty medium combined with a perfect gas. Typically, the issues related to shock propagation arise in high-pressure or low-temperature conditions. Consequently, the non-ideal medium model provides a more accurate explanation for the nonlinear wave problem compared to the ideal gas medium model. The prominence and significance of non-ideal gas filled with dust particle effects in theoretical and experimental investigations are greatly enhanced with the growing strength of shock waves. Vishwakarma and Nath [10] investigated the issue of a piston in motion inside a combination of non-ideal gas and dusty particles. They analyzed the magnitude of the non-ideal and dust characteristics in the flow field. Chadha and Jena [11] examine the impact of dusty and non-ideal variables on the various modes of wave propagation and their resulting discontinuities at the wavefronts in a plane, employing a singular-surface technique. Shah and Singh [12] utilized the Lie invariance group analysis approach to study and describe the collapse of implosion of nonlinear discontinuities in a non-ideal dust-filled reactive gas. In their study, Husain *et al.* [13] conducted a detailed analysis of the issue of strong shock proliferation in a dust-loaded van der Waals (vdW) gas. Subsequently, in another study, Husain *et al.* [14] further investigated the impact of dust and non-ideal parameters on the radius of shock waves and the total energy of weak shock proliferation in a non-ideal dusty medium. Chaturvedi *et al.* [15] investigated the evolution of shock waves along specific paths in polytropic reactive gas dynamics, particularly focusing on the impact of dust particles. It demonstrated how the reactive gas parameter affects the development of both compressive and expansion waves in reactive gas flows.

All of the previous studies have examined the flow of gas and particles without taking into account the magnetic field. However, it is widely recognized that most problems related to shock waves occur under high-pressure or low-temperature conditions. Therefore, in recent decades, researchers have been interested in studying the dynamics of shock waves in the presence of a magnetic field. Therefore, in recent years, numerous authors have conducted several research involving the propagation of shock waves in a two-phase medium consisting of a mixture of gas and dust-laden particles, while considering the influence of magnetic field effects. In a recent study, Sahu [16] derived solutions for the unsteady, self-similar flow field that occurs when a spherical shock encounters a moving piston in a dust loaded gas. Dust is comprised of minuscule solid particles that are constantly dispersed inside a non-ideal gas. The medium is exposed to gravitational and magnetic fields while being irradiated with monochromatic radiation. Under the influence of a magnetic field (transverse) and dust particles, Gupta and Singh [17] examined the evolutionary behavior of weakly nonlinear waves moving through an unstable 1-dimensional flow of a compressible, perfectly conducting fluid. According to their research, an axial magnetic field speeds up the expansion of a sawtooth wave (also known as a half N-wave) compared to when the field is not there. For a non-ideal dusty gas flow that is one-dimensional, isentropic, and completely conducting, Pang [18] discovered the precise solutions to the Riemann problem in the presence of a transverse magnetic field and also gave the behavior and expression of the wave curves.

The objective of this study is to derive a mathematical solution for the problem of the spread of a non-planar magnetic blast wave. The problem is described by a set of quasilinear hyperbolic differential equations in a mixture of simplified vdW gas with dust particles. The study aims to analyze the effects of dust particles and non-ideal parameters on flow profiles, shock path, and total energy of the blast waves in the presence of a magnetic (transverse) field. To do the analysis mentioned above, it is assumed that the gaseous medium is a limited type of vdW gas, and the dust particles are spherical form, small-sized, amorphous, and evenly dispersed throughout the mixture.

## 16.2 Fundamental Set of Equations

A basic set of Euler's conservation equations is considered to describe the one-dimensional compressible non-planar adiabatic flow motion of a dust-filled simplified vdW gas with magnetic field effects in the absence of radiation and viscosity, given as {see Courant and Friedrichs [19] and Whitham [20]}

$$\varrho_t + u\varrho_x + \varrho u_x + j\varrho u x^{-1} = 0, \tag{16.1}$$

$$\varrho(u_t + uu_x) + p_x + h_x = 0, \tag{16.2}$$

$$E_t + uE_x - \frac{p}{\varrho^2}(\varrho_t + u\varrho_x) = 0, \tag{16.3}$$

$$h_t + uh_x - B^2(\varrho_t + u\varrho_x) = 0. \tag{16.4}$$

Here, $x, t > 0$ are independent variables representing the position-time coordinates, $\varrho(x, t)$ is the fluid density, $p(x, t)$ is the pressure, $u(x, t)$ is the flow velocity, $h(x, t)$ is the magnetic pressure defined as $h = H^2\nu/2$, with H and $\nu$, respectively, as the transverse magnetic field and magnetic permeability. The letter $B = \left(\frac{2h}{\varrho}\right)^{1/2}$ denotes Alfven speed. Here we see, cylindrically symmetric flow with $j = 1$ and spherically symmetric flow with $j = 2$.

The entity $E$ denotes the internal energy of the vdW gas (simplified form) filled with tiny solid dust particles at per unit mass defined as

$$E = \frac{\{1-b\varrho(1-k_p)\}p(1-z)}{\varrho(\Gamma-1)} \tag{16.5}$$

and supplemented with the constitutive equation of state

$$p = \frac{(1-k_p)}{(1-Z)\{1-b\varrho(1-k_p)\}}\varrho\Re T, \tag{16.6}$$

where $k_p$ and $Z$ denotes fractions $k_p = m_{sp}/m_g$ and $Z = V_{sp}/V_g$; the quantities $m_{sp}$, $m_g$, $V_{sp}$, and $V_g$ respectively, defines the total mass of dusty particles and gas-mixture, and volumetric extension of the dusty particles and gas-mixture. Physical quantities $\Re$ and $T$ denote the gas (universal) constant and absolute temperature, respectively. 'Γ' denotes the Gruneisen coefficient defined as $\Gamma = \gamma(1+\lambda\beta)/(1+\lambda\beta\gamma)$ where $\gamma = c_p/c_v$ $\lambda = k_p/(1 - k_p)$, $\beta = c_{sp}/c_p$; here $c_p$, $c_v$ defines gas-mixture specific heat at constant pressure and volume, and $c_{sp}$ stands for specific heat of solid particles. The constant $b$ defines the vdW repulsion parameter.

Under inspection of Equations 16.3 and 16.5, the conservation of energy equation takes the form

$$p_t + u p_x - \mathcal{A}^2(\varrho_t + u\varrho_x) = 0, \tag{16.7}$$

where $\mathcal{A} = \left(\frac{\{\Gamma - b(1-k_p)\varrho Z\}p}{(1-Z)\{1-b\varrho(1-k_p)\}\varrho}\right)^{\frac{1}{2}}$ represents equilibrium sound speed.

Assuming the blast wave is propagating with speed $\frac{dR}{dt} = \mathcal{V}$, where $\mathbb{R} = \mathbb{R}(t)$ is the shock-location at time $t$. The flow variables in the undisturbed region is given by

$$u_* = 0,\ p_* = \text{constant},\ \varrho = \varrho_*(x) \text{ and } h_* = h_*(x).$$

In the case of physical flow variables, the values in the undisturbed zone are denoted by the numerical subscript '*'. A power law governs the relationship between the undisturbed region's density profile $\varrho_*$ and the shock radius are given as

$$\varrho_* = \varrho_a \mathbb{R}^{\mathfrak{S}},$$

where $\varrho_a$ and $\mathfrak{S}$ are constants. The analysis of the constant $\mathfrak{S}$ will be carried out in later sections.

The expression of total energy (addition of kinetic and internal energies) driven by the blast wave in a dust-filled vdW gas with the magnetic field is given as

$$\mathcal{E}_{\mathcal{J}} = 4\pi \int_0^{\mathbb{R}} \left( \frac{1}{2}\varrho u^2 + \frac{(1-Z)(1-b(1-k_p)\varrho)}{(\Gamma-1)} p + h \right) x^{j}\, \mathbf{d}x. \tag{16.8}$$

## 16.3 Rankine-Hugoniot Jump Boundary Conditions for Strong Shocks

The fundamental set of compressible non-ideal dusty magneto-gas-dynamics (MGD) equations for planar flow (Equations 16.1–16.4) can be reformulated into the following form (conservation) to obtain the shock jump boundary conditions:

$$\varrho_t + (\varrho u)_x = -\frac{j\varrho u}{x}, \tag{16.9}$$

$$(\varrho u)_t + (\varrho u^2 + p + h)_x = -\frac{j\varrho u^2}{x}, \tag{16.10}$$

$$\left(\frac{p(1-Z)(1-b\varrho(1-k_p))}{(\Gamma-1)} + \frac{\varrho u^2}{2} + h\right)_t + \left(\frac{((\Gamma-Z)-b\varrho(1-k_p)(1-Z))\,pu}{(\Gamma-1)} + \frac{\varrho u^3}{2} + 2uh\right)_x = -\frac{j\varrho u^3}{2x} - \frac{jpu}{x}\frac{((\Gamma-Z)-b\varrho(1-k_p)(1-Z))}{(\Gamma-1)} - \frac{2hju}{x} \tag{16.11}$$

$$\left((h)^{1/2}\right)_t + \left(u(h)^{1/2}\right)_x = -\frac{(h)^{1/2} ju}{x}. \tag{16.12}$$

Now, using the conservation form, we can derive the strong shock jump boundary conditions that connect the dynamics of the ahead and behind blast waves across the shock front $\mathbb{R}(t)$ as follows:

$$\varrho u - \varrho_* u_* = V(\varrho - \varrho_*), \tag{16.13}$$

$$(\varrho u^2 + p + h) - (\varrho_* u_*^2 + p_* + h_*) = V(\varrho u - \varrho_* u_*), \tag{16.14}$$

$$\left(\frac{((\Gamma-Z)-b\varrho(1-k_p)(1-Z))\,pu}{(\Gamma-1)} + \frac{\varrho u^3}{2} + 2uh\right) - \left(\frac{((\Gamma-Z_*)-\bar{b}(1-k_p)(1-Z_*))\,p_* u_*}{(\Gamma-1)} + \frac{\varrho_* u_*^3}{2} + 2u_* h_*\right) = V\left(\left(\frac{p(1-Z)\{1-b\varrho(1-k_p)\}}{(\Gamma-1)} + \frac{\varrho u^2}{2} + h\right) - \left(\frac{(1-Z_*)\{1-\bar{b}(1-k_p)\}\,p_*}{(\Gamma-1)} + \frac{\varrho_* u_*^2}{2} + h_*\right)\right) \tag{16.15}$$

$$(u(h)^{1/2}) - (u_*(h_*)^{1/2}) = V((h)^{1/2} - (h_*)^{1/2}). \tag{16.16}$$

Here $\bar{b} = b\varrho_*$ and $Z_* = \dfrac{k_p}{(1-k_p)G+k_p}$; G signifies the ratio of density of dust particles and gas-mixture.

The following equation which is quadratic in nature is found in terms of the density flow parameter $\varrho(x,t)$ using strong shock conditions (Equations 16.13–16.16).

$$2\bar{b}(1-k_p)Z_*\varrho^2 - [(\Gamma-1) + 2\{Z_* + \bar{b}(1-k_p)\}]\varrho\varrho_* + (\Gamma+1)\varrho_*^2 = 0, \tag{16.17}$$

where $\mathcal{A}_* = \left[\dfrac{(\Gamma-\bar{b}(1-k_p)Z_*)p_*}{(1-Z_*)(1-\bar{b}(1-k_p))\varrho_*}\right]^{1/2}.$

To derive the Rankine-Hugoniot (RH) jump conditions for the thermodynamic flow variables $\varrho$, $u$, $p$, and $h$, one must simplify quadratic Equation 16.17 along with Equations (conditions) 16.13, 16.14, and 16.16. So, we have

$$\varrho = \frac{1}{4\bar{b}(1-k_p)Z_*}\left(Q + \sqrt{Q^2 - 8\bar{b}(1-k_p)(\Gamma+1)Z_*}\right)\varrho_*, \tag{16.18}$$

$$u = \left(1 - \frac{4\bar{b}(1-k_p)Z_*}{Q + \sqrt{Q^2 - 8\bar{b}(1-k_p)(\Gamma+1)Z_*}}\right)V, \tag{16.19}$$

$$p = \left[\left\{1 - \frac{4\bar{b}(1-k_p)Z_*}{Q + \sqrt{Q^2 - 8\bar{b}(1-k_p)(\Gamma+1)Z_*}}\right\} + \frac{C_*}{2}\left\{1 - \left(\frac{Q + \sqrt{Q^2 - 8\bar{b}(1-k_p)(\Gamma+1)Z_*}}{4\bar{b}(1-k_p)Z_*}\right)^2\right\}\right]\varrho_* V^2, \tag{16.20}$$

$$h = \frac{C_*}{2}\left(\frac{Q + \sqrt{Q^2 - 8\bar{b}(1-k_p)(\Gamma+1)Z_*}}{4\bar{b}(1-k_p)Z_*}\right)^2 \varrho_* V^2. \tag{16.21}$$

where $Q = \Gamma - 1 + 2\{Z_* + \bar{b}(1-k_p)\}$ and the entity $C_* = \frac{2h_*}{\varrho_* V^2}$ termed as the shock Cowling number.

## 16.4 Closed Form Solution for Strong Shocks

In this section, we pursue the precise solution of the basic set of Euler's Equations 16.1–16.4 in a vdW gas (simplified form) including dust particles affected by a magnetic field by applying the simple analytical method suggested by Murata {see Reference [21]}.

Considering the above given context, we used shock boundary conditions from Equations 16.18–16.21 to derive explicit relations for the flow characteristics pressure and transverse magnetic field pressure just behind the shock front, using the physical variables density and velocity, stated as

$$\mathcal{p}=\left[\frac{2(4\bar{b}(1-k_{\mathcal{p}})Z_*)^2-\mathcal{C}_*\left(\mathcal{Q}+\sqrt{\mathcal{Q}^2-8\bar{b}(1-k_{\mathcal{p}})(\Gamma+1)\,Z_*}\right)\left(\mathcal{Q}+\sqrt{\mathcal{Q}^2-8\bar{b}(1-k_{\mathcal{p}})(\Gamma+1)\,Z_*}+4\bar{b}(1-k_{\mathcal{p}})Z_*\right)}{8\bar{b}(1-k_{\mathcal{p}})\,Z_*\left(\mathcal{Q}+\sqrt{\mathcal{Q}^2-8\bar{b}(1-k_{\mathcal{p}})(\Gamma+1)Z_*}-4\bar{b}(1-k_{\mathcal{p}})Z_*\right)}\right]\varrho_*u^2, \tag{16.22}$$

$$\hbar=\frac{\mathcal{C}_*\left(\mathcal{Q}+\sqrt{\mathcal{Q}^2-8\bar{b}(1-k_{\mathcal{p}})(\Gamma+1)Z_*}\right)^3}{8\bar{b}(1-k_{\mathcal{p}})Z_*\left[\left(\mathcal{Q}+\sqrt{\mathcal{Q}^2-8\bar{b}(1-k_{\mathcal{p}})(\Gamma+1)Z_*}\right)-4\bar{b}(1-k_{\mathcal{p}})Z_*\right]^2}\varrho_*u^2 \tag{16.23}$$

By considering relations (16.22) and (16.23), we can rewrite the basic governing equations (16.2)–(16.4) as

$$u_t+uu_x-\left[\frac{\{4\bar{b}(1-k_{\mathcal{p}})Z_*\}^2\{(2+\mathcal{C}_*)\mathfrak{T}-8\bar{b}(1-k_{\mathcal{p}})Z_*\}}{8\bar{b}(1-k_{\mathcal{p}})Z_*\{\mathfrak{T}-4\bar{b}(1-k_{\mathcal{p}})Z_*\}^2}\right]\frac{u}{\varrho}\left\{\frac{\partial\varrho}{\varrho t}-\varrho\frac{\partial u}{\partial x}+\frac{j\varrho u}{x}\right\}=0, \tag{16.24}$$

$$u_t+uu_x+\frac{\mathfrak{B}}{8\bar{b}(1-k_{\mathcal{p}})z_*}u\left(\frac{\partial u}{\partial x}+\frac{ju}{x}\right)=0, \tag{16.25}$$

$$u_t+uu_x+\frac{u}{2}\left(\frac{\partial u}{\partial x}+\frac{ju}{x}\right)=0, \tag{16.26}$$

where $\mathfrak{T}$ and $\mathfrak{B}$ are given as

$$\mathfrak{T}=\mathcal{Q}+\sqrt{\mathcal{Q}^2-8\bar{b}(1-k_{\mathcal{p}})(\Gamma+1)Z_*},$$

$$\mathfrak{B}=\frac{[(\Gamma-1)\{4\bar{b}(1-k_{\mathcal{p}})Z_*\}+\{Z_*+\bar{b}(1-k_{\mathcal{p}})\}\mathfrak{T}-(\mathfrak{T}^2/2)]}{\left\{1-\dfrac{\mathfrak{T}}{4\bar{b}(1-k_{\mathcal{p}})}\right\}.\left\{1-\dfrac{\mathfrak{T}}{4Z_*}\right\}}.$$

The following equation is obtained by performing integration after entering Equations 16.22 and 16.23.

$$\mathfrak{E}(t)=\varrho(u)\left(\frac{4(\mathfrak{T}-4\bar{b}(1-k_{\mathcal{p}})Z_*)-\mathcal{D}}{2(\mathfrak{T}-4\bar{b}(1-k_{\mathcal{p}})Z_*)}\right)_{(x)}-\left(\frac{j\mathcal{D}}{2(\mathfrak{T}-4\bar{b}(1-k_{\mathcal{p}})Z_*)}\right) \tag{16.27}$$

where $\mathfrak{E}(t)$ is time-function and observed constant, $\mathcal{D}$ is expressed as,

$$\mathcal{D} = \frac{\mathfrak{B}\left(1-\frac{\mathfrak{T}}{4\bar{b}(1-k_p)Z_*}\right)^2}{\left(1+\frac{\mathfrak{C}_*\mathfrak{T}}{2(\mathfrak{T}-4\bar{b}(1-k_p)Z_*}\right)}.$$

Looking at Equation 16.27, we can see that the basic set of conservation equations, Equation 16.1, may be expressed as

$$\left(\frac{1}{u}\frac{\partial u}{\partial t}-2\frac{ju}{x}\right)+\left(1-\frac{\mathcal{D}}{2\{\mathfrak{T}-4\bar{b}(1-k_p)Z_*\}}\right)\left(\frac{1}{u}\frac{\partial u}{\partial t}+\frac{\partial u}{\partial x}+\frac{ju}{x}\right)-\frac{1}{\mathfrak{E}}\frac{d\mathfrak{E}}{dt}=0, \quad (16.28)$$

On solving above Equations 16.25 and 16.28, we obtain

$$u=-\frac{16\bar{b}(1-k_p)Z_*\{\mathfrak{T}-4\bar{b}(1-\mathbf{k}_p)Z_*\}}{[(j+1)\{2\mathfrak{B}+8\bar{b}(1-k_p)Z_*\}\{\mathfrak{T}-4\bar{b}(1-k_p)Z_*\}-j\mathcal{D}\{\mathcal{B}-8\bar{b}(1-k_p)Z_*\}-\mathcal{D}\mathfrak{B}]}\frac{x}{\mathfrak{E}}\frac{\boldsymbol{d}\mathfrak{E}}{\boldsymbol{dt}}.$$

(16.29)

When we plug the velocity profile value from Equation 16.29 into Equation 16.28, we obtain

$$\mathfrak{E}(t)=\mathfrak{E}_*(t)^{-\left(\frac{4(j+1)\mathfrak{B}+4\bar{b}(1-k_p)Z_*)-(\mathfrak{T}-\mathbf{4}\bar{\boldsymbol{b}}(\mathbf{1}-\boldsymbol{k}_p)Z_*)-j\mathfrak{D}(\mathfrak{B}-8\bar{b}(1-k_p)Z_*)-\mathfrak{B}\mathfrak{D}}{2(\mathfrak{T}-4\bar{b}(1-k_p)Z_*)(8\bar{b}(1-k_p)Z_*+(j+1)\mathfrak{B})}\right)},$$

(16.30)

where $\mathfrak{E}_*$ is an arbitrary constant.

By utilizing the assistance of weak shock jump (RH) boundary conditions ((Equations 16.18 and 16.19), we can get the explicit formulation for the shock radius $\mathbb{R}(t)$ and the constant $\mathfrak{S}$ as

$$\mathbb{R}=(t)^{\left(\frac{8\bar{b}(1-k_p)Z_*\mathfrak{T}}{(8\bar{b}(1-k_p)Z_*+(j+1)\mathfrak{B})(\mathfrak{T}-4\bar{b}(1-k_p)Z_*)}\right)}, \quad (16.31)$$

$$\mathfrak{S}=\left(\frac{((1-j)(1-4\bar{b}(1-k_p)Z_*)\mathfrak{T}-2)(1-4\bar{b}(1-k_p)Z_*)\mathfrak{T}+2\bar{b}(1-k_p)Z_*(j+1)\mathcal{D}}{(\mathfrak{T}-4\bar{b}(1-k_p)Z_*)\mathfrak{T}}\right).$$

(16.32)

The problem of blast wave proliferation in a real gas flow affected by a magnetic field and subject to boundary conditions (Equations 16.18–16.21) is finally solved via a closed-form method, which is as follows,

$$\varrho=\frac{\mathfrak{E}_*(x)^{\left(\frac{(j+1)\mathcal{D}-4(\mathfrak{T}-4\bar{b}(1-k_p)Z_*)}{2(\mathfrak{T}-4\bar{b}(1-k_p)Z_*)}\right)}(t)^{\left(\frac{4\bar{b}(1-k_p)Z_*(2(1-j)(\mathfrak{T}-4\bar{b}(1-k_p)Z_*)-(j+1)\mathcal{D})}{(\mathfrak{T}-4\bar{b}(1-k_p)Z_*)(8\bar{b}(1-k_p)Z_*+(j+1)\mathfrak{B})}\right)}}{\left(\frac{8\bar{b}(1-k_p)Z_*}{(8\bar{b}(1-k_p)Z_*+(j+1)\mathfrak{B})}\right)\left(\frac{4(\mathfrak{T}-4\bar{b}(1-k_p)Z_*)-\mathcal{D}}{2(\mathfrak{T}-4\bar{b}(1-k_p)Z_*)}\right)},$$

(16.33)

$$u=\frac{8\bar{b}(1-k_p)Z_*}{(8\bar{b}(1-k_p)Z_*\,(j+1)\mathfrak{B})}\frac{x}{t},\tag{16.34}$$

$$p=\mathfrak{E}_*\left(\frac{2(4\bar{b}(1-k_p)Z_*)^2-C_*\mathfrak{T}(\mathfrak{T}+4\bar{b}(1-k_p)Z_*)}{(8\bar{b}(1-k_p)Z_*(\mathfrak{T}-4\bar{b}(1-k_p)Z_*)}\right)\left(\frac{8\bar{b}(1-k_p)Z_*}{(8\bar{b}(1-k_p)Z_*+(j+1)\mathfrak{B})}\right)^{\frac{\mathcal{D}}{2(\mathfrak{T}-4\bar{b}(1-k_p)Z_*)}}$$
$$(x)^{\frac{(j+1)\mathcal{D}}{2(\mathfrak{T}-4\bar{b}(1-k_p)Z_*)}}(t)^{\left(\frac{-2(j+1)((\mathfrak{T}-4\bar{b}(1-k_p)Z_*)(4\bar{b}(1-k_p)Z_*+\mathfrak{B})+2\bar{b}(1-k_p)Z_*\mathfrak{D})}{(\mathfrak{T}-4\bar{b}(1-k_p)Z_*)(8\bar{b}(1-k_p)Z_*+(j+1)\mathfrak{B})}\right)},$$

(16.35)

$$h=\frac{C_*\mathfrak{E}_*\mathfrak{T}^3}{8\bar{b}(1-k_p)Z_*(\mathfrak{T}-4\bar{b}(1-k_p)Z_*)^2}\left(\frac{8\bar{b}(1-k_p)Z_*}{8\bar{b}(1-k_p)Z_*+(j+1)\mathfrak{B})}\right)^{\frac{\mathcal{D}}{2(\mathfrak{T}-4\bar{b}(1-k_p)Z_*)}}(x)^{\frac{(j+1)\mathcal{D}}{2(\mathfrak{T}-4\bar{b}(1-k_p)Z_*)}}$$
$$(t)^{\left(\frac{-2(j+1)((\mathfrak{T}-4\bar{b}(1-k_p)Z_*)(4\bar{b}(1-k_p)Z_*+\mathfrak{B})+2\bar{b}(1-k_p)Z_*\mathfrak{D})}{(\mathfrak{T}-4\bar{b}(1-k_p)Z_*)(8\bar{b}(1-k_p)Z_*+(j+1)\mathfrak{B})}\right)},$$

(16.36)

With the help of the above values of the thermodynamic flow characteristics, the expression of the total energy (by summing the kinetic and internal energy) driven out by the blast wave in a dust-filled, simplified real gas influenced by the magnetic field, which reveals the time-position dependence, is rendered as follows,

$$\varepsilon_{\mathcal{T}}=\varepsilon_*(t)^{\frac{8\bar{b}(1-k_p)Z_*(j+1)(\mathcal{D}+2(\mathfrak{T}-4\bar{b}(1-k_p)Z_*))\mathfrak{T}-2(j+1)(8\bar{b}(1-k_p)Z_*(\mathfrak{T}-4\bar{b}(1-k_p)Z_*)\mathfrak{D}+2(\mathfrak{B}+4\bar{b}(1-k_p)Z_*)(\mathfrak{T}-4\bar{b}(1-k_p)Z_*)^2)}{2(8\bar{b}(1-k_p)Z_*+(j+1)\mathfrak{B})(\mathfrak{T}-4\bar{b}(1-k_p)Z_*)^2}},$$

(16.37)

where

$$\varepsilon_*=\frac{4\pi\mathfrak{E}_*\left(\begin{array}{c}16\bar{b}(1-k_p)Z_*(\Gamma-1)(4\bar{b}(1-k_p)Z_*(\mathfrak{T}-4\bar{b}(1-k_p)Z_*)^2+C_*\mathfrak{T}_*^3)\\+(\mathfrak{T}-4Z_*)(\mathfrak{T}-4\bar{b}(1-k_p))(\mathfrak{T}-4\bar{b}(1-k_p)Z_*)(2(4\bar{b}(1-k_p)Z_*)^2-(\mathfrak{T}+4\bar{b}(1-k_p)Z_*)C_*\mathfrak{T})\end{array}\right)}{(j+1)(\mathcal{D}+2(\mathfrak{T}-4\bar{b}(1-k_p)Z_*))(8\bar{b}(1-k_p)Z_*)^2(\mathfrak{T}-4\bar{b}(1-k_p)Z_*)\left(\frac{8\bar{b}(1-k_p)Z_*}{(8\bar{b}(1-k_p)Z_*+(j+1)\mathfrak{B})}\right)^{\frac{\mathcal{D}}{2(\mathfrak{T}-4\bar{b}(1-k_p)Z_*)}}}$$

## 16.5 Results and Conclusions

Explicit expressions of the thermodynamic flow variables and total energy that reveal the dependence of the independent space-time variables are set out in Equations 16.33 to 16.37. The obtained precise solution (Equations 16.33–16.37), along with the closed form expression of shock radius (path) given in Equation 16.31, depicts the influence of adiabatic exponent, volume, and mass concentration of dust-loaded particles, the ratio of the density of gas and particles, parameter of repulsion, and Shock-Cowling number (however, the velocity profile and shock-radius do not influence by the magnetic field which is evident from Equations 16.34 and 16.31, respectively). Moreover, this indicates that the blast wave proliferation process in a non-ideal dusty flow is different from that of non-linear waves propagating in an ideal gas or in a simplified non-ideal gas. Furthermore, it is worth noting that either non-magnetic effect or in the absence of non-ideal parameters with the presence of a magnetizing effect; the findings

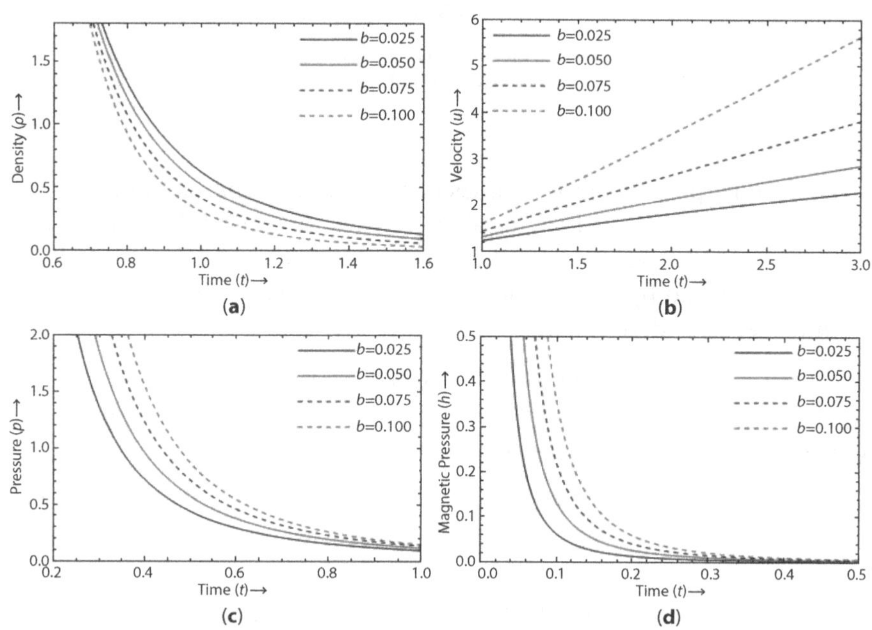


**Figure 16.1** Behavior of the solution profiles of the flow variables for cylindrically symmetric flow (a) density ($\rho$), (b) velocity ($u$), (c) pressure ($p$), and (d) magnetic pressure ($h$) for various values of $b$ with $\gamma = 1.4$ (solid line), $\gamma = 1.67$ (dashed line); $k = 0.3$; $G = 10$; $\beta = 0.5$; $C_{*} = 0.01$.

from this inquiry are in reasonable agreement with those from earlier investigations, see Murata [21] and Husain *et al.* [22, 23].

In the dust-loaded non-ideal magneto-gas-dynamical flow, the solution profiles of physical flow quantities such as density, velocity, pressure, and magnetic pressure at the shock location, $\chi(t)$ are shown in Figure 16.1 (a–d). The profiles also show how the vdW excluded co-volume ($b$) changes with respect to time $t$, impacting the blast wave proliferation problem. Observing Figure 16.1 (a–d) makes it clear that, when the value of the excluded volume '$b$' increases, the behavior of the mass concentration ratio $k_p$ with magnetic field effects generally rises (see Figure 16.1 (b–d)), but the density profile noticeably falls (see Figure 16.1 (a)).

The energy's behavior with respect to the change in $b$ and a constant $k_p$ as a function of time $t$ is shown in Figure 16.2 (a). As we approach infinity, the energy profile of the cylindrical shock waves decreases exponentially with time ($t$), as shown in Figure 16.2 (a), and a general drop in the solution profile of energy is indicated by an increasing value of the parameter of non-idealness ($b$). Figure 16.2 (b) shows the energy's behavior as a function of time $t$ as $k_p$ varies with a fixed $b$. Figure 16.2 (b) shows that as the mass concentration ratio $k_p$ increases, the energy profile values decrease dramatically.

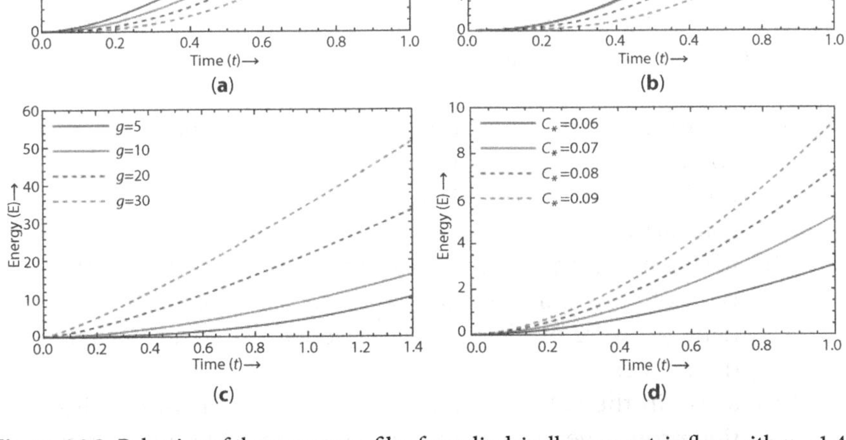


**Figure 16.2** Behavior of the energy profiles for cylindrically symmetric flow with $\gamma = 1.4$ (solid line), $\gamma = 1.67$ (dashed line).

The behavior of the energy with variation in $g$ with fixed $C_*$; versus the time, $t$ is illustrated in Figure 16.2 (c). As we approach infinity, the energy profile of the cylindrical shock waves increases exponentially with time ($t$), as seen in Figure 16.2 (c), and a larger value of $g$ indicates a general rise in the solution energy profile. Figure 16.2 (d) shows the energy's behavior as a function of time $t$ as a function of the change in $C_*$ with a constant $g$. As we approach infinity, the energy profile of the cylindrical shock waves increases exponentially with time ($t$), as seen in Figure 16.2 (d), and a general rise in the solution profile of energy is indicated by a rising value of the parameter $C_*$.

The impact of a simplified vdW gas containing dust on the movement of non-planar blast waves in a magnetic field is the subject of an analytical study in this chapter. On the basis of a straightforward analytical method initiated by Murata [21], a closed-form precise solution for physical flow variables, shock path, and total energy driven by strong discontinuity has been obtained. In order to reach the precise solution of the study, we build a correlation by utilizing robust shock conditions, which define the magnetic pressure and pressure profiles in terms of density and velocity, respectively. The derived mathematical equations, specifically Equations 16.21 and 16.32–16.37, clearly demonstrate the impact of various dusty parameters. These parameters include the mass concentration ratio ($k$), the ratio of species density of the gas, the density of dusty particles ($G$), the ratio of specific heat of dusty particles to the specific heat of the gas at constant pressure ($\beta$), and the parameter of non-idealness ($b$).

It is worth pointing out that explicit solutions comprise the effect of the magnetic field, however velocity and radius of shock do not show any significant influence of the magnetic field. From the analytical expressions of flow variables, the following conclusions has been made which closely aligns with theoretical expectations.

(i) for fixed $k_p$ and varying b, with the rise in the values of co-volume b, the pressure, magnetic pressure and energy profiles leads to rise while density and velocity behaves contrarily.

(ii) for fixed b and varying $k_p$, with the rise in the values of the ratio of mass concentration $k_p$, velocity, desity and magnetic pressure profiles leads to rise while pressure profile decelerates.

(iii) a rise in the values of adiabatic exponent, pressure solution profiles get to rise, and the solution profiles of density, velocity and magnetic pressure behaves reversely.

## References


1. Carrier, G.F., *J. Fluid Mech.*, **4**, 376–382, 1958.
2. Soo, S.L., *Ind. Eng. Chem. Fundam.*, **4**, 426–433, 1965.
3. Marble, F.E., *Annu. Rev. Fluid Mech.*, **2**, 397–446, 1970.
4. Pai, S.-I., *Two-Phase Flows*, K. Oswatitsch (Ed.), Vieweg+Teubner Verlag, Wiesbaden, 1977.
5. Rudinger, G., *Fundamentals of Gas Particle Flow*, Elsevier Scientific Publishing Company, Amsterdam, 1980.
6. Elperin, T., Ben-Dor, G., Igra, O., *Int. J. Heat Fluid Flow*, **8**, 303–312, 1987.
7. Hamad, H. and El-Fayes, F., *Acta Mech.*, **119**, 53–63, 1996.
8. Nath, G., *Res. Astron. Astrophys.*, **10**, 445, 2010.
9. Singh, D. and Arora, R., *Math. Methods Appl. Sci.*, **45**, 5149–5164, 2022.
10. Vishwakarma, J.P. and Nath, G., *Commun. Nonlinear Sci. Numer. Simul.*, **17**, 2382–2393, 2012.
11. Chadha, M. and Jena, J., *Comput. Appl. Math.*, **34**, 729–739, 2015.
12. Shah, S. and Singh, R., *J. Math. Phys.*, **61**, 033506, 2020.
13. Husain, A., Singh, V.K., Haider, S.A., *Int. J. Eng. Adv. Technol.*, **9**, 2490–2494, 2020.
14. Husain, A., Aftab Haider, S., Singh, V.K., *Mater. Today Proc.*, **62**, 1613–1619, 2022.
15. Chaturvedi, R.K., Srivastava, S.K., Singh, L.P., Koteswararao, N.V., Makrariya, A., *Chin. J. Phys.*, 2023.
16. Sahu, P.K., *Eur. Phys. J. Plus*, **136**, 452, 2021.
17. Gupta, P. and Singh, L.P., *Chin. J. Phys.*, **77**, 1912–1926, 2022.
18. Pang, Y., Ge, J., Liu, Z., Hu, M., 701–717 23, 2022.
19. Courant, R. and Friedrichs, K.O., *Supersonic Flow and Shock Waves*, Interscience Publishers, 1948.
20. Whitham, G.B., *Linear and Nonlinear Waves*, Wiley, New York, 1974.
21. Murata, S., *Chaos, Solitons Fractals*, **28**, 327–330, 2006.
22. Husain, A., Haider, S.A., Singh, V.K., *Eng. Technol. Appl. Sci. Res.*, **12**, 9503–9509, 2022.
23. Husain, A., Haider, S.A., Singh, V.K., Analytical Study of Quasi-One-Dimensional Cylindrical Weak Shock Wave Problem Under the Action of Magnetic Field at Stellar Surface, in: *Mathematical Modeling, Computational Intelligence Techniques and Renewable Energy*, M. Sahni, J.M. Merigó, W. Hussain, E. León-Castro, R.K. Verma, R. Sahni (Eds.), pp. 449–462, Springer Nature Singapore, Singapore, 2023.

## References

[illegible]

17

# The Effect of Casson Fluid Flow on a Stagnation Point Over a Porous Stretching Sheet with Thermal Radiation

**Wajeeha K.[1], Sushma M. N.[1], U.S. Mahabaleshwar[1], Mahesh R.[1] and Dhananjay Yadav[2]***

[1]*Department of Studies in Mathematics, Davangere University, Shivagangothri, Davangere, India*

[2]*Department of Mathematical & Physical Sciences, University of Nizwa, Nizwa, Oman*

***Abstract***

The current work investigates the influence of magnetite ($Fe_3O_4$) nanoparticles mixed with water ($H_2O$) on a Casson fluid with a stagnation point on a porous stretched sheet. A realistic contemporary nanofluid model is used to assess the new model's formulation to change the thermophysical properties such as viscosity and thermal conductivity that are involved. With this new model, we can examine precisely how the expansion of velocity and temperature profiles is influenced by the size of the nanoparticles and the direction of heat transfer (heating or cooling, for example). To construct a mathematical model, we utilize the governing equations with momentum equations, and energy equations, which are transferred into nonlinear ordinary differential equations. Using the appropriate similarity transformations. This obtained system of nonlinear ordinary differential equations is solved numerically using the shooting method, and the results for preeminent physical parameters on velocity and temperature profile are analyzed with the help of graphs.

***Keywords*:** Iron oxide, Casson fluid, porous medium, thermal radiation, stagnation point

**Corresponding author*: dhananjayadav@gmail.com
U.S. Mahabaleshwar: ORCID: 0000-0002-6542-0490
Dhananjay Yadav: ORCID: 0000-0001-8404-2053

Ashwani Kumar, Nitesh Dutt and Mukesh Kumar Awasthi (eds.) Heat Transfer Enhancement Techniques: Thermal Performance, Optimization and Applications, (375–390) © 2025 Scrivener Publishing LLC

## 17.1 Introduction

The Casson fluid was first developed by Casson [1] in 1959 and is well-known for its non-Newtonian behavior, which under some flow circumstances resembles an elastic solid and shows little reaction to minor shear stress. Because of its unique properties, Casson's model which he developed to explain the flow characteristics of pigment oil suspensions attracted the attention of academics. Shear stresses are mostly determined by the Casson fluid material characteristics. Its applicability as a blood rheological model is especially notable. Researchers frequently use a two-fluid method when modelling blood, considering erythrocytes as a Casson fluid. Anusha *et al.* [2] investigated the dynamic behavior of the Casson fluid in an unsteady flow across a permeable flat plate. Meanwhile, taking into account thermal radiation and heat source/sink characteristics, Mahabaleshwar *et al.* [3] investigated the two-dimensional flow of a graphene-water-Casson nanofluid on a stretching/shrinking sheet. In an alternative investigation, Vaidya *et al.* [4] examined how heat transfer affects non-Newtonian fluid flow via a porous axisymmetric tube while taking slip effects into account. A solution for the flow of a non-Newtonian fluid through a porous shrinking surface in the presence of radiation was developed by Bhattacharyya *et al.* [5] with a focus on wall temperature.

A stagnation point in fluid flow refers to a specific location where the velocity of the fluid becomes zero due to the obstruction or change in geometry. At this point, the pressure is typically higher, creating a stagnation or high-pressure region. Optimizing thermal management requires a thorough understanding of heat transfer at stagnation points, particularly in cooling systems where stagnation points are purposefully placed at leading edges to maximize convective heat transfer and boost efficiency. A large amount of work has been done on stagnation-point flows with Newtonian and/or non-Newtonian fluids; this may be seen in References [6–14]. As an illustration, Reference [6] looked at the wedge flow of a viscous fluid with a wall that was nonlinearly expanding and contracting. For some nonlinearity exponents, they found dual solutions. Kolomenskiy and Moffatt's [7] unstable similarity solutions added instability to free stream flow. Further research on mixed convection flows and magnetohydrodynamics was conducted in other works, such as References [8–14], which provided insights into the variables influencing stagnation-point flow. This degree of investigation, demonstrated by References [9, 10], not only advances basic comprehension but also provides practical solutions for optimizing heat transfer and cooling efficiency in diverse scenarios.

In fluid mechanics, a material or medium with interconnected spaces or pores is called "porous." The fluids that can flow through these gaps have an impact on the material's permeability and transport properties. Porosity and porous media were concepts initially developed in the 1800s by French engineer Henry Darcy. As a result of his studies, Darcy came up with the mathematical formula known as "Darcy's law," which defines fluid flow through porous media. Porous materials and media have several applications in a wide range of industries. Geological awareness of porous rocks and soils is essential to comprehending groundwater circulation, oil and gas reserves, and geological formations. In filtering, porous materials are also used to remove contaminants and pollutants.

When it comes to improving heat and mass transport in engineering, porous materials are essential. For instance, they improve heat exchangers' fluid contact surface, which raises efficiency. They also help species move around and react in catalytic reactors and electrochemical systems. Properties like porosity, permeability, and tortuosity have a big impact on how fluid flows through various kinds of material. Developing novel, optimally performing materials and models for fluid mechanics applications is the aim of continuing research. Lin *et al.* [15] and Rashid *et al.* [16] examined a copper-water nanofluid on a porous disc inside a Marangoni boundary layer, highlighting the influence of particle shape. Mahabaleshwar *et al.* [17–19] investigated the effects of Newtonian and non-Newtonian fluids on porous media, and the effects of magnetohydrodynamics (MHD) on Rivlin-Ericksen liquid flow sliding over porous surfaces were illustrated by Vishalakshi *et al.* [20, 21]. Sneha *et al.* [22] investigated the impact of Walters' liquid B flow on porous media, whereas Animasaun *et al.* [23, 24] looked at dust particle motion in air and chemical reactions in an alumina-water nanofluid. Researchers Ammani *et al.* [25], Sastry [26], and Arifin *et al.* [27] looked into numerical solutions that took skin friction and Nusselt number into account. As a whole, this extensive study improves our knowledge and optimization of porous materials in many fluid mechanics applications.

Thermal radiation has a broad frequency range that includes radio waves and X-rays. It is caused by the movement of charged particles, particularly electrons, within a substance as a result of its temperature. It has a wide range of uses, including heating and energy systems such as boilers, heaters, radiators, and thermal power plants. Thermal radiation detection in thermography is used for imaging in applications like electrical inspections and building diagnostics. The study of thermal radiation from celestial bodies is essential to astronomy in order to comprehend temperature, composition, and energy production. Utilizing the sun's thermal radiation, solar energy powers electricity-generating devices like concentrated solar power (CSP). Notable studies highlight the effect of heat radiation on bioconversion in MHD non-homogeneous flows of Maxwell nanofluids across vertical surfaces, such as Wakif's [28] numerical simulations. The thermodynamics of stable two-dimensional Williamson non-homogeneous flows was studied by Eswaramoorthi *et al.* [29] by integrating modified Fourier's and Fick's laws.

Nanofluids, seen as a promising alternative for conventional heat transfer methods, have garnered significant interest among academics aiming to enhance heat transfer efficiency. The exploration of nanofluids, particularly their stagnation-point flow, has led to unique discoveries. In this study, we investigate nanofluids composed of water ($H_2O$) and $Fe_3O_4$ nanoparticles. The $Fe_3O_4$, or magnetite, a highly magnetic iron oxide, traces its use back to ancient civilizations and finds modern applications as nanoparticles in medicine, environmental remediation, electronics, catalysis, and energy storage. Because of the unclear diffusion variables connected to thermophoresis and Brownian effects, single-phase modelling for nanofluids may be better despite the frequently recommended two-phase modelling. This approach ensures a more rigorous physical basis, as emphasized in Reference [30].

In order to particularly target the Casson fluid flow on porous stretching sheet with the stagnation point of $Fe_3O_4/H_2O$ nanofluids and the thermal radiation impact is incorporated in the heat transfer, a modern and realistic nanofluid model is used in this research. The model was created in Reference [31]. The $Fe_3O_4/H_2O$ nanofluid, which is regarded as a potential nanofluid for thermal applications in the relevant sectors, is one of the practical nanofluids that this model has empirical information about. Due to this, there are some experimental studies of this particular class of nanofluids in the literature. These studies mostly focus on the convective heat transfer in pipes and tubes in both laminar and turbulent regimes (see, for example, References [32–36]).

## 17.2 Mathematical Formulation

Consider a steady and incompressible Casson fluid with the stagnation-point flow of $Fe_3O_4$ nanoparticle mixed with $H_2O$ over a porous stretching sheet presented using the model that was suggested in Reference [31].

Governing equations:

$$\frac{\partial u}{\partial x}+\frac{\partial v}{\partial y}=0 \tag{17.1}$$

$$u\frac{\partial u}{\partial x}+v\frac{\partial u}{\partial y}=u_e\frac{d}{dx}(u_e)+\left(1+\frac{1}{\wedge}\right)\frac{\partial^2 u}{\partial y^2}-\frac{v_{nf}}{k}(u_e-u) \tag{17.2}$$

$$(\rho C_p)nf\left(u\frac{\partial T}{\partial x}+v\frac{\partial T}{\partial y}\right)=\frac{\partial}{\partial y}\left(\left(K_{nf}+\frac{16\sigma T_\infty^3}{3k^*}\right)\frac{dT}{\partial y}\right) \tag{17.3}$$

The associated boundary conditions are:

$$\begin{aligned} &u(x,0)=ax,\ v(x,0)=-\sqrt{cv_{bf}}\,Q, T(x,0)=T_W \\ &u(x,\infty)=cx, T(x,\infty)=T_\infty \end{aligned} \tag{17.4}$$

The relationships between viscosity and thermal conductivity, as shown in Reference [31], cover a wide range of nanofluids. Nonetheless, it is important to recognize some limits. First, the correlations are found based on particular base fluids that have different concentrations of nanoparticles. The correct ranges for temperature, nanoparticle diameter, and volumetric proportion of nanoparticles should also be taken into consideration. The application of this new model demands the incorporation of valid intervals for both associated thermal conductivity and viscosity. In this case, the temperature varies from 294 K to 323 K, and the chosen nanoparticle size falls between 20 nm and 140 nm. The range of 0.00 to 0.03 is the maximum volumetric fraction of nanoparticles. To use the suggested model properly, it is essential to take these parameters into account. Recognizing that the base fluid has a freezing point of 273 K is also critical. As a result, it is possible to represent the thermophysical characteristics of the nanofluids as described in Reference [31].

$$\mathrm{Re}=\frac{2\rho_{bf}k_{Bo}T}{\pi\mu_{bf}^2 d_{np}}$$

$$\rho_{nf}=\phi\rho_{np}+(1-\phi)\rho_{bf}$$

$$\mu_{nf} = \mu_{bf}\left(1-34.87\left(\frac{d_{np}}{d_{bf}}\right)^{-0.3}\phi^{1.03}\right)^{-1}$$

$$d_{bf} = \frac{1}{10}\left(\frac{6M}{N_A\pi\rho_{bfo}}\right)$$

$$k_{nf}\left(T\right) = k_{bf}\left(1+4.4\,\mathrm{Re}^{0.4}\,pr^{0.66}\left(\frac{T}{T_{fr}}\right)^{10}\left(\frac{-k_{np}}{k_{bf}}\right)^{0.03}\phi^{0.66}\right)$$

$$C\rho_{nf} = \frac{\phi\left(\rho C_p\right)_{np}+\left(1-\phi\right)\left(\rho C_p\right)_{bf}}{\rho_{nf}}$$

$$k_{nf} = k_{bf}\left(1+AT^{10.4}\right)$$

$$A = 4.4\left(\frac{2\rho_{bf}k_{Bo}}{\pi\mu_{bf}^{2}d_{np}}\right)^{0.4}P_r^{0.66}\left(\frac{1}{T_{fr}}\right)^{10}\left(\frac{k_{np}}{k_{bf}}\right)^{0.03}\phi^{0.66}$$

By introducing the similarity transformation as (refer to [36]):

$$\eta = \sqrt{\frac{c}{\vartheta_{bf}}}y,\quad \theta = \left(\frac{T-T_\infty}{T_W-T_\infty}\right),\quad \frac{u}{u_e} = f'(\eta), \tag{17.5}$$

Upon substitution into Equations 17.1–17.5, we get

$$\frac{A_1}{A_2}\left(1+\frac{1}{\Lambda}\right)f'''(\eta)-\frac{A_1}{A_2}Da^{-1}\left(1-f'(\eta)\right)+f(\eta)f''(\eta)-f'(\eta)^2+1=0 \tag{17.6}$$

$$\left(\frac{k_{nf}}{k_f}+Nr\right)\theta''(\eta)+\Pr\frac{\left(\rho C_p\right)_{nf}}{\left(\rho C_p\right)_f}f(\eta)\theta'(\eta)=0 \tag{17.7}$$

In Equation 17.9, $\Delta T_o = T_w - T_\infty$. The similarity boundary conditions become:

$$f(0)=Q,\ f'(0)=\frac{a}{c}=p,\ f'(\infty)=1 \tag{17.8}$$

$$\theta(0)=1,\ \theta(\infty)=0 \tag{17.9}$$

## 17.3 Result and Discussion

This work looks into how the concentration of the nanoparticles works with 25-nm diameter on effect of Casson nanofluid flow on a porous stretching sheet with a combination of thermal radiation. The $Fe_3O_4$ nanoparticles are considered with $H_2O$ as a base fluid. The altered nonlinear ordinary differential equations (ODEs) are solved numerical by employing the shooting method using bvp4c algorithm differential equation method. The Prandtl number is fixed for the basic fluid (Pr = 6.2), and the impacts of the physical parameters are explored with the use of graphics that are provided below. The thermophysical properties of $Fe_3O_4$ and base fluid $H_2O$ is presented in the Table 17.2.1.

The variation of Casson parameter ($\beta$) on an axial velocity profile is shown in Figure 17.1. From this figure, we came to know that the velocity

**Table 17.2.1** Thermophysical properties of $Fe_3O_4$ and water [36].

| **Properties** | **$Fe_3O_4$ (magnetite)** | **Base fluid water ($H_2O$)** |
|---|---|---|
| $\rho(kg/m^3)$ | 5180 | 997.1 |
| $\mu(Ns/m^2)$ | — | $89 \times 10^{-5}$ |
| $C_p(J/kg,\ k)$ | 670 | 0.6134 |
| $k(W/m,k)$ | 60.4 | 0.613 |

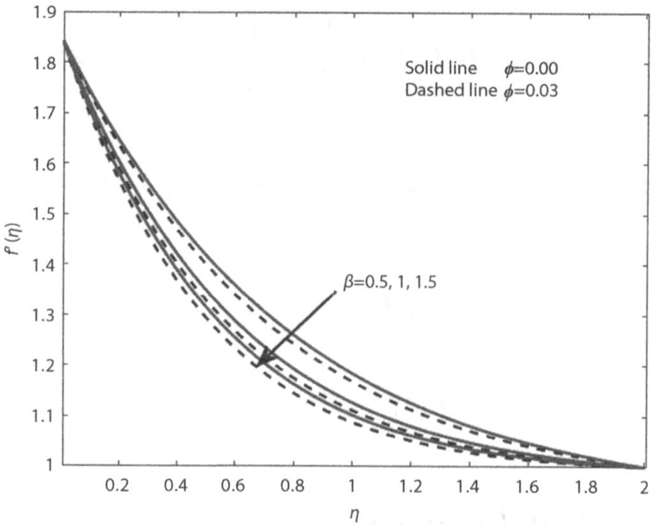


**Figure 17.1** *f*/*n* v/s ($\eta$) for various values of $\beta$ (Casson parameter).

boundary layer decreases by increasing the Casson parameter ($\beta$) for both cases when the concentrations of the nanoparticle $Fe_3o_4$ are at 0.00 and 0.03, and the diameter of the nanoparticle is 25 nm. Furthermore, we observe that with the nanoparticle volume concentration at 0.03, the velocity boundary layer decreases faster as compared to without nanoparticle volume concentration.

The variation of Casson parameter ($\beta$) on a temperature profile is shown in Figure 17.2. From this figure, we came to know that the thermal boundary layer decreases by increasing the Casson parameter ($\beta$) for both cases when the concentrations of the nanoparticle $Fe_3o_4$ are at 0.00 and 0.03, and the diameter of the nanoparticle is 25 nm. Furthermore, we observe that with the nanoparticle volume concentration at 0.03, the thermal boundary layer decreases faster as compared to without nanoparticle volume concentration.

The Variation of inverse Darcy number ($Da^{-1}$) on an axial velocity profile is shown in Figure 17.3. From this figure, we came to know that the velocity boundary layer decreases by increasing the porous parameter ($Da^{-1}$) for both cases when the concentrations of the nanoparticle $Fe_3o_4$ are at 0.00 and 0.03, and the diameter of the nanoparticle is 25 nm. Furthermore, we observe that with the nanoparticle volume concentration at 0.03, the velocity boundary layer decreases faster compared to without nanoparticle volume concentration.

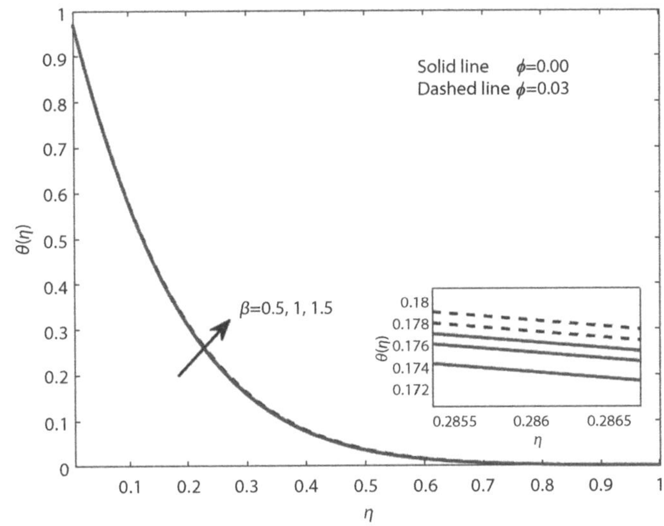


**Figure 17.2** $\theta(\eta)$ *v*/*s* (*n*) for various values of $\beta$ (Casson parameter).

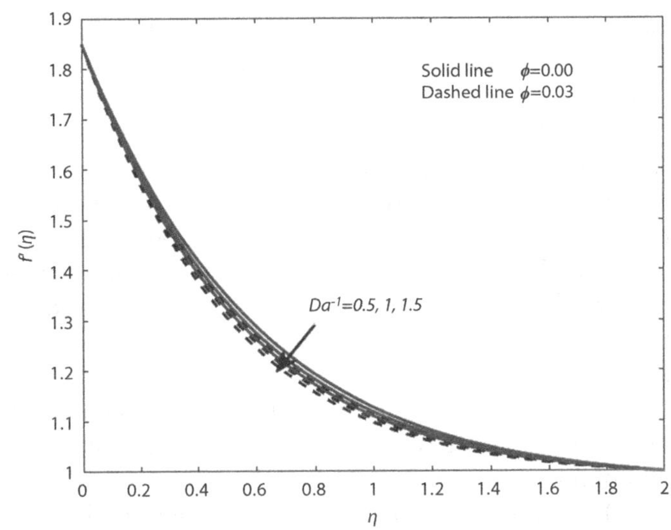


**Figure 17.3** $f'(\eta)$ *v*/*s* (*n*) for rising value of $Da^{-1}$ (porous parameter).

The variation of inverse Darcy number ($Da^{-1}$) on a temperature profile is shown in Figure 17.4. From this figure, we came to know that the thermal boundary layer decreases by increasing the inverse Darcy number ($Da^{-1}$) for both cases when the concentrations of the nanoparticle $Fe_3o_4$ are at 0.00 and 0.03, and the diameter of the nanoparticle is 25 nm. Furthermore, we

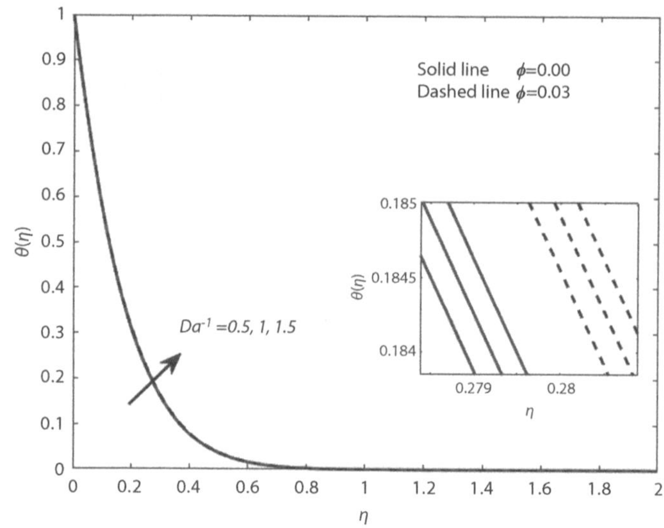


**Figure 17.4** $\theta(\eta)$ *v/s* $(\eta)$ for rising value of $Da^{-1}$.

observe that with the nanoparticle volume concentration at 0.03, the thermal boundary layer decreases faster compared to without nanoparticle volume concentration.

The variation of the suction-injection parameter ($V_C$) on an axial velocity profile is shown in Figure 17.5. From this figure, we came to know that the velocity boundary layer decreases by increasing the suction-injection parameter ($V_C$) for both cases when the concentrations of the nanoparticle $Fe_3o_4$ are at 0.00 and 0.03, and the diameter of the nanoparticle is 25 nm. Furthermore, we observe that with the nanoparticle volume concentration at 0.03, the velocity boundary layer decreases faster compared to without nanoparticle volume concentration.

The variation of the suction-injection parameter ($V_C$) on a temperature profile is shown in Figure 17.6. From this figure, we came to know that the thermal boundary layer decreases by increasing the suction-injection parameter ($V_C$) for both cases when the concentrations of the nanoparticle $Fe_3o_4$ are at 0.00 and 0.03, and the diameter of the nanoparticle is 25 nm. Furthermore, we observe that with the nanoparticle volume concentration at 0.03, the thermal boundary layer decreases faster compared to without nanoparticle volume concentration.

The variation of the thermal radiation parameter ($N_r$) on a temperature profile is shown in Figure 17.7. From this figure, we came to know that the thermal boundary layer decreases by increasing the thermal radiation

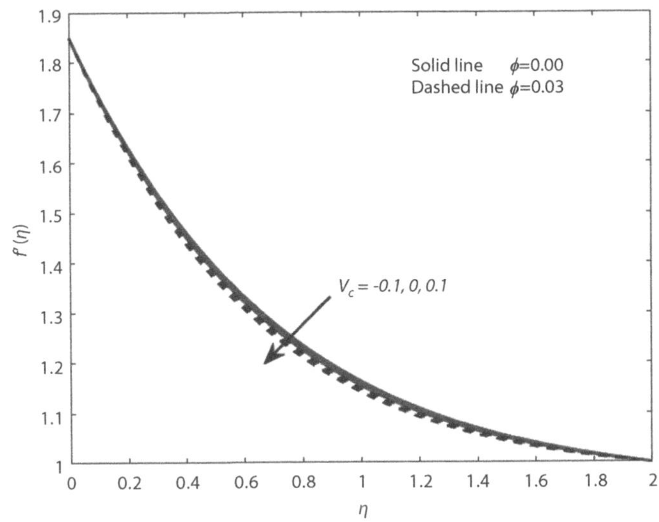


**Figure 17.5** $f'(\eta)$ *v/s* $(\eta)$ for rising value of $V_C$ (suction-injection).

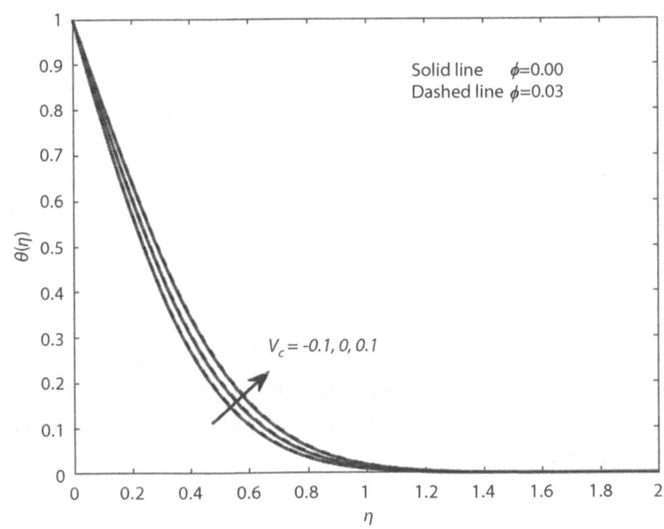


**Figure 17.6** $\theta(\eta)$ *v/s* $(\eta)$ for rising value of $V_C$ (suction-injection).

parameter ($N_r$) for both cases when the concentrations of the nanoparticle $Fe_3o_4$ are at 0.00 and 0.03, and the diameter of the nanoparticle is 25 nm. Furthermore, we observe that with the nanoparticle volume concentration at 0.03, the thermal boundary layer decreases faster compared to without nanoparticles volume concentration.

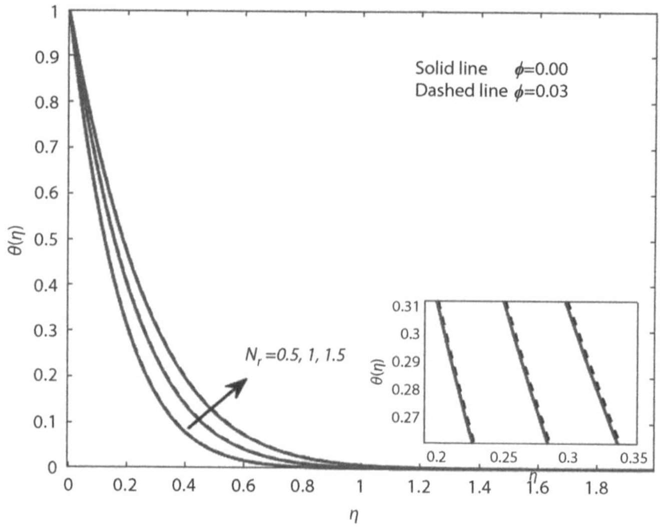


**Figure 17.7** $\theta(\eta)$ *v/s* $(\eta)$ various values of $N_r$.

## 17.4 Conclusion

The objective of this work is to quantitatively examine the effects of $Fe_3O_4/H_2O$ nanofluid Casson fluid flow at a stagnation point on a porous stretched wall. The study makes use of a realistic nanofluid model and the shooting technique using the bvp4c algorithm. In this novel model, the impacts of nanoparticle size (diameter 25 nm), heat transfer direction, and thermal radiation on velocity and temperature profiles are explored. The findings show that the temperature profile rises but the velocity profile falls as the Casson parameter, inverse Darcy number, stretching parameter, and mass transpiration increase. This clarifies the ways in which these variables affect the heat transfer and flow properties when there is a volume percentage of nanoparticles present.

## References

1. Casson, N., A flow equation for pigment-oil suspensions of the printing ink type, 1959.
2. Anusha, T., Huang, H.-N., Mahabaleshwar, U.S., Two dimensional unsteady stagnation point flow of Casson hybrid nanofluid over a permeable flat surface and heat transfer analysis with radiation. *J. Taiwan Inst. Chem. Eng.*, 127, 79–91, 2021, https://doi.org/10.1016/j.jtice.2021.08.014.

3. Mahabaleshwar, U.S., Aly, E.H., Vishalakshi, A.B., MHD and Thermal Radiation Flow of Graphene Casson Nanofluid Stretching/Shrinking Sheet. *Int. J. Appl. Comput. Math.*, 8, 113, 2022, https://doi.org/10.1007/s40819-022-01300-w.
4. Vaidya, H., Rajashekhar, C., Manjunatha, G., Prasad, K.V., Effects of Heat Transfer on Peristaltic Transport of a Bingham Fluid through an Inclined Tube with Different Wave Forms. *Defect Diffus. Forum*, 392, 158–177, 2019, https://doi.org/10.4028/www.scientific.net/DDF.392.158.
5. Bhattacharyya, K., Uddin, M.S., Layek, G.C., Exact solution for thermal boundary layer in Casson fluid flow over permeable shrinking sheet with variable wall temperature and thermal radiation. *Alexandria Eng. J.*, 55, 1703–1712, 2016, https://doi.org/10.1016/j.aej.2016.03.010.
6. Bachok, N. and Ishak, A., Similarity solutions for the stagnationpoint flow and heat transfer over a nonlinearly stretching/shrinking sheet. *Sains Malays.*, 40, 11, 1297–1300, 2011.
7. Kolomenskiy, D. and Moffatt, H.K., Similarity solutions for unsteady stagnation point flow. *J. Fluid Mech.*, 711, 394–410, 2012, https://doi.org/10.1017/jfm.2012.39.
8. Seddighi Chaharborj, S., Ismail, F., Gheisari, Y., Seddighi Chaharborj, R., Abu Bakar, M.R., Abdul Majid, Z., Lie group analysis and similarity solutions for mixed convection boundary layers in the stagnation-point flow toward a stretching vertical sheet. *Abstr. Appl. Anal.*, 2013, 269420, 2013, https://doi.org/10.1155/2013/26942 0.
9. Farooq, M., ul ainAnzar, Q., Hayat, T., Ijaz Khan, M., Anjum, A., Local similar solution of MHD stagnation point flow in Carreau fluid over a non-linear stretched surface with double stratified medium. *Results Phys.*, 7, 3078–3089, 2017, https://doi.org/10.1016/j.rinp.2017.08.019.
10. Subba, R., Gorla, R., Dakappagari, V., Pop, I., Boundary layer flow at a three-dimensional stagnation point in power-law nonNewtonian fluids. *Int. J. Heat Fluid Flow*, 14, 408–412, 1993, https://doi.org/10.1016/0142-727X(93)90015-F.
11. Ahmad, M., Sajid, M., Hayat, T., Ahmad, I., On numerical and approximate solutions for stagnation point flow involving third order fluid. *AIP Adv.*, 5, 067138, 2015, https://doi.org/10.1063/1.4922878.
12. Bhattacharyya, K., Boundary layer stagnation-point flow of casson fluid and heat transfer towards a shrinking/stretching sheet. *Front. Heat Mass Transf. (FHMT)*, 4, 023003, 2013.
13. Hayat, T., Farooq, M., Alsaedi, A., Iqbal, Z., Melting heat transfer in the stagnation point flow of Powell–Eyring fluid. *J. Thermophys. Heat Transf.*, 27, 4, 761–766, 2013.
14. Asin, M.H.M., Ishak, A., Pop, I., MHD stagnation-point flow and heat transfer with effects of viscous dissipation, Joule heating and partial velocity slip. *Sci. Rep.*, 5, 17848, 2015, https://doi.org/10.1038/srep17848.
15. Lin, Y. and Zheng, L., Marangoni boundary layer flow and heat transfer of copper-water nanofluid over a porous medium disk. *AIP Adv.*, 5, 10, 107225, 2015, doi:10.1063/1.4934932.

16. Rashid, U., Baleanu, D., Liang, H., Abbas, M., Iqbal, A., Rahman, J., Marangoni Boundary Layer Flow and Heat Transfer of Graphene–Water Nanofluid with Particle Shape Effects. *Processes*, 8, 9, 1120, 2020, https://doi.org/10.3390/pr8091120.
17. Mahabaleshwar, U.S., Nagaraju, K.R., Sheremet, M.A., Baleanu, D., Lorenzini, E., Mass transpiration on Newtonian flow over a porous stretching/shrinking sheet with slip. *Chin. J. Phys.*, 63, 130–137, 2020, https://doi.org/10.1016/j.cjph.2019.11.016.
18. Mahabaleshwar, U.S., Vinay Kumar, P.N., Nagaraju, K.R., Bognár, G., Ravichandra Nayakar., S.N., A New Exact Solution for the Flow of a Fluid through Porous Media for a Variety of Boundary Conditions. *Fluids*, 4, 2019, https://doi.org/10.3390/fluids4030125.
19. Mahabaleshwar, U.S., Vishalakshi, A.B., Azese, M.N., The role of Brinkmann ratio on non-Newtonian fluid flow due to a porous shrinking/stretching sheet with heat transfer. *Eur. J. Mech. B. Fluids*, 92, 153–165, 2022.
20. Vishalakshi, A.B., Mahabaleshwar, U.S., Sheikhnejad, Y., Impact of MHD and mass transpiration on Rivlin–Ericksen liquid flow over a stretching sheet in a porous media with thermal communication. *Transp. Porous Med.*, 142, 353–381, 2022.
21. Vishalakshi, A.B., Mahabaleshwar, U.S., Sarris, I.E., An MHD fluid flow over a porous stretching/shrinking sheet with slips and mass transpiration. *Micromachines*, 13, 116, 2022.
22. Sneha, K.N., Mahabaleshwar, U.S., Sheikhnejad, Y., Heat and mass transfer of Walters' liquid B flow over a porous stretching/shrinking plate with mass transpiration and slip. *Transp. Porous Med.*, 142, 383–405, 2022.
23. Animasaun, I.L., 47nm alumina–water nanofluid flow within boundary layer formed on upper horizontal surface of paraboloid of revolution in the presence of quartic autocatalysis chemical reaction. *Alexandria Eng. J.*, 2016, S1110016816300862–. doi:10.1016/j.aej.2016.04.030.
24. Animasaun, I.L., Koriko, O.K., Mahanthesh, B., Dogonchi, A.S., A Note on the Significance of Quartic Autocatalysis Chemical Reaction on the Motion of Air Conveying Dust Particles. *Z. Naturforsch. A*, 74, 10, 879–904, 2019, doi: 10.1515/zna-2019-0180.
25. Baby, A., Manjunatha, S., Jayanthy, S., Gireesha, B., Archana, M., Effect of Variable Viscosity on Marangoni Convective Boundary Layer Flow of Nanofluid in the Presence of Mixed Convection. *J. Nanofluids*, 8, 845–851, 2019, 10.1166/jon.2019.1637.
26. Sastry, D. R. V. S. R. K., Thermosolutal MHD Marangoni Convective Flow of a Nanofluid Past a Flat Plate with Viscous Dissipation and Radiation Effects. *WSEAS Trans. Math.*, 15, 271–279, 2016.
27. Arifin, N.M., Nazar, R., Pop, I., Non-isobaric Marangoni boundary layer flow for Cu, Al2O3 and TiO2 nanoparticles in a water based fluid. *Meccanica*, 46, 833–843, 2011, https://doi.org/10.1007/s11012-010-9344-6.

28. Wakif, A., Numerical Inspection of Two-Dimensional MHD Mixed Bioconvective Flows of Radiating Maxwell Nanofluids Nearby a Convectively Heated Vertical Surface. *Waves Random Complex Media*, 1–22, 2023, https://doi.org/10.1080/17455030.2023.2179853.
29. Eswaramoorthi, S., Loganathan, K., Faisal, M., Botmart, T., Shah, N.A., Analytical and numerical investigation of Darcy-Forchheimer flow of a non-linear-radiative non-Newtonian fluid over a Riga plate with entropy optimization. *Ain Shams Eng. J.*, 14, 101887, 2023, https://doi.org/10.1016/j.asej.2022.101887.
30. Buongiorno, J., Convective transport in nanofluids. *J. Heat Transfer*, 128, 240–250, 2006, https://doi.org/10.1115/1.2150834.
31. Corcione, M., Empirical correlating equations for predicting the effective thermal conductivity and dynamic viscosity of nanofluids. *Energy Convers. Manage.*, 52, 789–793, 2011.
32. Li, Q. and Xuan, Y., Convective heat transfer and flow characteristics of Cu-water nanofluid. *Sci. China Ser. E-Technol. Sci.*, 45, 408, 2002, https://doi.org/10.1360/02ye9047.
33. Khoshvaght-Aliabadi, M., Hormozi, F., Zamzamian, A., Selfsimilar analysis of fluid flow, heat, and mass transfer at orthogonal nanofluid impingement onto a flat surface. *Heat Mass Transfer*, 51, 423, 2015, https://doi.org/10.1007/s00231-014-1422-1.
34. El-Maghlany, W.M., Hanafy, A.A., Hassan, A.A., El-Magid, M.A., Experimental study of Cu–water nanofluid heat transfer and pressure drop in a horizontal double-tube heat exchanger. *Exp. Therm Fluid Sci.*, 78, 100–111, 2016, https://doi.org/10.1016/j.expthermflusci.2016.05.015.
35. Khoshvaght-Aliabadi, M. and Alizadeh, A., An experimental study of Cu–water nanofluid flow inside serpentine tubes with variable straight-section lengths. *Exp. Therm Fluid Sci.*, 61, 1–11, 2015, https://doi.org/10.1016/j.expthermflusci.2014.09.014.
36. Shekaramiz, M., Fathi, S., Ataabadi, H.A., Kazemi-Varnamkhasti, H., Toghraie, D., MHD nanofluid free convection inside the wavy triangular cavity considering periodic temperature boundary condition and velocity slip mechanisms. *Int. J. Therm. Sci.*, 170, 2021, https://doi.org/10.1016/j.ijthermalsci.2021.107179.

# 18

# Emerging Trends in Smart Green Building Technologies

**Gongutri Borah**

*Faculty of Paramedical Sciences, Assam down town University, Sankar Madhab Road, Panikhaiti, Guwahati, Assam, India*

***Abstract***

The chapter explores the ever-changing landscape of intelligent green building technologies, shedding light on transformative trends reshaping the architecture and construction industry. Given the global environmental challenges, there is an imperative need for integrating intelligent and sustainable solutions within the built environment. The chapter conducts a thorough examination of recent innovations, methodologies, and practices driving the evolution of smart green buildings. The discussion commences by framing the urgency for environmentally conscious structures in response to climate change and resource depletion. It then traverses through various dimensions of smart green buildings, encompassing energy-efficient designs, advanced materials, and responsive systems. The utilization of Internet of things (IoT) devices, artificial intelligence, and data analytics are explored as catalysts for optimizing energy consumption, improving occupant comfort, and enabling predictive maintenance. Additionally, the chapter scrutinizes the pivotal role of renewable energy sources and energy storage solutions in establishing net-zero energy buildings. Case studies and real-world examples are presented to illustrate the successful implementation of these technologies across diverse architectural contexts. The socio-economic impact of smart green buildings is also investigated, highlighting their potential contribution to creating sustainable and resilient communities. The discourse extends beyond physical construction aspects, addressing the significance of smart green building policies, regulations, and certifications. The chapter concludes by envisioning the future trajectory of smart green building technologies, foreseeing their continued integration into urban planning and the construction industry. By providing a comprehensive overview of the latest advancements, this chapter serves

*Email*: gongutri28@gmail.com; ORCID: 0000-0002-9366-4388

Ashwani Kumar, Nitesh Dutt and Mukesh Kumar Awasthi (eds.) Heat Transfer Enhancement Techniques: Thermal Performance, Optimization and Applications, (391–416) © 2025 Scrivener Publishing LLC

as a valuable resource for architects, engineers, policymakers, and researchers navigating the evolving landscape of sustainable and intelligent building practices.

***Keywords:*** Green building, environment, design, urban, artificial intelligence

## 18.1 Introduction

Smart green buildings represent a paradigm shift in contemporary architectural and construction practices, emphasizing sustainable and intelligent design principles. These buildings integrate advanced technologies to efficiently utilize resources, reduce environmental impact, and enhance occupant well-being. According to Alwaer and Clements-Croome [1], smart green buildings are characterized by a holistic approach that incorporates environmental, economic, and social considerations into the design and operation processes. The integration of technologies such as Internet of things (IoT) devices, building automation systems, and renewable energy sources enables these buildings to actively respond to environmental conditions, optimize energy consumption, and create comfortable living or working environments [2]. One defining feature of smart green buildings is their focus on energy efficiency. Through innovative architectural designs and the integration of advanced materials, these buildings aim to minimize energy consumption and environmental impact. For instance, passive design strategies, including optimized building orientation, natural ventilation, and the use of thermal mass, contribute to reducing the need for mechanical heating and cooling systems. Additionally, smart green buildings often incorporate renewable energy sources such as solar panels and wind turbines to generate clean energy on-site, aligning with the principles of sustainability and resource efficiency. Furthermore, the concept of occupant-centric design is a key characteristic of smart green buildings. These structures prioritize the health and well-being of occupants by integrating technologies that monitor and improve indoor environmental quality. Smart lighting systems, advanced heating ventilation and air conditioning (HVAC) controls, and real-time air quality monitoring contribute to creating comfortable and healthy indoor environments [3]. The integration of such technologies not only aligns with the principles of green building design but also reflects the growing recognition of the impact of the built environment on human health and productivity. In essence, smart green buildings embody a symbiotic relationship between sustainable practices and cutting-edge technologies, fostering a more resilient and environmentally conscious built environment.

In the contemporary context, the imperative for sustainable and intelligent building solutions is driven by the urgent need to address global environmental challenges. The construction and operation of buildings contribute significantly to carbon emissions, resource depletion, and environmental degradation [4]. According to recent reports [5], buildings are responsible for approximately 40% of global energy consumption and 36% of total greenhouse gas emissions. The rising awareness of climate change and its adverse effects underscores the critical role that sustainable building practices play in mitigating environmental impact and fostering a more resilient and adaptive built environment.

The necessity for sustainable building solutions is intricately tied to the objective of attaining energy efficiency and diminishing reliance on finite resources. The depletion of conventional energy sources and the fluctuation of energy prices require a transition towards buildings that not only consume reduced energy but also produce and utilize renewable energy [6]. Sustainable building solutions underscore the amalgamation of energy-efficient technologies, renewable energy sources, and advanced building materials to develop structures that actively play a role in a low-carbon future [7]. This imperative aligns with worldwide initiatives like the United Nations Sustainable Development Goals (SDGs), emphasizing responsible consumption and production patterns, and the shift to clean and affordable energy sources.

Additionally, the necessity for intelligent building solutions stems from the escalating complexity of urban environments and the demand for resilient and adaptable structures. Intelligent buildings employ state-of-the-art technologies such as the IoT, artificial intelligence (AI), and data analytics to optimize operational efficiency, enhance occupant comfort, and provide real-time monitoring and control [8]. The integration of intelligent systems not only enhances building performance but also contributes to the development of smart cities capable of addressing dynamic challenges such as population growth, urbanization, and climate change.

## 18.2 Environmental Challenges and the Need for Innovation

The construction sector plays a crucial role in exacerbating environmental concerns, primarily due to its substantial energy use, resource exploitation, and waste production. Construction processes are energy-intensive, and the extraction, manufacturing, and transportation of building materials contribute to greenhouse gas emissions and other environmental consequences [9]. As highlighted in various reports [10], the construction industry significantly contributes to global energy consumption and carbon emissions, particularly through the production of cement, a key construction material that is a major source of carbon dioxide emissions. The extraction and processing of raw materials like steel and concrete also contribute to environmental harm and habitat loss [11]. Moreover, the construction industry serves as a substantial consumer of natural resources, leading to resource depletion and environmental stress. The demand for raw materials such as timber, minerals, and water puts significant pressure on ecosystems, contributing to issues like deforestation, soil erosion, and water scarcity [12]. The extraction of non-renewable resources like sand and gravel for construction activities has been linked to habitat destruction and alterations in river ecosystems [13]. Additionally, the improper disposal of construction waste, including demolition debris and unused materials, presents a challenge, adding to landfills and environmental pollution [14].

Furthermore, the construction industry's impact on environmental issues extends into the operational phase of the built environment, as buildings consume substantial energy for heating, cooling, and lighting. Inefficient building designs and outdated technologies contribute to high energy consumption, resulting in increased carbon emissions [15]. To address these challenges, sustainable construction practices, incorporating energy-efficient designs, environmentally friendly materials, and renewable energy sources, are essential for reducing the environmental impact of the construction industry and fostering a more sustainable and resilient built environment.

## 18.3 The Urgency for Adopting Smart Green Building Technologies

The pressing need to embrace intelligent green building technologies is highlighted by the increasing global challenges posed by climate change, resource depletion, and the imperative to transition towards sustainable development. Buildings play a significant role in energy consumption and greenhouse gas emissions, contributing substantially to environmental impact during both the construction and operation phases (Figure 18.1). Reports indicate that the building sector is accountable for nearly 30% of global energy consumption and approximately one-third of global greenhouse gas emissions [2]. Urgency to address these environmental impacts necessitates a swift transition to smart green building technologies, seamlessly integrating sustainability and intelligence into the fabric of the built environment. Smart green building technologies play a vital role in optimizing resource utilization and minimizing environmental impact. Traditional buildings often lag in terms of energy efficiency, occupant comfort, and adaptability to changing environmental conditions. The incorporation of smart technologies, such as building automation systems and IoT devices, facilitates real-time monitoring and control of building operations, resulting in heightened energy efficiency and reduced environmental impact [16]. These technologies contribute to the development of responsive and adaptive building systems capable of dynamically adjusting to varying energy demands, weather conditions, and occupant preferences. The urgency to adopt smart green building technologies is also fueled by the imperative to achieve global sustainability goals. Initiatives like the United Nations SDGs stress responsible consumption, climate action, and the creation of sustainable cities and communities [17]. Smart green buildings align with these goals by promoting energy conservation, integrating renewable energy, and managing resources intelligently, thereby contributing to the broader agenda of sustainable development [18].

Furthermore, the urgency for adopting smart green building technologies is accentuated by economic considerations. The long-term cost savings associated with energy-efficient designs, predictive maintenance, and sustainable practices present a compelling case for the adoption of these technologies. The initial investments in smart green building technologies are often outweighed by significant returns on investment through reduced energy costs, increased building efficiency, and overall improved operational performance.

## 18.4 Innovative Architectural Designs That Prioritize Energy Efficiency

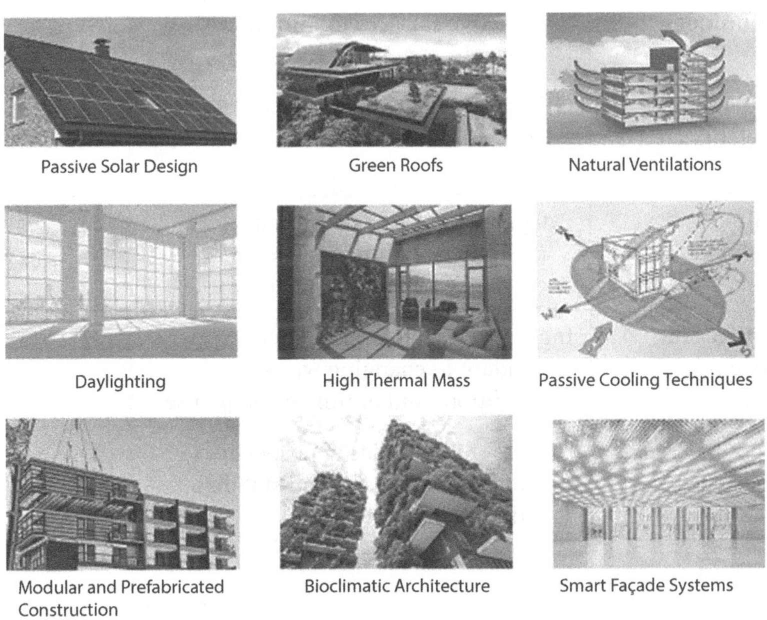


**Figure 18.1** Example of energy-efficient buildings.

**Table 18.1** Design elements details.

| Design element | Key points | Scientific references |
|---|---|---|
| Passive solar design | Utilizes the sun's energy for heating and lighting, optimizing solar gain, and reducing reliance on mechanical heating. | [19] |
| Green roofs | Integrates living vegetation on rooftops, providing insulation, and reducing heat absorption. | [20] |
| Natural ventilation | Utilizes natural air currents for cooling, reducing the need for mechanical cooling systems. | [21] |
| Daylighting | Maximizes natural light to reduce artificial lighting, employing skylights, and reflective surfaces. | [22] |
| High thermal mass | Uses materials with high thermal mass to stabilize indoor temperatures and reduce heating/cooling needs. | [23] |
| Passive cooling techniques | Implements features like overhangs and reflective surfaces to minimize solar heat gain. | [24] |
| Modular and prefabricated construction | Off-site manufacturing reduces waste and optimizes material usage. | [25] |
| Bioclimatic architecture | Adapts designs to local climate conditions, optimizing energy efficiency and occupant comfort. | [26] |
| Smart façade systems | Incorporates dynamic elements to adapt to changing weather conditions and optimize energy use. | [27] |
| Earth-sheltered construction | Utilizes soil's insulating properties to reduce heat transfer and stabilize indoor temperatures. | [28] |

## 18.5 Passive Design Principles and Their Impact on Building Performance

Passive design principles are fundamental in the development of energy-efficient buildings, utilizing natural elements and characteristics to optimize comfort and minimize dependence on active mechanical systems (Table 18.1). These principles involve various strategies, such as orientation, insulation, shading, and natural ventilation, focusing on utilizing the inherent properties of the building and its surroundings. The impact of passive design on building performance is substantial, influencing energy consumption, thermal comfort, and overall sustainability. A key passive design principle is proper building orientation, maximizing sunlight exposure for heating and minimizing heat gain during warmer periods. Research underscores the significance of orientation in improving energy efficiency. For example, a study explores the influence of building orientation on energy consumption, revealing that well-oriented buildings can significantly reduce heating and cooling needs, resulting in considerable energy savings [29]. Insulation serves as another critical passive design element regulating heat transfer to maintain comfortable indoor temperatures. Adequate insulation minimizes heat loss in colder seasons and heat gain in warmer periods. Research also investigates the energy performance of insulated buildings, highlighting a positive correlation between insulation levels and energy efficiency [30]. The study emphasizes the role of insulation in reducing energy consumption and enhancing overall building performance.

Shading strategies, such as overhangs, awnings, or vegetation, play a crucial role in passive design by preventing excessive solar heat gain and glare. A scientific study [15] examines the impact of shading on building performance, emphasizing its potential to improve thermal comfort and reduce the need for air conditioning. Shading strategies effectively mitigate the adverse effects of solar radiation, contributing to energy savings and occupant well-being. Natural ventilation, a passive design principle, utilizes natural air movement to cool indoor spaces. Well-designed ventilation strategies can decrease reliance on mechanical cooling systems, resulting in energy savings. The impact of natural ventilation on indoor air quality and thermal comfort is explored, highlighting its potential to enhance building performance by providing a sustainable and energy-efficient solution. In summary, passive design principles significantly shape building performance by optimizing energy efficiency, improving thermal comfort, and promoting sustainability [31]. The integration of these principles is crucial for creating buildings that respond to their environments while minimizing their ecological footprint.

## 18.6 Exploration of Eco-Friendly and Sustainable Construction Materials

The investigation into eco-friendly and sustainable construction materials is a crucial aspect of addressing the environmental impact of the construction industry. As the demand for buildings continues to grow, the choice of materials becomes pivotal in minimizing resource depletion, reducing energy consumption, and mitigating the overall ecological footprint. This exploration encompasses diverse alternatives that are environmentally responsible and have a lower impact on ecosystems and human health. One noteworthy eco-friendly material is bamboo, recognized as a rapidly renewable resource that offers excellent strength and versatility. A study assesses the feasibility of incorporating bamboo into the building industry, emphasizing its potential as a sustainable alternative to traditional timber. With its fast growth rate and minimal environmental impact, bamboo emerges as an attractive option for construction practices [2].

Recycled steel stands out as another significant material in sustainable construction, as it lessens the demand for virgin steel production and reduces energy consumption. Reports delve into the embodied energy of common and alternative building materials, underlining the energy-saving benefits associated with incorporating recycled steel into construction projects. This approach not only addresses resource scarcity but also contributes to the reduction of greenhouse gas emissions linked to steel production [32]. Hempcrete, a bio-composite material made from hemp fibers, lime, and water, is gaining attention as a sustainable construction material, offering a more environmentally friendly choice compared to traditional concrete. The use of recycled plastic in construction materials presents an innovative approach to curbing plastic waste and fostering a circular economy. Researchers discuss the potential of mixed recycled aggregates from construction and demolition wastes for concrete production. This not only tackles the issue of plastic pollution but also diminishes the demand for virgin materials in construction applications [33]. Fly ash concrete represents another sustainable option that utilizes a byproduct of coal combustion. Scientists explore the performance characteristics of high-volume Class F fly ash concrete, showcasing its potential to reduce the carbon footprint of construction. Incorporating fly ash into concrete formulations helps manage industrial waste while enhancing the material's strength and durability [34]. Reclaimed wood offers a classic yet effective approach to sustainable construction. Salvaged or deconstructed wood from existing structures reduces the need for new timber and minimizes the environmental impact associated with logging. Researchers delve into the characterization and utilization of wood ash from biomass combustion plants, highlighting the importance of considering waste byproducts in sustainable construction practices. Thus, the exploration of eco-friendly and sustainable construction materials encompasses various alternatives aligned with the principles of environmental stewardship. Scientific research plays a crucial role in assessing the viability and environmental benefits of these materials, guiding the construction industry toward more sustainable and resilient practices [35].

## 18.7 Case Studies Showcasing the Use of Advanced Materials in Real-World Projects

### 18.7.1 Case Study: The Edge, Amsterdam—Innovative Use of Recycled Materials

The Edge, a sustainable office building in Amsterdam, incorporates advanced materials such as recycled steel and concrete. The façade features recycled aluminum and glass, contributing to its energy-efficient design. The building's sustainability aspects are highlighted by various reports [36], which assesses the environmental impact of The Edge, emphasizing the use of recycled materials in construction.

### 18.7.2 Case Study: One Central Park, Sydney—Living Façade with Green Wall Technology

One Central Park, a mixed-use development in Sydney, showcases the use of innovative materials, including green walls composed of living plants. The building employs advanced irrigation and plant selection technologies to create a sustainable and aesthetically pleasing façade. A study [37] explores the ecological performance of living walls, providing insights into the environmental benefits and challenges associated with such advanced materials.

### 18.7.3 Case Study: Bosco Verticale, Milan—Vertical Forest Towers

Bosco Verticale, or Vertical Forest, in Milan features two residential towers covered with trees and plants, incorporating advanced materials in its construction. The buildings aim to improve air quality and energy efficiency. A scientific review [38] assesses the sustainable features of Bosco Verticale, emphasizing the use of advanced materials and their impact on building performance.

### 18.7.4 Case Study: The Crystal, London—Sustainable Building with Photovoltaic Integration

The Crystal in London is a sustainable building that integrates advanced materials, including low-emissivity glass and photovoltaic (PV) panels. These materials contribute to energy efficiency and renewable energy generation. A study [39] evaluates the energy performance of The Crystal, emphasizing the impact of advanced materials on the building's sustainability.

### 18.7.5 Case Study: The Edge, New York—High-Performance Façade

The Edge in New York exemplifies the use of advanced materials in its high-performance façade, incorporating dynamic shading systems and triple-glazed windows. The building aims for energy efficiency and occupant comfort. Scientific research investigates the thermal performance of dynamic shading devices, providing insights into the effectiveness of these advanced materials in real-world applications [40].

### 18.7.6 Case Study: King Abdullah Petroleum Studies and Research Center (KAPSARC), Riyadh—Advanced Building Envelope

The KAPSARC in Riyadh incorporates advanced materials in its building envelope, utilizing double-skin façades and innovative shading systems. These materials contribute to energy savings and thermal comfort. Reports also examines the sustainable features of KAPSARC, emphasizing the role of advanced materials in achieving energy-efficient building performance [41].

These case studies exemplify the practical application of advanced materials in real-world construction projects, showcasing their impact on sustainability, energy efficiency, and overall building performance. The referenced scientific studies provide in-depth analyses and assessments of these projects, offering valuable insights for future sustainable construction endeavors.

## 18.8 The Intersection of IoT and Smart Green Buildings

The intersection of the IoT and smart green buildings represents a paradigm shift in the construction industry, offering unprecedented opportunities for enhanced sustainability, energy efficiency, and occupant well-being. Internet-of-things technology involves the integration of interconnected sensors, actuators, and communication devices, enabling real-time data collection, analysis, and decision-making. When applied to smart green buildings, IoT transforms traditional structures into intelligent systems capable of responding dynamically to environmental changes and user behaviors. This integration optimizes resource utilization, reduces energy consumption, and contributes to the creation of environmentally responsible built environments [42]. Scientific studies highlight the multifaceted impact of IoT on smart green buildings. A study discuss the role of IoT in building automation and energy management, emphasizing the potential for real-time monitoring and control of various building systems. The study underscores the importance of leveraging IoT-enabled devices to enhance energy efficiency, reduce operational costs, and improve overall building performance [43]. The integration of IoT technologies allows for a more holistic and data-driven approach to building management, fostering a dynamic and responsive environment.

The use of IoT in smart green buildings goes beyond energy efficiency and extends to occupant comfort and well-being. Smart sensors can monitor indoor air quality, lighting conditions, and temperature, providing real-time feedback for adjustments that optimize the indoor environment for health and productivity [44]. A research explores the recent advances in data-driven optimization methods for building performance analysis, emphasizing the potential for IoT to enhance the quality of life within smart green buildings. Furthermore, the synergy between IoT and smart green buildings introduces the concept of predictive maintenance. By continuously monitoring building systems and equipment through IoT-enabled sensors, potential issues can be identified and addressed proactively, minimizing downtime and reducing the environmental impact associated with unexpected failures [42]. This approach aligns with the principles of sustainability by promoting resource efficiency and extending the lifespan of building components, ultimately contributing to a more resilient and environmentally conscious built environment.

## 18.9 Artificial Intelligence in Smart Buildings

Artificial Intelligence (AI) is transforming the operational dynamics of smart buildings, optimizing energy efficiency, occupant comfort, and overall performance. In the context of smart buildings, AI involves the integration of machine learning (ML) algorithms and data analytics, enabling buildings to learn, adapt, and make informed decisions based on real-time data. This revolutionary technology has the potential to significantly enhance the sustainability and intelligence of built environments. Predictive analytics is a crucial aspect of AI in smart buildings. Machine learning algorithms analyze historical data, weather patterns, and occupant behavior to predict future energy demands and optimize HVAC systems accordingly. This predictive approach leads to more efficient energy consumption, thereby reducing operational costs and environmental impact. A study explores the application of AI-based predictive control in building HVAC systems, emphasizing the potential for significant energy savings [40]. Another important application of AI in smart buildings is the implementation of intelligent building management systems. These systems use AI algorithms to monitor and control various building components, such as lighting, heating, and ventilation, in real-time. By continuously learning and adapting to changing conditions, AI-driven building management systems optimize energy usage and enhance overall performance. Researchers also investigate the integration of AI in building management systems, highlighting the benefits in terms of energy efficiency and occupant satisfaction. Artificial intelligence plays a crucial role in occupant-centric applications within smart buildings as well. Personalized climate control, lighting preferences, and space utilization patterns can be learned and adapted by AI algorithms, contributing to enhanced occupant comfort and well-being [44]. Various reports explore the implementation of AI in personalized comfort control, emphasizing the potential for AI to tailor the indoor environment to individual preferences while optimizing energy efficiency. Furthermore, the continuous advancement of AI technologies, including the integration of ML, natural language processing, and computer vision, expands the capabilities of smart buildings. These advancements enable buildings not only to respond to immediate environmental conditions but also to proactively identify potential issues, make autonomous decisions, and continuously improve operational efficiency. The scientific exploration of AI in smart buildings is rapidly expanding, reflecting the increasing recognition of its potential to revolutionize the sustainable and intelligent operation of built environments [45].

## 18.10 The Role of Solar and Wind Energy in Achieving Net-Zero Energy Buildings

Solar and wind energy play vital roles in achieving net-zero energy buildings, a concept emphasizing that a building's total energy consumption equals the renewable energy it generates on-site. These renewable sources significantly reduce dependence on conventional energy grids, minimize greenhouse gas emissions, and advance the sustainability of built environments. Photovoltaic (PV) solar panels are a prevalent technology for capturing solar energy, converting sunlight into electricity, while wind turbines capture wind energy and transform it into electrical power. Their integration into the built environment is crucial for achieving energy autonomy and environmental sustainability [46]. When strategically deployed on building rooftops or integrated into façades, PV solar panels enable on-site electricity generation from sunlight. Scientists highlight the importance of solar energy in achieving net-zero energy buildings, showcasing the potential of PV systems in meeting a significant portion of a building's energy needs. Additionally, advancements in building-integrated photovoltaics (BIPV) contribute to architectural aesthetics and energy efficiency, ensuring a seamless integration of solar energy solutions into the building envelope [47]. Wind energy complements solar power, offering a consistent energy source even when sunlight is limited. Small-scale wind turbines, designed for urban environments, can be integrated into building designs to harness wind energy. A study explores the integration of small wind turbines into residential buildings, demonstrating their potential to contribute to on-site renewable energy generation and facilitate the transition towards net-zero energy buildings. Achieving net-zero energy buildings often involves implementing energy storage systems to store excess energy generated during peak times [48]. Battery technologies, such as lithium-ion batteries, enable efficient storage and utilization of surplus energy during periods of low renewable energy production. Scientists also review advancements in lithium-ion battery technologies, emphasizing their role in enhancing the reliability and resilience of net-zero energy systems. Thus, the integration of solar and wind energy technologies is instrumental in realizing net-zero energy buildings, fostering a sustainable and resilient built environment. Ongoing scientific research explores new technologies, materials, and strategies to optimize the performance of renewable energy systems, providing valuable insights for the widespread adoption of net-zero energy building practices [49].

## 18.11 Energy Storage Solutions in Buildings for Balancing Intermittent Renewable Sources

Energy storage solutions within buildings play a crucial role in addressing the intermittent nature of renewable energy sources like solar and wind. These storage systems are vital for balancing the irregular energy production associated with renewables, storing surplus energy during periods of high generation and releasing it during peak demand or when renewable sources are inactive. An exemplary instance of energy storage in buildings involves the utilization of batteries, particularly lithium-ion batteries, known for their high energy density, efficiency, and relatively extended lifespan. Scholars investigate the integration of lithium-ion batteries for energy storage in buildings, underscoring their effectiveness in enhancing the dependability and stability of renewable energy systems.

Pumped hydro storage stands out as another energy storage solution suitable for incorporation into buildings or nearby terrains. This method involves using excess energy to pump water to a higher elevation during surplus periods, releasing the stored water to generate electricity during high-demand periods. A study examines the potential of small-scale pumped hydro storage systems for residential buildings, emphasizing their role in improving the flexibility and grid integration of intermittent renewable energy sources.

Additionally, thermal energy storage offers an innovative solution by storing excess energy in the form of heat, which can later be converted to electricity or directly used for heating. Phase change materials (PCMs) prove particularly effective in this context, capable of storing and releasing significant energy during phase transitions. Researchers explore the use of PCM-enhanced thermal energy storage in buildings, demonstrating its potential to enhance the performance of renewable energy systems by providing a reliable and efficient means of energy storage. Therefore, energy storage solutions in buildings are vital for mitigating the intermittent nature of renewable energy sources, contributing to the stability and reliability of the overall energy system. Ongoing scientific research explores various storage technologies, their integration into building structures, and their impact on optimizing the utilization of intermittent renewable energy sources [50].

## 18.12 Technologies Enhancing Occupant Well-Being and Productivity

Technological advancements in building design and management are increasingly focused on enhancing occupant well-being and productivity. These innovations span various aspects, from indoor air quality to lighting and acoustics. The integration of such technologies contributes to creating healthier and more productive indoor environments. One critical area is the implementation of smart lighting systems that adjust to circadian rhythms. Human-centric lighting, which mimics natural sunlight patterns, has been shown to positively impact mood, alertness, and overall well-being. A study explores the effects of dynamic lighting on sleep and cognitive performance, emphasizing the potential of lighting systems to enhance occupant well-being and productivity [51].

Indoor air quality (IAQ) monitoring and management technologies also play a crucial role in ensuring healthy environments. Advanced sensors can detect pollutants and regulate ventilation systems accordingly. Studies, such as the one by Wargocki *et al.*, highlight the impact of indoor air quality on cognitive function and productivity, emphasizing the importance of effective IAQ management in buildings. Incorporating biophilic design elements, which connect occupants with nature, has gained traction for its positive effects on well-being and productivity. Green walls, indoor plants, and natural materials contribute to a more visually appealing and stress-reducing environment [52]. Scientists explore the impact of biophilic design on workplace well-being and productivity, showcasing its potential to create more engaging and satisfying spaces. The use of wearable devices and personalized comfort control systems is another innovative approach. Wearables can monitor individual health metrics, while personalized control systems allow occupants to tailor their immediate environment, such as adjusting temperature and lighting. A study investigates the implementation of personalized comfort control using wearable devices, highlighting the potential for improving occupant satisfaction and well-being. Technological solutions also address acoustics, which significantly impact occupant comfort and productivity. Sound masking systems and acoustic panels are examples of technologies used to manage noise levels and create more pleasant indoor environments. Scientists examines the impact of noise on cognitive performance, emphasizing the need for effective acoustic solutions to enhance productivity in workplaces. Thus, the integration of technologies in building design and management is essential for promoting occupant well-being and productivity. Scientific research continually explores the efficacy of these technologies, providing valuable insights into their effects on cognitive function, emotional well-being, and overall satisfaction in indoor environments [53].

## 18.13 The Impact of a Human-Centric Approach on Building Design

A human-centric approach to building design prioritizes the well-being, comfort, and productivity of occupants. This design philosophy emphasizes understanding and accommodating the needs, preferences, and behaviors of the people who will inhabit the space. Scientific research underscores the positive impact of a human-centric approach on building design, demonstrating its potential to enhance physical and mental health, as well as overall satisfaction. One key aspect of a human-centric approach is the incorporation of natural elements into the built environment. Biophilic design, which integrates nature-inspired elements, has been shown to reduce stress, improve cognitive function, and enhance overall well-being. Researchers explores the impact of biophilic design on workplace well-being, emphasizing the importance of incorporating natural elements, such as plants and daylight, into building environments. Lighting is a critical factor in human-centric design, influencing circadian rhythms, mood, and overall comfort. Dynamic lighting systems that mimic the natural progression of sunlight can positively impact occupants [54].

Acoustic considerations are also central to a human-centric approach. Noise levels, sound quality, and privacy significantly impact occupant comfort and concentration. Various reports emphasize the importance of acoustic design in promoting well-being and reducing stress in indoor environments. Furthermore, ergonomic considerations in furniture and spatial layouts contribute to a human-centric design that prioritizes comfort and functionality. Scientific research also explores the impact of ergonomic design on occupant comfort and productivity, emphasizing the need for tailored solutions to accommodate diverse user needs. Thus, a human-centric approach to building design is essential for creating environments that support the well-being and satisfaction of occupants. Scientific research continues to validate the positive impact of incorporating natural elements, optimizing lighting conditions, addressing acoustic concerns, and considering ergonomic factors in building design, promoting healthier and more productive indoor environments [55].

## 18.14 Smart Green Building Policies and Certifications

Green building standards and certifications play a crucial role in advancing sustainable and environmentally responsible practices within the construction industry. These certifications offer directives and standards for the design, construction, and operation of buildings to minimize environmental impact. One of the globally acknowledged standards is the Leadership in Energy and Environmental Design (LEED), established by the U.S. Green Building Council (USGBC). The LEED assesses buildings according to criteria like energy efficiency, water conservation, indoor environmental quality, and the use of sustainable materials.

The Building Research Establishment Environmental Assessment Method (BREEAM), originating in the United Kingdom, is another influential green building certification. The BREEAM assesses buildings across categories such as energy, health and well-being, water usage, and materials, providing a comprehensive evaluation of a building's environmental performance. Scientific studies have investigated the environmental implications of BREEAM-certified buildings, confirming their positive contributions to sustainability and energy efficiency. The Living Building Challenge (LBC), developed by the International Living Future Institute, sets a high bar for regenerative design. The LBC certification focuses on achieving net-positive energy and water use, as well as the use of sustainable and socially responsible materials [56].

In addition to these global certifications, countries often have their own standards tailored to regional contexts. For example, in Germany, the Passive House Standard focuses on achieving ultra-low energy buildings through rigorous design and construction principles. The scientific literature has extensively analyzed the energy performance and environmental benefits of buildings meeting the Passive House Standard, showcasing its significance in promoting energy-efficient construction practices. Thus, existing green building standards and certifications, such as LEED, BREEAM, LBC, and the Passive House Standard, contribute significantly to shaping the sustainability landscape in the construction industry. These certifications offer holistic frameworks that address various aspects of building design and operation, and ongoing scientific research continues to validate their positive impact on environmental and resource sustainability [57].

## 18.15 The Influence of Regulations on Industry Adoption of Smart Green Technologies

Regulations play a pivotal role in shaping the adoption of smart green technologies in various industries. Government policies and regulations can act as drivers or barriers, influencing the extent to which businesses integrate sustainable and intelligent technologies into their operations. Scientific research has explored the impact of regulations on industry adoption, providing insights into the dynamics of this relationship. Government mandates and regulations often set the stage for the widespread adoption of smart green technologies. For example, regulations promoting energy efficiency and emissions reduction can incentivize industries to invest in technologies such as smart sensors, automation, and renewable energy solutions [58].

Conversely, regulatory barriers or uncertainties can hinder the adoption of smart green technologies. Ambiguous policies or inconsistent regulations may create challenges for industries seeking to invest in innovative solutions. Researcher explores the impact of regulatory uncertainties on the adoption of sustainable technologies in the construction industry, emphasizing the need for clear and stable regulatory frameworks to support industry initiatives. In some cases, industry-specific regulations may drive the adoption of smart green technologies. For instance, the transportation sector is increasingly influenced by regulations aimed at reducing carbon emissions. Studies examine the impact of regulatory requirements on the adoption of electric vehicles (EVs) in the automotive industry, illustrating how regulations can steer industries toward environmentally friendly technologies. Furthermore, international agreements and standards can shape the global landscape for smart green technologies. The Paris Agreement, for instance, sets targets for countries to reduce carbon emissions, influencing industries worldwide to adopt technologies that contribute to sustainability goals [59].

Hence, the adoption of smart green technologies in diverse industries is markedly impacted by regulations. Whether by encouraging certain practices or imposing challenges, the regulatory framework plays a vital role in molding the conduct of enterprises. Ongoing scientific research consistently offers valuable perspectives on the intricate relationship between regulations and the integration of sustainable and intelligent technologies. This, in turn, establishes a basis for informed policy formulation and industry-driven initiatives.

## 18.16 Speculations on the Future Trajectory of Smart Green Building Technologies

Speculating on the future trajectory of smart green building technologies involves envisioning a range of innovations and advancements that may shape the built environment. While the precise direction is challenging to predict, emerging trends and research offer insights into potential developments. Scientific references provide a foundation for these speculations, highlighting ongoing discussions and areas of exploration within the field.

### 18.16.1 Incorporation of Artificial Intelligence (AI) and Machine Learning (ML)

The amalgamation of AI and ML is anticipated to have a pivotal impact on the prospective development of intelligent and environmentally-friendly building technologies. Systems propelled by AI can enhance building operations through perpetual learning and adjustment to occupant behavior, environmental factors, and energy consumption trends. Scholars explore the utilization of AI in the optimization of building energy management systems, suggesting its capability to augment the effectiveness and sustainability of intelligent buildings.

### 18.16.2 Advanced Building Materials and Energy Harvesting

Innovations in building materials, such as self-healing concrete, smart windows, and energy-harvesting surfaces, may become more prevalent. These materials can contribute to improved energy efficiency and sustainability. A study explores the potential of energy-harvesting technologies in buildings, discussing the integration of piezoelectric and thermoelectric materials for capturing ambient energy.

### 18.16.3 IoT-Driven Smart Building Ecosystems

The IoT is expected to further evolve, creating interconnected smart building ecosystems. Internet-of-things sensors and devices can enhance occupant comfort, optimize energy use, and enable predictive maintenance. Researchers discusses the role of IoT in smart buildings, emphasizing its potential for real-time monitoring, data analytics, and creating responsive environments.

### 18.16.4 Circular Economy and Sustainable Construction Practices

The future of smart green building technologies may see a stronger emphasis on circular economy principles and sustainable construction practices. This involves designing buildings for disassembly, recycling materials, and minimizing waste. A study explores the integration of circular economy principles in the construction industry, emphasizing the potential for reducing environmental impacts.

### 18.16.5 Human-Centric Design and Well-Being Focus

The future trajectory may involve a heightened focus on human-centric design and occupant well-being. Technologies that prioritize indoor air quality, circadian lighting, and personalized comfort control may become more prevalent. A study investigates the impact of indoor environmental quality on occupant well-being, emphasizing the importance of designing smart buildings with a holistic approach to human health.

Thus, the future of smart green building technologies is likely to be shaped by a convergence of AI, advanced materials, IoT ecosystems, sustainable practices, and a strong emphasis on occupant well-being. While these speculations provide a glimpse into potential trends, ongoing scientific research and technological advancements will ultimately determine the trajectory of smart green buildings in the coming years [60].

## 18.17 Conclusion: Socio-Economic Impact and Community Resilience

The implementation of smart green buildings can have profound socio-economic impacts and contribute to community resilience. From an economic standpoint, the construction and operation of smart green buildings create job opportunities in areas such as sustainable design, renewable energy installation, and building automation. The growth of the green building sector stimulates economic activity and supports the development of a skilled workforce. Additionally, the increased energy efficiency and resource conservation associated with smart green buildings can prime to cost savings for both occupants as well as the community, fostering economic stability and resilience. Community resilience is enhanced through the social and environmental benefits of smart green buildings. The incorporation of sustainable practices, such as green spaces, energy-efficient technologies, and eco-friendly materials, contributes to improved public health and well-being. Research by Upadhyay *et al.* (2017) highlights the positive correlation between green building features and health outcomes, emphasizing the potential for smart green buildings to create healthier and more resilient communities. Furthermore, the application of green energy sources in smart green buildings contributes to community energy independence and reduces vulnerability to external energy supply disruptions, enhancing overall resilience in the face of environmental and economic challenges. The socio-economic impact of smart green buildings extends beyond their physical structures, influencing the broader community through educational and awareness initiatives. The adoption of sustainable building practices often involves community engagement, education, and collaboration. For instance, smart green buildings can serve as educational hubs, demonstrating the benefits of sustainable living and providing resources for residents to adopt eco-friendly practices. This educational aspect contributes to building a more environmentally conscious and socially responsible community, further strengthening the resilience of the society as a whole.

In conclusion, the socio-economic impact and community resilience of smart green buildings are interconnected, creating a cycle of positive influence. These buildings not only provide economic opportunities and cost savings but also foster healthier communities, contribute to environmental sustainability, and promote education and awareness. The integration of smart green buildings into communities reflects a holistic approach to development that considers economic, social, and environmental dimensions, ultimately contributing to the longstanding resilience and welfare of the public.

## Acknowledgement

GB thanks Chancellor Assam Down Town University for the support and inspiration.

# References


1. Alwaer, H. and Clements-Croome, D., Key performance indicators (KPIs) and priority setting in using the multi-attribute approach for assessing sustainable intelligent buildings. *Build. Environ.*, 42, 7, 2729–2739, 2007.
2. Lu, Y., Wu, D., Wang, Y., Jin, Y., A review of the recent advances in data-driven optimization methods for building performance analysis. *Renew. Sustain. Energy Rev.*, 74, 73–86, 2017.
3. Lam, K.P., Lau, S.S.Y., Goggins, G., Chan, E.H.W., Bai, Y., Impact of indoor environmental quality on occupant well-being and comfort: A review of the literature. *Int. J. Sustain. Built Environ.*, 7, 2, 214–228, 2018.
4. Hossain, M.A., An, V., Zhu, Y., Skitmore, M., Embodied carbon in construction materials: A review. *Renew. Sustain. Energy Rev.*, 127, 109896, 2020.
5. Sartori, I. and Hestnes, A.G., Energy use in the life cycle of conventional and low-energy buildings: A review article. *Energy Build.*, 39, 3, 249–257, 2007.
6. Kibert, C.J., *Sustainable construction: Green building design and delivery*, 2nd ed., John Wiley & Sons, 2008. United Nations, Transforming our world: The 2030 Agenda for Sustainable Development. 2015, Retrieved from https://sdgs.un.org/2030agenda.
7. Weis, G., Bilec, M.M., Sullivan, J.L., Evaluating the environmental impact of various construction waste disposal approaches: A case study of the Greater Pittsburgh Region. *J. Cleaner Prod.*, 140, 1021–1030, 2017.
8. Cabeza, L.F., Pérez, G., Castell, A., Barreneche, C., Building technologies and the energy challenge. *Energy Build.*, 76, 829–835, 2014.
9. Villoria Sáez, P., Environmental impact assessment of concrete structures in buildings: A review. *J. Cleaner Prod.*, 180, 720–734, 2018.
10. Xu, X., Cao, S., Zhang, L., Chen, X., Global ecological and economic consequences of changes in hydrological cycle and human water use in China. *Water Res.*, 165, 114981, 2019.
11. Rinaldi, M., Sanches, F.O., Tavares, C.R.G., de Campos, T.M., The environmental impact of sand exploitation in rural areas of Birbhum District, West Bengal, India. *J. Environ. Manage.*, 188, 456–465, 2017.
12. Tam, V.W.Y. and Tam, C.M., Waste reduction through incentives for source separation of construction waste: A case study. *Waste Manage. (Oxford)*, 28, 9, 1652–1660, 2008.
13. Zhang, J., Zuo, J., Zillante, G., Zhao, Z.Y., Green building research trend: A bibliometric analysis. *Build. Environ.*, 114, 244–254, 2016.
14. Santamouris, M., Sfakianaki, A., Pavlou, K., Onoufriou, A., Alexandri, E., Using cool pavements as a mitigation strategy to fight urban heat island—A promising but challenging approach. *J. Build. Perform. Simul.*, 10, 4, 381–396, 2017.

15. Zhang, T., Hong, T., Lin, B., A comprehensive review of a two decades long building cooling load estimation methods. *Energy Build.*, 173, 116–129, 2018.
16. Hong, T., Yan, D., D'Oca, S., Chen, C., Occupant behavior models: A critical review of implementation in energy simulation programs. *Build. Environ.*, 132, 21–33, 2018.
17. Szokolay, S.V., *Introduction to architectural science: The basis of sustainable design*, Architectural Press, 2004.
18. D'Oca, S. and Hong, T., A quantitative comparison of the performance of four building simulation programs (BSP)—EnergyPlus, DeST, DOE-2.1E, and IDA-ICE—against the ASHRAE 140 empirical method. *Build. Environ.*, 47, 58–70, 2012.
19. Morrison, G. and Szokolay, S.V., Solar access to existing buildings. *Archit. Sci. Rev.*, 45, 3, 209–216, 2002.
20. Getter, K.L., Rowe, D.B., Robertson, G.P., Sedum spp. as a proxy for assessing changes in green roof technology. *Environ. Sci. Technol.*, 41, 11, 3899–3904, 2007.
21. Etheridge, D. and Sandberg, M., Ventilation and indoor air quality in new houses—final report, Swedish Council for Building Research, 1996.
22. Reinhart, C.F. and Walkenhorst, O., Validation of dynamic radiance-based daylight simulations for a test office with external blinds. *Energy Build.*, 33, 7, 683–697, 2001.
23. Givoni, B., *Passive and low energy cooling of buildings*, CRC Press, 1994.
24. Heschong, L. and Wright, R.L., Daylighting impacts on human performance in school. *J. Illum. Eng. Soc.*, 31, 2, 101–114, 2002.
25. Smith, A. and Slaughter, S., Prefabricated housing: A sustainable alternative for the production of affordable homes. *Struct. Surv.*, 27, 4, 302–313, 2009.
26. Santamouris, M., *Energy and climate in the urban built environment*, James & James, 1999.
27. Wienold, J. and Wagner, A., Advanced facades: Climate-adaptive building shells—The current state of research. *Energy Build.*, 107, 316–330, 2015.
28. Geiger, D. and Pfluger, R., Earth-sheltered building: Thermal performance of historical structures. *Energy Procedia*, 121, 99–106, 2016.
29. Santamouris, M., Asimakopoulos, D.N., Papaglastra, M., On the impact of urban climate on the energy consumption of buildings. *Sol. Energy*, 70, 3, 201–216, 2001.
30. Awbi, H.B., *Ventilation of buildings*, Taylor & Francis, 2003.
31. Ahuja, P., Yadav, V., Thakur, S., Internet of Things (IoT): A review of enabling technologies, challenges, and open research issues. *J. Comput. Theor. Nanosci.*, 14, 1, 104–117, 2017.
32. Chen, X., Zhao, R., Hong, T., A review of predictive building control. *Renew. Sustain. Energy Rev.*, 103, 442–454, 2019.

33. Lund, H., Mathiesen, B.V., Christensen, P., Energy system analysis of marginal electricity supply in consequential LCA. *Int. J. Life Cycle Assess.*, 20, 6, 849–860, 2015.
34. Lai, C.L., Wu, Y.S., Chen, J.C., Liu, Y.T., Life cycle assessment of building-integrated photovoltaics. *Energy Build.*, 227, 110403, 2020.
35. Carlini, E.M., Lira, M.S., Loures, L.F.M., Performance analysis of wind turbines integrated into the building. *Renew. Energy*, 116, 769–781, 2018.
36. Rincón-Mora, G.A. and Li, J., On lifetime and performance of lithium-ion batteries. *J. Power Sources*, 360, 243–249, 2017.
37. Zhang, Y., Wang, Y., Fang, Z., Advanced battery technologies for building-scale stationary electrical energy storage. *Energy Storage Mater.*, 16, 237–261, 2019.
38. Komor, P., Reiter, U., Hauer, A., Ofner, H., Small-scale pumped hydro storage for residential buildings. *Appl. Energy*, 239, 238–249, 2019.
39. Su, Y., Darkwa, J., Kokogiannakis, G., A review of phase change materials integrated in building applications. *Renew. Sustain. Energy Rev.*, 79, 1375–1397, 2017.
40. Viola, A.U., James, L.M., Schlangen, L.J., Dijk, D.J., Blue-enriched white light in the workplace improves self-reported alertness, performance, and sleep quality. *Scand. J. Work Environ. Health*, 34, 4, 297–306, 2008.
41. Wargocki, P., Wyon, D.P., Baik, Y.K., Clausen, G., Fanger, P.O., Perceived air quality, sick building syndrome (SBS) symptoms and productivity in an office with two different pollution loads. *Indoor Air*, 15, 2, 76–87, 2015.
42. Browning, W.D., Ryan, C.O., Clancy, J.O., *14 Patterns of Biophilic Design*, Terrapin Bright Green, 2014.
43. Zhang, Y., Xiao, F., Lu, Y. *et al.*, Artificial Intelligence in Personalized Comfort Control for Sustainable Buildings: A Review. *Energies*, 13, 15, 3807, 2020.
44. Klatte, M., Bergström, K., Lachmann, T., Does noise affect learning? A short review on noise effects on cognitive performance in children. *Front. Psychol.*, 1, 1–6, 2010.
45. Jahncke, H., Hygge, S., Halin, N., Green, A.M., Dimberg, K., Open-plan office noise: Cognitive performance and restoration. *J. Environ. Psychol.*, 31, 4, 373–382, 2011.
46. Hedge, A., Effects of an adjustable chair and keyboard stand on spinal angles and discomfort. *Hum. Factors*, 46, 1, 130–147, 2004.
47. Ardente, F., Beccali, M., Cellura, M., Mistretta, M., Energy performance assessment in LEED-certified buildings: Case study of an Italian green office. *Energy Build.*, 117, 126–139, 2016.
48. Shu, L., Wang, Y., Guo, S., An environmental assessment of BREEAM certification in the UK. *Build. Environ.*, 122, 258–269, 2017.
49. Oh, S.H., Kim, Y.H., Lee, J.K., Kim, J.T., Assessing the environmental impact of a building by applying the living building challenge standard. *J. Cleaner Prod.*, 156, 509–520, 2017.

50. Feist, W., Schnieders, J., Dorer, V., Cost-optimal energy performance requirements for buildings in the European Union: An approach to achieve the 2020 target. *Energy Build.*, 49, 2–8, 2012.
51. Janda, K.B., Zhang, T., Hong, T., Adoption of green building technologies: The role of policies, human capital, and innovation diffusion dynamics. *Sustain. Cities Soc.*, 28, 339–346, 2017.
52. Van der Heijden, J., Boons, F., van Sark, W., The role of policy in the innovation game: Testing the systemic innovation framework with a case study of the transition towards sustainable construction. *Technol. Forecasting Social Change*, 118, 288–305, 2017.
53. Husein, A.R., Azis, A., Warshawsky, D., Government policies and the adoption of electric vehicles: An exploratory study. *Energy Rep.*, 6, 1173–1182, 2020.
54. Nauclér, T., Azar, C., Lindgren, K., Industry decarbonization: The role of international cooperation in the cement sector. *Energy Policy*, 130, 27–36, 2019.
55. Nistor, I., Ioan, D., Baetica, R., Artificial intelligence for optimal energy management in smart buildings. *Energies*, 14, 3, 604, 2021.
56. Hasan, N., Yaqoob, I., Ahmedy, I. *et al.*, Enabling technologies for the internet of health things. *IEEE Access*, 8, 27889–27911, 2020.
57. Bello, O., Zeadally, S., Badra, M., Internet of Things (IoT) for smart energy management in buildings: A review. *Energy Build.*, 211, 109762, 2020.
58. Dastjerdi, A.V. and Fortino, G., Fog computing in the context of Internet of Things: A survey. *IEEE Internet Things J.*, 6, 2, 899–918, 2019.
59. Yang, X., Han, X., Liu, J., Zhao, X., A comprehensive review on indoor environmental quality: Integration and improvement. *Build. Environ.*, 185, 107285, 2021.
60. Upadhyay, A., Nieuwenhuijsen, M.J., Mohee, R., Health benefits and costs of green building design and construction. *J. Environ. Manage.*, 195, Pt 1, 86–92, 2017, https://doi.org/10.1016/j.jenvman.2017.03.051.

# About the Editors

**Ashwani Kumar, PhD** is a senior lecturer teaching mechanical engineering in the Technical Education Department at Uttar Pradesh, Kanpur, India. He has more than 13 years of research, academic, and administrative experience and serves as a guest editor and editorial board member for eight international journals, as well as a review board member for 20 international journals. Being an academician and researcher, he has authored four book series, authored or co-authored over 30 books, and published over 100 research articles and two patents.

**Nitesh Dutt, PhD** is an assistant professor in the Department of Mechanical Engineering, College of Engineering Roorkee, Uttarakhand, India. He has more than seven years of teaching experience. and has published more than 11 research articles in international journals and conferences. His main areas of research include nuclear engineering, heat and mass transfer, thermodynamics, fluid mechanics, refrigeration and air conditioning, and computational fluid dynamics.

**Mukesh Kumar Awasthi, PhD** is an assistant professor in the Department of Mathematics at Babasaheb Bhimrao Ambedkar University, Lucknow. He has published more than 115 research publications across various mediums, including national and international journals and conferences, as well as eight books. In addition to attending many symposia, workshops, and conferences in mathematics and fluid mechanics, he has received funding for a project researching a nonlinear study of the interface in a multilayer fluid system.

# About the Editors

# Index

1D, 2D, and 3D thermal models, 273

Accelerated rate calorimetry (ARC), 277
Active cooling with earth-air heat exchangers, 137–138
Adaptive formulations, 113
Adaptive traits, 116
Advanced building envelope, 400
Advanced building materials and energy harvesting, 409
Advanced heat sink channel, 164
Advancements, 287
Aesthetic appeal, 115
Aging, 271
Aging tests, 284
Air cooling, 286
Ambient temperature, 271
Ancient wisdom & technology, 116
Antibacterial applications, 335
Anti-fungal properties, 114
Anti-inflammatory applications, 336
Arduino, 257
Artificial intelligence in smart buildings, 402
Artificial roughness-based turbulence, 25
Augmentation of solar, geothermal, and Earth-air heat exchanger in sustainable buildings, 119–121

Baffles, 235, 236, 249
Battery components, 284
Battery design, 287
Battery integrity, 285
Battery lifespan, 284
Battery management systems (BMS), 278
Battery pack failures, 285
Battery performance, 284
Battery safety, 274
Benefits of Earth-air heat exchanger (EAHE) systems, 140–141
Biodegradability, 114
Biodegradability & well-being, 114, 115
Biodegradable paints, 113
Biomass thermal energy, 41
Biosensing, 335
Blasius friction equation, 176
Blast waves, 360
Blood flow, 332
Boundary layer, 235, 243

Calorimetry, 271
Capacity, 274
Capacity retention, 284
Casson fluid, 313, 376
Casson fluid parameters, 345
Casson nanofluid, 382
Casson parameter, 321, 382
Catastrophic failure, 275
Cattaneo–Christov heat flux, 332
Ceiling design, 108
Cell design, 271
Challenges in integrating renewable systems, 143–144

Characterization methods, 286
Charge cycles, 272
Charging behavior, 275
Charging rates, 279
Chemical properties, 113
Chemical reaction parameter, 326
Chemical reactions, 276
Climate and geographic considerations, 146–147
Closed-loop and open-loop systems, 131
Comfort & sustainability, 108
Comfort factors, 114
Comparison among desiccant and VCR cooling, 99
Compound method, 163
Computational models, 283
Concentrated solar power, 54
Concentration gradients, 273
Concentration polarization, 273
Concrete & climate, 108
Concrete alternatives, 108
Conductive additives, 286
Construction processes, 393
Continuity equation, 169
Controlled cycling tests, 284
Cooling characteristics, 108
Cooling mechanisms, 275
Copper-inserted diamond fin heat sink, 171
Cost, 110
Cost efficiency, 108
Cost implications of renewable integration, 145
Counter flow double pipe heat exchanger (CD-DPHE), 203, 205, 212
Cow dung paint, 106, 107
Critical conditions, 276
Crush tests, 277
Cultural & ethical alignment, 111
Current rate, 271
Current state of renewable energy technologies, 122–123
Customization & diversity, 116

Darcy's law, 377
Data-driven method, 301
Degradation mechanisms, 271
Degradation rates, 284
Deintercalation, 272
Density, 17
Desiccant (recirculation mode), 93
Desiccant system (ventilation mode), 92
Desiccant-assisted dehumidification and cooling system, 94
Design elements details, 396
Design optimization, 272
Design requirements, 281
Diamond shape heat exchanger sink, 186
Differential scanning calorimetry (DSC), 277
Dimpled tubes, 77
Direct-use geothermal systems, 130
Discharge cycles, 272
Discharging behavior, 275
Discharging limits, 279
District heating and cooling, 41
Dittus Boelter equation, 175
Double pipe heat exchanger (DPHE), 193, 195, 201, 205
Drug delivery, 335
Drying, 233, 235, 248
Durability, 110, 270
Durability & longevity, 115
Durability & protection, 111
Dust-filled gas, 360
Dynamic control in EAHE systems, 139

EAHE, 224, 225
EAHE performance evaluation, 227
Earth-air heat exchangers: passive and active cooling, 135–136
Earth-to-air heat exchanger, 223, 224
Eco-conscious solutions, 107
Eco-conscious spaces, 111
Eco-friendly characteristics, 113

Eco-friendly finishes, 113
Eco-friendly options, 112
Efficiency, 256
EIS, 297
Electrochemical characteristics, 283
Electrochemical impedance spectroscopy (EIS), 283
Electrochemical models, 274
Electrochemical reactions, 272
Electrode cracking, 271
Electrode degradation, 273
Electrode designs, 286
Electrode materials, 272
Electrode–electrolyte interface, 273
Electrodes, 273
Electrohydrodynamics, 78
Electrolyte decomposition, 271, 273
Electrolytes, 284
Electro-passive heat transfer enhancement techniques, 78
Energy conservation, 106
Energy consumption, 106
Energy efficiency, 111
Energy efficiency and environmental impact of solar technologies, 127–128
Energy equation, 170
Energy storage, 270
Energy storage solutions, 404
Energy storage systems, 287
Energy-efficient buildings, 41
Enhanced geothermal systems (EGS), 133
Enhanced indoor comfort, 114
Environmental & health benefits, 108
Environmental adaptation, 113
Environmental challenges, 393
Environmental consciousness, 111
Environmental impact, 109, 114
Environmental impact of geothermal systems, 130–131
Environmental sustainability, 107
External damage, 285

Feature extraction, 302
Financial incentives for geothermal systems, 132
Fins and extended surfaces, 7
Fluid flow, 235, 236, 238, 241
Frequency analysis, 283
Friction factor, 233, 235, 236, 238, 240, 241, 242, 243, 244, 245, 246, 247, 248
Future trends in integrated renewable energy systems, 150

Geothermal energy, 40
Geothermal energy in building systems, 128–130
Geothermal power plants, 132–133
Gold nanoparticles, 332
Grashof numeral, 346
Green building standards, 407
Grid independency test, 238, 239, 234
Ground source heat pumps (GSHP), 130
Ground temperature stability, 137

Hamilton-Crosser model, 341
Handling procedures, 286
Harnessing subsurface heat, 130–132
Health & sustainability, 116
Health benefits, 106
Healthier environment, 115
Heat capacity of nanofluids, 19
Heat conduction, 273
Heat dissipation, 271
Heat exchangers, 13
Heat flow, 277
Heat generation, 271
Heat pipes, 22
Heat pump systems, 41
Heat release rates, 277
Heat transfer, 2, 235, 236, 238, 240, 241, 243, 245, 247, 283
Heat transfer enhancement method, 163
Heat transfer enhancement techniques, 3
Heat transfer mechanisms, 284

Heat transfer rate, 194, 204
Heat transport fluid, 202
Heating mechanisms, 275
Heat-related issues, 283
Helical microfins tubes, 69
Herringbone tube, 72
High temperatures, 275
High thermal conductivity components, 286
Holistic approach, 107
Holistic comfort, 111
Human well-being, 107
Human-centric approach, 406
HVAC, 63, 64, 223, 224
HVAC&R, 64
Hydrochlorofluorocarbon (HCFC), 230

Imaging, 335
Impact of policy and regulations on renewable integration, 151–152
Impact studies, 116
Impedance, 297
Impedance changes, 283
Improper handling, 285
Indoor air quality, 106, 405
Industrial applications of geothermal energy, 133–134
Infrared (IR) thermography, 283
Ingredients, 109
Innovative architectural designs, 395
Innovative formulations, 116
Input features, 304
In-situ measurements, 277
Integration of EAHE systems with HVAC, 139
Integration of geothermal systems into building design, 131–132
Integration of solar PV panels, 124
Internal resistance, 275
Inverse Darcy number, 382
Ion mobility, 275
IoT-driven smart building ecosystems, 410
IR radiation & sunlight, 111

Jet impingement, 22

K-e model, 169

LIB durability, 285
Life cycle, 276
Liquid cooling, 286
Lithium-ion batteries (LIBs), 270, 271, 272, 273, 274, 275, 276, 277, 278, 279, 280, 281, 282
Lithium-ion intercalation, 272
Living spaces, 106, 107, 111, 112
Longevity, 271
Low temperatures, 275
Low VOC content, 113
Low VOC formulas, 111

Magnetic field, 362
Magnetic iron oxide, 378
Magneto-gas-dynamics (MGD), 364
Maintenance requirements, 145–146
Manufacturing guidelines, 281
Market demand, 111
Mass flow rate, 196, 217
Materials innovation, 286
MATLAB, 214
Maxwell model, 342
Mean ambient temperature, 314
Mechanical aspects, 276
Microchannel diamond fin heat sink, 165
Microchannel heat exchanger sink, 160
Microchannel heat sink method, 164
Microchannels, 13
Microcontroller, 258
Model validation, 299
Modern innovation, 111
Mold resistance, 114
Momentum equation, 170
Multi objective type optimization (MOTOT), 207
Multiphysics coupling, 273
Multiphysics simulations, 273
MXene nanomaterials, 334

Nano bloodstream, 348
Nanofluids, 16, 333, 378
Nanomedicine applications, 334
Nanostructured materials, 286
Natural composition, 114
Natural composition impact, 115
Natural decomposition, 113
Natural heat convection in nanofluids, 19
Need for vedic paints, 111
Neural interfaces, 336
Next-generation collector designs, 55
Non-load bearing applications, 108
Non-toxic formulations, 112
Non-toxic safety benefits, 115
Novel applications, 287
Novel materials, 286
Nusselt number, 233, 236, 238, 240, 241, 242, 244, 245, 352
NYCC, 299

Occupant well-being, 116
Ohmic heating, 272
Operating conditions, 279
Operational conditions, 276
Operational safety, 282
Optimization, 195
Overcharge tests, 277

Paint Impact, 106
Parameter identification, 298
Particle swarm optimization (PSO), 193, 195, 197, 198, 208, 209
Passive cooling with Earth-air heat exchangers, 137
Passive design principles, 397
Passive heat transfer enhancement techniques, 66
Passive method, 163
Perfect gas, 361
Performance, 109, 271
Phase change material (PCMs), 5, 11, 22, 225, 286
Photothermal therapy, 335
Photovoltaic, 221
Physical impact, 285
Porosity, 21
Porous, 377
Porous media, 20
Porous spinning disk, 314
Positive energy, 113
Power output, 275
Power series method, 361
Practical implementation techniques for EAHE systems, 141–142
Prakritik certification, 108
Prandtl number, 382
Preventative measures, 285
Prime colors, 107
Principles and functionality of Earth-air heat exchangers (EAHE), 136–138
Pulmonary physiology pressure, 332
Puncture tests, 277

Quasilinear hyperbolic differential equation, 362

Radiation parameter, 325
Rankine-Hugoniot (RH) jump conditions, 364
Rapid response plans, 280
Reaction kinetics, 276
Recycling & reuse, 111
Reduced maintenance, 115
Reduction of VOCs, 114
Reflective coatings, 113
Regulatory and permitting challenges, 147–148
Renewable energy, 270
Renewable thermal energy systems, 53
Reynolds number, 233, 235, 237, 238, 242, 244, 247
Ribs, 235, 236, 249
Risk minimization, 280
RLS, 298

Role in sustainable living, 116
Roughened microchannel HeatSink, 183
Roughness, 233, 235, 236, 237, 238, 241, 242, 247
Runaway triggers, 284

Safety & health benefits, 114
Safety management strategies, 286
Safety mechanisms, 277
Safety protocols, 285
Safety tests, 276
Schmidt number, 325
Seasonal thermal energy storage (STES), 143
Semi-cylindrical tubes, 233, 236, 237, 238, 241, 243, 245, 247, 248
Shell powder use, 108
Short-circuit tests, 277
Side reactions, 273
Significance of renewable energy in sustainable buildings, 121
Single objective type optimization, 195
Slip parameter, 321
Smart building integration, 111, 112
Smart buildings, 106, 107, 108, 109, 110, 111, 112, 113, 114, 115, 116
Smart green building technologies, 394
Smart green buildings, 392
Solar air heater, 233, 235, 249
Solar augmentation strategies, 123–124
Solar panel, 259
Solar shingles and solar paint, 125
Solar thermal collectors, 124–125
Solar thermal energy, 40
Solar tracking, 257
Sound & radiation resistance, 108
Specialized finishes, 113
Stagnation point, 376
Standards and regulations, 5
State of charge, 271
State of Health, 293
Straight microfins, 69
Stretching sheet, 378
Sub-colors integration, 107
Suction-injection parameter, 384
Supernova remnants, 360
Surfactants, 68
Sustainable building, 221
Sustainable contributions, 115
Sustainable design, 106
Sustainable development goals, 42
Sustainable formulations, 116
Sustainable ingredients, 112
Synergies among solar, geothermal, and Earth-air systems, 142–143
Synthetic vs Prakritik, 106
System compatibility issues, 144

Technological advancements in renewable systems, 148
Technological integration, 116
Temperature, 275
Temperature distribution, 278, 283
Temperature monitoring, 279
Temperature performance, 285
Temperature regulation, 113, 286
Temperature rise, 276
Temperature spikes, 271
Temperature thresholds, 277, 284
Thermal & breathability, 115
Thermal abuse, 277
Thermal analysis techniques, 285
Thermal behavior, 270
Thermal capacitances, 273
Thermal characterization, 270, 285
Thermal conductivity, 284
Thermal considerations, 287
Thermal efficiency, 235, 236, 248
Thermal energy storage, 41, 55, 404
Thermal gradients, 283
Thermal imaging, 271
Thermal inertia in EAHE systems, 138
Thermal insulation, 106, 286
Thermal management, 275
Thermal management systems, 275
Thermal modeling, 271, 285

Thermal models, 278
Thermal performance, 333
Thermal properties, 108, 270
Thermal protection algorithms, 279
Thermal radiation, 377
Thermal radiation parameter, 384
Thermal regulation, 272
Thermal resistances, 273
Thermal runaway, 274
Thermal runaway mechanisms, 278
Thermal runaway modeling, 276
Thermal stability, 275, 277
Thermal stress prevention, 279
Thermal-related challenges, 285
Thermal-related degradation, 270
Tissue engineering, 336
Traditional cow dung paint, 107
Tri-hybrid nanofluid, 313
Twisted tapes insert tube, 74
Types of vedic paints, 112

Use of solar heat in regenerating the desiccant wheel, 95

van der Waals gas, 361
Vedic paint, 106, 107, 108, 109, 110, 111, 112, 113, 114, 115, 116
Vedic paint ethos, 108
Vedic paint in buildings, 107
Vedic principles, 107
Vedic vs synthetic, 108
Vertical Forest Towers, 400
Vibration, 285
Viscosity, 18
Visual comfort, 115
Vortex generators, 23

Waste heat recovery, 41
Wired coils, 76

Zero/Low VOCs, 112

# Also of Interest

## Check out these other related titles from Scrivener Publishing

*CHEMICAL PROCESS ENGINEERING VOLUME 1: Design, Analysis, Simulation, Integration, and Problem Solving with Microsoft Excel-UniSim Software for Chemical Engineers Computation, Physical Property, Fluid Flow, Equipment & Instrument Sizing, Pumps & Compressors, Mass Transfer,* by A. Kayode Coker and Rahmat Sotudeh-Gharebagh, ISBN 9781119510185. Written by one of the most prolific and respected chemical engineers in the world and his co-author, also a well-known and respected engineer, this two-volume set is the "new standard" in the industry, offering engineers and students alike the most up-do-date, comprehensive, and state-of-the-art coverage of processes and best practices in the field today.

*CHEMICAL PROCESS ENGINEERING VOLUME 2: Design, Analysis, Simulation, Integration, and Problem Solving with Microsoft Excel-UniSim Software for Chemical Engineers, Heat Transfer and Integration, Process Safety, Chemical Kinetics and Reactor Design, Engineering Economics, Optimization,* by A. Kayode Coker and Rahmat Sotudeh-Gharebagh, ISBN: 9781119853992. Written by one of the most prolific and respected chemical engineers in the world and his co-author, also a well-known and respected engineer, this two-volume set is the "new standard" in the industry, offering engineers and students alike the most up-do-date, comprehensive, and state-of-the-art coverage of processes and best practices in the field today.

*Liquid Biofuels: Fundamentals, Characterization, and Applications,* Edited by Krushna Prasad Shadangi, ISBN: 9781119791980. From the selection and pretreatment of raw materials to design of reactors, methods of conversion, selection of process parameters, optimization, and production of various types of biofuels to the industrial applications for the technology, this is the most up-to-date and comprehensive coverage of liquid biofuels for engineers and students.

*PETROLEUM REFINING DESIGN AND APPLICATIONS HANDBOOK VOLUME 1,* by A. Kayode Coker, ISBN: 9781118233696. The most comprehensive and up-to-date coverage of the advances of petroleum refining designs and applications, written by one of the world's most well-known process engineers, this is a must-have for any chemical, process, or petroleum engineer.

*PETROLEUM REFINING DESIGN AND APPLICATIONS HANDBOOK VOLUME 2: Rules of Thumb, Process Planning, Scheduling, and Flowsheet Design, Process Piping Design, Pumps, Compressors, and Process Safety Incidents,* by A. Kayode Coker, ISBN: 9781119476412. The second of a three-volume set of the most comprehensive and up-to-date coverage of the advances of petroleum refining designs and applications, written by one of the world's most well-known process engineers, this is a must-have for any chemical, process, or petroleum engineer.

*PETROLEUM REFINING DESIGN AND APPLICATIONS HANDBOOK VOLUME 3: Mechanical Separation, Distillation, Packed Towers, Liquid-Liquid Extraction, Heat Transfer and Process Safety Incidents,* by A. Kayode Coker, ISBN: 9781119794868. The third volume of a four-volume set of the most comprehensive and up-to-date coverage of the advances of petroleum refining designs and applications, written by one of the world's most well-known process engineers, this is a must-have for any chemical, process, or petroleum engineer.

*PETROLEUM REFINING DESIGN AND APPLICATIONS HANDBOOK VOLUME 4: Heat Transfer, Pinch Analysis and Process Safety Incidents,* by A. Kayode Coker, ISBN: 9781119827528. The fourth volume of a multi-volume set of the most comprehensive and up-to-date coverage of the advances of petroleum refining designs and applications, written by one of the world's most well-known process engineers, this is a must-have for any chemical, process, or petroleum engineer.

Printed in the USA/Agawam, MA
September 16, 2026

910661.147